£19.99

FUNDAMENTALS OF LIGHT SOURCES AND LASERS

FUNDAMENTALS OF LIGHT SOURCES AND LASERS

Mark Csele

WILEY-INTERSCIENCE

A JOHN WILEY & SONS, INC., PUBLICATION

Copyright © 2004 by John Wiley & Sons, Inc. All rights reserved.

Published by John Wiley & Sons, Inc., Hoboken, New Jersey.
Published simultaneously in Canada.

No part of this publication may be reproduced, stored in a retrieval system, or transmitted in any form or by any means, electronic, mechanical, photocopying, recording, scanning, or otherwise, except as permitted under Section 107 or 108 of the 1976 United States Copyright Act, without either the prior written permission of the Publisher, or authorization through payment of the appropriate per-copy fee to the Copyright Clearance Center, Inc., 222 Rosewood Drive, Danvers, MA 01923, 978-750-8400, fax 978-646-8600, or on the web at www.copyright.com. Requests to the Publisher for permission should be addressed to the Permissions Department, John Wiley & Sons, Inc., 111 River Street, Hoboken, NJ 07030, (201) 748-6011, fax (201) 748-6008.

Limit of Liability/Disclaimer of Warranty: While the publisher and author have used their best efforts in preparing this book, they make no representations or warranties with respect to the accuracy or completeness of the contents of this book and specifically disclaim any implied warranties of merchantability or fitness for a particular purpose. No warranty may be created or extended by sales representatives or written sales materials. The advice and strategies contained herein may not be suitable for your situation. You should consult with a professional where appropriate. Neither the publisher nor author shall be liable for any loss of profit or any other commercial damages, including but not limited to special, incidental, consequential, or other damages.

For general information on our other products and services please contact our Customer Care Department within the U.S. at 877-762-2974, outside the U.S. at 317-572-3993 or fax 317-572-4002.

Wiley also publishes its books in a variety of electronic formats. Some content that appears in print, however, may not be available in electronic format.

Library of Congress Cataloging-in-Publication Data:

Csele, Mark.
   Fundamentals of light sources and lasers/Mark Csele.
     p. cm.
   "A Wiley-Interscience publication."
   Includes bibliographical references and index.
   ISBN 0-471-47660-9 (cloth : acid-free paper)
   1. Light sources. 2. Lasers. I. Title.

QC355.3.C74 2004
621.36'6--dc22
                              2004040908

Printed in the United States of America

10 9 8 7 6 5 4 3 2 1

*To my parents
for fostering and encouraging
my interest in science*

CONTENTS

Preface   xiii

1. **Light and Blackbody Emission**   1

   1.1   Emission of Thermal Light   1
   1.2   Electromagnetic Spectrum   2
   1.3   Blackbody Radiation and the Stefan–Boltzmann Law   2
   1.4   Wein's Law   4
   1.5   Cavity Radiation and Cavity Modes   6
   1.6   Quantum Nature of Light   9
   1.7   Electromagnetic Spectrum Revisited   10
   1.8   Absorption and Emission Processes   10
   1.9   Boltzmann Distribution and Thermal Equilibrium   13
   1.10   Quantum View of Blackbody Radiation   14
   1.11   Blackbodies at Various Temperatures   15
   1.12   Applications   17
   1.13   Absorption and Color   18
   1.14   Efficiency of Light Sources   18
           Problems   19

2. **Atomic Emission**   21

   2.1   Line Spectra   21
   2.2   Spectroscope   22
   2.3   Einstein and Planck: $E = h\nu$ 26
   2.4   Photoelectric Effect   27
   2.5   Atomic Models and Light Emission   28
   2.6   Franck–Hertz Experiment   31
   2.7   Spontaneous Emission and Level Lifetime   34
   2.8   Fluorescence   35
   2.9   Semiconductor Devices   37
   2.10   Light-Emitting Diodes   44
           Problems   48

3. **Quantum Mechanics**   49

   3.1   Limitations of the Bohr Model   50
   3.2   Wave Properties of Particles (Duality)   50

	3.3	Evidence of Wave Properties in Electrons	52
	3.4	Wavefunctions and the Particle-in-a-Box Model	53
	3.5	Reconciling Classical and Quantum Mechanics	55
	3.6	Angular Momentum in Quantum States	56
	3.7	Spectroscopic Notation and Electron Configuration	57
	3.8	Energy Levels Described by Orbital Angular Momentum	60
	3.9	Magnetic Quantum Numbers	62
	3.10	Direct Evidence of Momentum: The Stern–Gerlach Experiment	63
	3.11	Electron Spin	65
	3.12	Summary of Quantum Numbers	67
	3.13	Example of Quantum Numbers: The Sodium Spectrum	69
	3.14	Multiple Electrons: The Mercury Spectrum	71
	3.15	Energy Levels and Transitions in Gas Lasers	72
	3.16	Molecular Energy Levels	73
	3.17	Infrared Spectroscopy Applications	77
		Problems	79
4.	**Lasing Processes**		**83**
	4.1	Characteristics of Coherent Light	84
	4.2	Boltzmann Distribution and Thermal Equilibrium	86
	4.3	Creating an Inversion	87
	4.4	Stimulated Emission	90
	4.5	Rate Equations and Criteria for Lasing	92
	4.6	Laser Gain	98
	4.7	Linewidth	101
	4.8	Thresholds for Lasing	104
	4.9	Calculating Threshold Gain	106
		Problems	113
5.	**Lasing Transitions and Gain**		**117**
	5.1	Selective Pumping	117
	5.2	Three- and Four-Level Lasers	119
	5.3	CW Lasing Action	124
	5.4	Thermal Population Effects	127
	5.5	Depopulation of Lower Energy Levels in Four-Level Lasers	128
	5.6	Rate Equation Analysis for Atomic Transitions	130
	5.7	Rate Equation Analysis for Three- and Four-Level Lasers	136
	5.8	Gain Revisited	143
	5.9	Saturation	146
	5.10	Required Pump Power and Efficiency	149

	5.11	Output Power	154
		Problems	156

6. Cavity Optics 159

	6.1	Requirements for a Resonator	159
	6.2	Gain and Loss in a Cavity	160
	6.3	Resonator as an Interferometer	162
	6.4	Longitudinal Modes	164
	6.5	Wavelength Selection in Multiline Lasers	166
	6.6	Single-Frequency Operation	169
	6.7	Characterization of a Resonator	174
	6.8	Gaussian Beam	176
	6.9	Resonator Stability	178
	6.10	Common Cavity Configurations	180
	6.11	Spatial Energy Distributions: Transverse Modes	185
	6.12	Limiting Modes	186
	6.13	Resonator Alignment: A Practical Approach	187
		Problems	190

7. Fast-Pulse Production 193

	7.1	Concept of Q-Switching	193
	7.2	Intracavity Switches	195
	7.3	Energy Storage in Laser Media	196
	7.4	Pulse Power and Energy	198
	7.5	Electrooptic Modulators	202
	7.6	Acoustooptic Modulators	206
	7.7	Cavity Dumping	211
	7.8	Modelocking	212
	7.9	Modelocking in the Frequency Domain	215
		Problems	217

8. Nonlinear Optics 219

	8.1	Linear and Nonlinear Phenomena	219
	8.2	Phase Matching	223
	8.3	Nonlinear Materials	227
	8.4	SHG Efficiency	229
	8.5	Sum and Difference Optical Mixing	230
	8.6	Higher-Order Nonlinear Effects	231
	8.7	Optical Parametric Oscillators	232
		Problems	233

9. Visible Gas Lasers — 235

9.1	Helium–Neon Lasers	235
9.2	Lasing Medium	236
9.3	Optics and Cavities	237
9.4	Laser Structure	239
9.5	HeNe Power Supplies	241
9.6	Output Characteristics	245
9.7	Applications	246
9.8	Ion Lasers	247
9.9	Lasing Medium	247
9.10	Optics and Cavities	251
9.11	Laser Structure	252
9.12	Power Supplies	256
9.13	Output Characteristics	258
9.14	Applications and Operation	259

10. UV Gas Lasers — 261

10.1	Nitrogen Lasers	261
10.2	Lasing Medium	262
10.3	Gain and Optics	264
10.4	Nitrogen Laser Structure	265
10.5	Output Characteristics	269
10.6	Applications and Practical Units	269
10.7	Excimer Lasers	270
10.8	Lasing Medium	271
10.9	Gain and Optics	274
10.10	Excimer Laser Structure	274
10.11	Applications	277
10.12	Practical and Commercial Units	278

11. Infrared Gas Lasers — 283

11.1	Carbon Dioxide Lasers	283
11.2	Lasing Medium	283
11.3	Optics and Cavities	285
11.4	Structure of a Longitudinal CO_2 Laser	286
11.5	Structure of a Transverse CO_2 Laser	289
11.6	Alternative Structures	290
11.7	Power Supplies	290
11.8	Output Characteristics	292
11.9	Applications	292
11.10	Far-IR Lasers	293

12. Solid-State Lasers — 295

12.1	Ruby Lasers	295
12.2	Lasing Medium	296
12.3	Optics and Cavities	297
12.4	Laser Structure	298
12.5	Power Supplies	299
12.6	Output Characteristics	300
12.7	Applications	301
12.8	YAG (Neodymium) Lasers	301
12.9	Lasing Medium	302
12.10	Optics and Cavities	302
12.11	Laser Structure	303
12.12	Power Supplies	306
12.13	Applications, Safety, and Maintenance	308
12.14	Fiber Amplifiers	309

13. Semiconductor Lasers — 313

13.1	Lasing Medium	313
13.2	Laser Structure	315
13.3	Optics	319
13.4	Power Supplies	320
13.5	Output Characteristics	321
13.6	Applications	324

14. Tunable Dye Lasers — 327

14.1	Lasing Medium	327
14.2	Laser Structure	330
14.3	Optics and Cavities	334
14.4	Output Characteristics	334
14.5	Applications	335

Index — 337

PREFACE

The field of photonics is enormously broad, covering everything from light sources to geometric and wave optics to fiber optics. Laser and light source technology is a subset of photonics whose importance is often underestimated. This book focuses on these technologies with a good degree of depth, without attempting to be overly broad and all-inclusive of various photonics concepts. For example, fiber optics is largely omitted in this book except when relevant, such as when fiber amplifiers are examined. Readers should find this book a refreshing mix of theory and practical examples, with enough mathematical detail to explain concepts and enable prediction of the behavior of devices (e.g., laser gain and loss) without the use of overwhelmingly complex calculus. Where possible, a graphical approach has been taken to explain concepts such as modelocking (in Chapter 7) which would otherwise require many pages of calculus to develop.

This book, targeted primarily to the scientist or engineer using the technology, offers the reader theory coupled with practical, real-world examples based on real laser systems. We begin with a look at the basics of light emission, including blackbody radiation and atomic emission, followed by an outline of quantum mechanics. For some readers this will be a basic review; however, the availability of background material alleviates the necessity to refer back constantly to a second (or third) book on the subject. Throughout the book, practical, solved examples founded on real-world laser systems allow direct application of concepts covered. Case studies in later chapters allow the reader to further apply concepts in the text to real-world laser systems.

The book is also ideal for students in an undergraduate course on lasers and light sources. Indeed, the original design was for a textbook for an applied degree course (actually, two courses) in laser engineering. Unlike many existing texts which cover this material in a single chapter, this book has depth, allowing the reader to delve into the intricacies. Chapter problems assist the reader by challenging him or her to make the jump between theory and reality. The book should serve well as a text for a single course in laser technology or two courses where a laboratory component is present. Introductory chapters on blackbody radiation, atomic emission, and quantum mechanics allow the book to be used without the requirement of a second or third book to cover these topics, which are often omitted in similar texts. It is assumed that students will already have a grasp of geometric and wave optics (including the concepts of interference and diffraction), as well as basic first-year physics, including kinematics.

Chapter 1 begins with a look at the most basic light source of all, blackbody radiation, and includes a look at standard applications such as incandescent lighting as well as newer applications such as far-IR viewers capable of "seeing" a human body hidden in a trunk! Chapter 2 is a look at atomic emission, in which we examine the nature and origins of emission of light from electrically excited gases as well as mechanisms such as fluorescence, with applications ranging from common fluorescent lamps to vacuum-fluorescent displays and colored neon tubes. The chapter concludes with a look at semiconductor light sources (LEDs). This chapter also includes atomic emission theory (such as the origins of line spectra) as well as practical details of spectroscopy, including operating principles of a spectroscope and examples of its use in identifying unknown gas samples, which serve to reinforce with practical applications the usefulness of the entire theory of atomic structure.

Investigation of both blackbody radiation as well as atomic emission light leads us on a path to quantum mechanics (in Chapter 3), vital to understanding the mechanisms responsible for light emission at an atomic level and later for understanding the origins of transitions responsible for laser emission in the ultraviolet, visible, and infrared regions of the spectrum. Although few books include this topic, it is vital to understanding emission spectra as well as basic laser processes, and so is included in the book for some readers as a review, for others as a new topic.

In Chapter 4 we begin with a fundamental look at lasers and lasing action. Aside from the basic processes, such as stimulated emission and rate equations governing lasing action, we also outline key laser mechanisms, such as pumping, the requirement for feedback (examined in detail in Chapter 6), and gain and loss in a real laser. Real-world examples are embedded within the chapter, such as noise in a fiber amplifier, which demonstrates the rate equations in action as well as details of an experiment in which the gain of a gas laser is measured by insertion of a variable loss into the optical cavity. In Chapter 5 we examine lasing transitions in detail, including selective pumping mechanisms and laser energy-level systems (three- and four-level lasers). Examples of transitions and energies in real laser systems are given along with theoretical examples, allowing the reader to compare how well theoretical models fit real laser systems.

In addition to expanding the concepts of gain and loss introduced in Chapter 4, in Chapter 6 we examine the laser resonator as an interferometer. The mathematical requirements for stability of a resonator and longitudinal and transverse modes in a real resonator are detailed. Wavelength selection mechanisms, including gratings, prisms, and etalons, are outlined in this chapter, with examples of applications in practical lasers, such as single-frequency tuning of a line in an argon laser.

Chapter 7 provides the reader with an introduction to techniques used to produce fast pulses such as Q-switching and modelocking. In the case of modelocking a graphical approach is used to illustrate how a pulse is formed from many simultaneous longitudinal modes. In Chapter 8 we cover nonlinear optics as they apply to lasers. Harmonic generation and optical parametric oscillators are examined.

The last six chapters of the book provide case studies allowing the reader to see the practical application of laser theory. The lasers chosen represent the vast majority of commercially available lasers, allowing the reader to relate the theory learned to practical lasers that he or she encounters in the laboratory or the manufacturing environment. In these chapters various lasers are outlined with respect to the lasing process involved (including quantum mechanics, energy levels, and transitions), details of the laser itself (lasing medium, cooling requirements), power sources for the laser, applications, and a survey of commercially available lasers of that type. Visible gas lasers, including helium–neon and ion lasers, are covered in Chapter 9. UV gas lasers such as nitrogen and excimer lasers are covered in Chapter 10, including details of the unique constraints on electrical pumping for these types of lasers. In Chapter 11 we examine infrared gas lasers focusing primarily on carbon dioxide and similar lasers using rotational and vibrational transitions.

Chapter 12 details common solid-state lasers, including YAG and ruby. Pump sources, including flashlamps, CW arc lamps, and semiconductor lasers, are examined as well as techniques such as nonlinear harmonic generation (often used with these lasers). Chapter 13 details the basics of semiconductor lasers, and Chapter 14 covers dye lasers which feature wide continuously tunable wavelength ranges and are often used in modelocking schemes to generate extremely short pulses of laser radiation. In each of these case study chapters, photographs and details have been included, allowing the reader to see the structure of each laser. Chapter 10, for example, includes numerous photographs of the various structures in a real excimer laser, including details of the electrodes and preionizers, the cooling system, and heat exchanger, as well as electrical components such as the energy storage capacitor and thyratron trigger. In each case a clear explanation is given guiding the reader to understanding the function of each critical component.

Adjunct material to this text, including in-depth discussions of spectroscopy (with color photos of example spectra), analysis and photographic details of real laser systems (such as real ion, carbon dioxide, and solid-state lasers), additional problems, and instructional materials (such as downloadable MPEG videos) covering practical details such as cavity resonator alignment may be found on the author's host web site at http://technology.niagarac.on.ca/lasers.

I must thank a number of people who have made this book possible. First, Dr. Marc Nantel of PRO (Photonics Research Ontario) for instigating this in the first place—it was his suggestion to write this book and he has been extremely supportive throughout, reviewing manuscripts and providing invaluable input. I would also like to thank colleague and fellow author Professor Roy Blake, who has helped coach me through the entire process; director of the CIT department at Niagara College, Leo Tiberi, who has been supportive throughout and has provided an environment in which creative thought flourishes; Dr. Johann Beda of PRO for reviewing the manuscript and for suggestions on material for the book; my counterpart at Algonquin College, Dr. Bob Weeks, for suggestions and illuminating discussions on excimer laser technology; and assistant editor Roseann Zappia of John Wiley & Sons. In addition to providing reviews that provided excellent feedback on the material, Roseann did a remarkable job of simplifying the entire publishing process and has gone out of her way to make this endeavor as painless as possible. I'd also like to thank my wife (who has become a widow to my laptop computer for the past year and a half) for her patience and support all along.

Finally, I would like to acknowledge not only Niagara College but especially my photonics students in various laser engineering courses for their review of my manuscripts (many under the guise of course notes) and providing invaluable input from a student's point of view. Many students provided critical insight into proofs and problems in this book which would only be possible from an undergraduate. In some cases they highlighted difficult concepts that required clarification.

I welcome the opportunity to hear from readers, especially those with suggestions for improving the book. Please feel free to e-mail me at mcsele@ieee.org. Since I get a large volume of e-mail (and spam), please refer to the book in the subject line.

MARK CSELE

CHAPTER 1

Light and Blackbody Emission

As a reader of this book, you are no doubt familiar with the basic properties of light, such as reflection and interference. This book deals with the *production* of light in its many forms: everything from incandescent lamps to lasers. In this chapter we examine the fundamental nature of light itself as well as one of the most basic sources of light: the blackbody radiator. Blackbody radiation, sometimes called *thermal light* since the ultimate power source for such light is heat, is still a useful concept and governs the workings of many practical light sources, such as the incandescent electric lamp. In later chapters we shall see that these concepts also form a base on which we shall develop many thermodynamic relationships which govern the operation of other light sources, such as lasers.

1.1 EMISSION OF THERMAL LIGHT

We have all undoubtedly encountered *thermal light* in the form of emission of light from a red-hot object such as an element on an electric stove. Other examples of such light are in the common incandescent electric lamp, in which electrical current flowing through a thin filament of tungsten metal heats it until it glows white-hot. The energy is supplied by an electrical current in what is called resistance heating, but it could just as well have been supplied by, say, a gas flame. In fact, the original incandescent lamp was developed in 1825 for use in surveying Ireland and was later used in lighthouses. The lamp worked by spraying a mixture of oxygen and alcohol (which burns incredibly hot) at a small piece of lime and igniting it. The lime was placed at the hottest part of the flame and heated until it glowed white-hot, emitting an immense quantity of light. It was the brightest form of artificial illumination at the time and was claimed to be 83 times as bright as conventional gas lights of the time. Improvements to the lamp were made by using a parabolic reflector behind the piece of lime concentrating the light. The lamp allowed the surveying of two mountain peaks over 66 miles apart and was later improved by using hydrogen and oxygen as fuel but eventually was superseded by the more convenient electric arc lamp. This light source did, however, find its way into theaters, where it was used as a

Fundamentals of Light Sources and Lasers, by Mark Csele
ISBN 0-471-47660-9 Copyright © 2004 John Wiley & Sons, Inc.

2 LIGHT AND BLACKBODY EMISSION

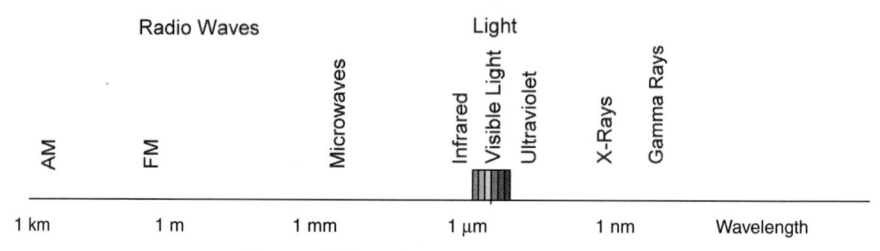

Figure 1.2.1 Electromagnetic spectrum.

spotlight which replaced the particularly dangerous open gas flames used at the time for illumination, and hence the term *limelight* was born.[1]

1.2 ELECTROMAGNETIC SPECTRUM

Anyone with a knowledge of basic physics knows that light can be viewed as a particle or as a wave, as we shall examine in this chapter. Regardless of the fact that it exhibits particlelike behavior, it surely does exhibit wavelike behavior, and light and all other forms of electromagnetic behavior are classified based on their wavelength. In the case of visible light, which is simply electromagnetic radiation visible to the human eye, the wavelength determines the color: Red light has a wavelength of about 650 nm and blue light has a wavelength of about 500 nm. Electromagnetic radiation includes all forms of radio waves, microwaves, infrared radiation, visible light, ultraviolet radiation, x-rays, and gamma rays. Figure 1.2.1 outlines the entire spectrum, including corresponding wavelengths.

1.3 BLACKBODY RADIATION AND THE STEFAN–BOLTZMANN LAW

Imagine a substance that absorbs all incident light, of all frequencies, shining on it. Such an object would reflect no light whatsoever and would appear to be completely black—hence the term *blackbody*. If the blackbody is now heated to the point where it glows (called *incandescence*), emissions from the object should, in theory, be as perfect as its absorption—one would logically expect it to emit light at all frequencies since it absorbs at all frequencies. In the 1850s a physicist named Kirchhoff, a pioneer in the use of spectroscopy as a tool for chemical analysis, observed that real substances absorb better at some frequencies than others. When heated, those substances emitted more light at those frequencies. The paradox spawned research into radiation in general and specifically, how radiation emitted from an object varied with temperature.

---

[1] An excellent historical description of the events surrounding the development of the limelight as well as the arc lamp is presented in the *Connections* series hosted by James Burke (BBC, 1978).

It was observed that the amount of radiation emitted from an object varied with the temperature of the object. The mathematical relationship for this dependence on temperature was established in 1879 by the physicist Josef Stefan, who showed that the total energy radiated by an object increased as the fourth power of the temperature of the object. All objects at a temperature above absolute zero (0 K) radiate energy, and when the temperature of an object is doubled, the total amount of energy radiated from the object will be 16 times as great!

In 1884, Ludwig Boltzmann completed the mathematical picture of a blackbody radiator, and the Stefan–Boltzmann law was developed, which allows calculation of the total energy integrated over a blackbody spectrum.

$$W = \sigma T^4 \tag{1.3.1}$$

where σ is the Stefan–Boltzmann constant (=5.67 × 10^{-8} W/m^2 · K^4) and T is the temperature in kelvin. This law applies, strictly speaking, to ideal blackbodies. For a nonideal blackbody radiator, a third term, called the *emissivity* of an object, is added, so the law becomes

$$W = e\sigma T^4 \tag{1.3.2}$$

where the emissivity of the object (e) is a measure of how well it radiates energy. An ideal blackbody has a value of 1; real objects have a value between 0 and 1.

Example 1.3.1 Use of the Stefan–Boltzmann Formula The Stefan–Boltzmann formula gives an answer in watts per square meter. This can be rearranged to give the power output of an object as

$$P = A\sigma T^4$$

where A is the surface area of the object. Now consider two objects: The first is a large (1-m^2) object at a relatively cool temperature of 300 K. The total power radiated from this (ideal) object is 459 W. Now consider a much smaller (10 cm^2 = 1 × 10^{-4} m^2) object at 3000 K. The total power radiated from this object is also 459 W. Although 10,000 times smaller, the object is also much hotter, so radiates a great deal of power.

Perhaps the most startling revelation from all this is that the 1-m^2 object at room temperature emits 459 W at all. This may seem like an enormous amount of energy, especially when compared to a 500-W floodlight; however, essentially all of this output is in the far-infrared region of the spectrum and is manifested as heat. The human body, similarly, emits a fair quantity of heat (hence the reason for large air conditioners in office buildings that house large quantities of bodies in a relatively confined space).

1.4 WEIN'S LAW

Aside from the fact that radiated energy increases with temperature, it was also observed that the wavelengths of radiation emitted from a heated object also change as an object is heated. Objects at relatively low temperatures such as 200°C (473 K) do not glow but do indeed emit something—namely, infrared radiation—felt by human beings as heat. As an object is heated to about 1000 K, the object is seen to glow a dull-red color. We call this object "red hot." As the temperature increases, the red color emitted becomes brighter, eventually becoming orange and then yellow. Finally, the temperature is high enough (like that of the sun at 6000 K) that the light emitted is essentially white (white hot).

In an attempt to model this behavior, physicists needed an ideal blackbody with which to experiment; however, in reality, none exists. In most cases a nonideal material radiated energy in a pattern of wavelengths, depending on the chemical nature of the material. The physicist Wilhelm Wein, in 1895, thought of a way to produce, essentially, a perfect blackbody in a cavity radiator. His thought was to produce a heated object with a tiny hole in it which opens to an enclosed cavity as in Figure 1.4.1. Light incident on the hole would enter the cavity and be absorbed by the irregular walls inside the cavity—essentially a perfect absorber. Light that did reflect from the inner walls of the cavity would eventually be absorbed by other surfaces that it hit after reflection, as evident in the figure. If the entire cavity is now placed in a furnace and heated to a certain temperature, the radiation emitted from the tiny hole is blackbody radiation.

Wein's studies of cavity radiation showed that regardless of the actual temperature of an object, the pattern of the emission spectrum always looked the same, with the amount of light emitted from a blackbody increasing as wavelengths became shorter, then peaking at a certain wavelength and decreased rapidly at yet still shorter wavelengths in a manner similar to that shown in Figure 1.4.2 (in this example for an object at a temperature of 3000 K). In the figure, a peak emission is evident around 950 nm in the near-infrared region of the spectrum. Emission is also seen throughout the visible region of the spectrum (from 400 through

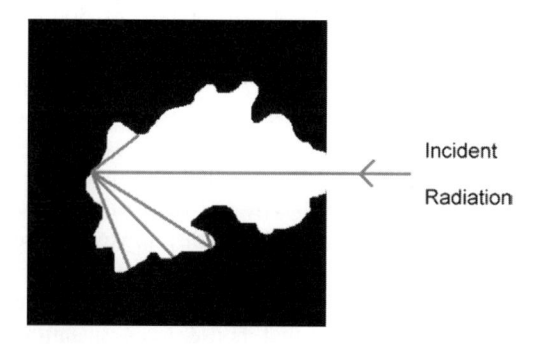

Figure 1.4.1 Cavity radiator absorbing incident radiation.

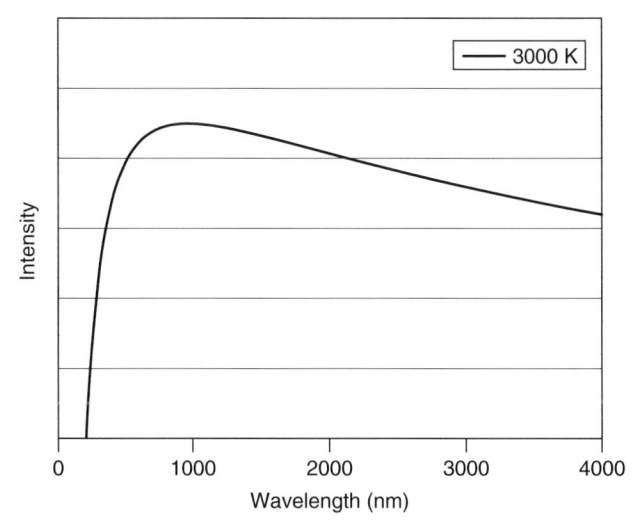

Figure 1.4.2 Typical blackbody output for an object at 3000 K.

700 nm), but more intensity is seen in the red than in the blue. It was observed that when the temperature was increased, the total amount of energy was increased according to the Stefan–Boltzmann law, and the wavelength of peak emission also became shorter. Wein's law predicts the wavelength of maximum emission as a function of object temperature:

$$\lambda_{max} T = 2.897 \times 10^{-3} \text{ m} \cdot \text{K} \tag{1.4.1}$$

where λ_{max} is the wavelength where the peak occurs and T is the temperature in kelvin. As an example, contrast the 3000 K object to the sun, which has essentially a blackbody temperature of 5270 K. The formula gives the peak at 550 nm in the yellow-green portion of the visible spectrum. This is indeed the most predominant wavelength in sunlight and is (presumably through evolutionary means) the wavelength of maximum sensitivity of the human eye.

Wein's law allows the prediction of temperature based on the wavelength of peak emission and may be used to estimate the temperature of objects such as hot molten steel as well as stars in outer space. A white star such as our sun has a temperature of around 5000 K, whereas a blue star is much hotter, with a temperature around 7000 K. Hotter stars are thought to consume themselves more quickly than cooler stars.

Both the Stefan–Boltzmann law and Wein's law may be verified experimentally using an incandescent lamp connected to a variable power source. As a practical thermal source an incandescent light does not exhibit perfect blackbody emission; however, it does follow the same general behavior as that of a blackbody source. Consider the actual spectral output of a 25-W incandescent light bulb measured

6 LIGHT AND BLACKBODY EMISSION

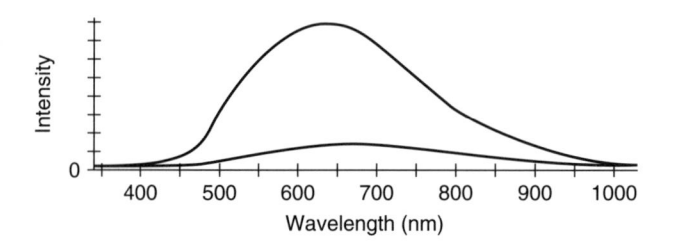

Figure 1.4.3 Output of an incandescent lamp.

both at full power and at 58% of full power (Figure 1.4.3). At full power the lamp exhibits maximum output around 625 nm, whereas at 58% of full power (the lower trace in the figure) the wavelength of peak emission shifts to around 660 nm, exhibiting the behavior predicted by Wein's law. The effect of filament temperature on intensity is also evident in the figure. Visually, light emitted from the lamp is observed to be more orange when the lamp is operated at the lower power (and hence, cooler filament) and quite white when operated at full power. We would expect white light to be consistent across the entire visible spectrum, having an intensity at 400 nm comparable to that at 600 nm, but this is clearly not the case here, and it may be surprising that this light which we call white is very rich in red and orange light (600 to 700 nm) and relatively weak in violet and blue light (400 to 500 nm). We shall revisit this idea later in the chapter.

1.5 CAVITY RADIATION AND CAVITY MODES

To derive a mathematical expression to describe the blackbody radiation curve (e.g., in Figure 1.4.2), many approaches were taken. As it is difficult to compute the behavior of a real blackbody, an easier avenue was found by Wein in the form of a cavity radiator, in which we assume that the object is a heated (isothermal) cavity with a hole in it from which light is emitted. Inside the enclosure the absorption of energy balances with emission. The cavity itself may be seen as a resonator in which standing waves are present. To simplify the problem further (to permit mathematical analysis), consider a cubical heated cavity of dimensions L (Figure 1.5.1).

A standing wave can be produced in any one of three dimensions, and standing waves of various wavelengths are possible as long as they fit inside the cavity (i.e., are an integral multiple of the dimensions of the cavity). The condition exists, then, that any electromagnetic wave (e.g., a light wave) inside the cavity must have a node at the walls of the cavity. As frequency increases, more nodes will fit inside the cavity, as evident in Figure 1.5.1. Rayleigh and Jeans showed mathematically that the number of modes per unit frequency per unit volume in such a cavity is

$$\frac{8\pi v^2}{c^3} \quad (1.5.1)$$

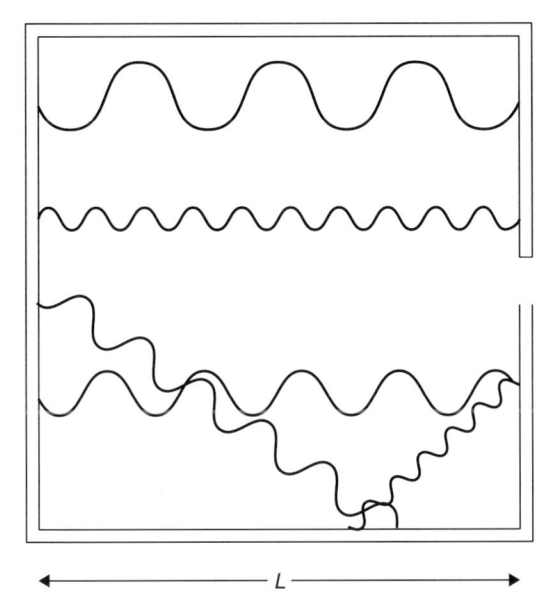

Figure 1.5.1 Modes in a heated cavity.

where v is the frequency of the mode and c is the speed of light. Rayleigh and Jeans went on to postulate that the probability of occupying any given mode is equal for all modes (in other words, all wavelengths are radiated with equal probability), an assumption from classical wave theory, and that the average energy per mode was kT (from Boltzmann statistics). The last assumption was made according to the classical prequantum (i.e., before the Rutherford atomic model) view of radiation, in which each atom with an orbiting electron was considered to be an oscillator continually emitting radiation with an average energy of kT. The resulting law formulated by Rayleigh and Jeans describes the intensity of a blackbody radiator as a function of temperature as follows:

$$\text{intensity} = \frac{8\pi v^2}{c^3} kT \tag{1.5.2}$$

The law works well at low frequencies (i.e., long wavelengths); however, at higher frequencies it predicts an intense ultraviolet (UV) output, which simply was not observed. For any given object, 16 times as much energy observed as red light should be emitted as violet light, and this simply does not happen. This law predicts the equilibrium intensity to be proportional to kTv^2, a result that may be arrived at using classical electromagnetic theory and which states that an oscillator will radiate energy at a rate proportional to v^2. This failure of the Rayleigh–Jeans law, dubbed the *UV catastrophe*, illustrates the problems with applying classical physics to the domain of light and why a new approach was needed.

8 LIGHT AND BLACKBODY EMISSION

A different approach to the problem was developed by Max Planck in 1899. His key assumption was that the energy of any oscillator at a frequency v could exist only in discrete (quantized) units of hv, where h was a constant (called *Planck's constant*). The fundamental difference in this approach from the classical approach was that modes were quantized and required an energy of hv to excite them (more on this in the next section). Upper modes, with higher energies, were hence less likely to be occupied than lower-energy modes.

Bose–Einstein statistics predict the average energy per mode to be the energy of the mode times the probability of that mode being occupied. Central to this idea was that the energy of the actual wave itself is quantized as $E = hv$. This is not a trivial result and is examined in further detail in Chapter 2, where experimental proof of this relation will be given. Multiplying this energy by the probability of a mode being occupied (the Bose–Einstein distribution function) gives the average energy per mode as

$$\frac{hv}{\exp(hv/kT) - 1} \tag{1.5.3}$$

where h is Planck's constant, v the frequency of the wave, k is Boltzmann's constant, and T is the temperature. When this is multiplied by the number of modes per unit frequency per unit volume, the same number that Rayleigh and Jeans computed, the resulting formula allows calculation of the intensity radiated at any given wavelength for any given temperature:

$$\frac{8\pi h v^3}{kT} \left[\exp\left(\frac{hv}{kT}\right) - 1 \right]^{-1} \tag{1.5.4}$$

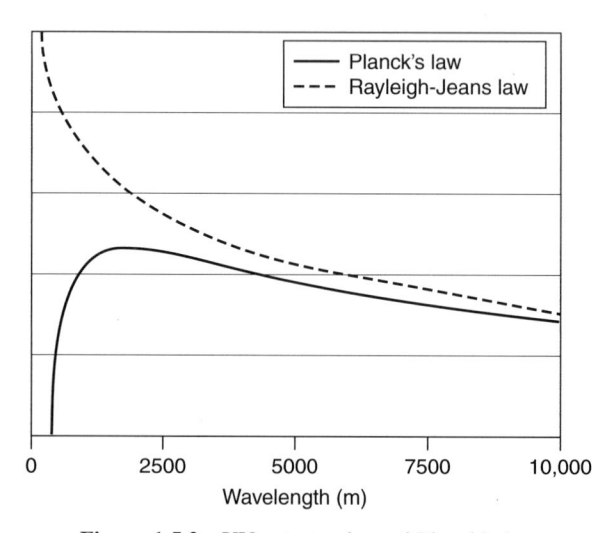

Figure 1.5.2 UV catastrophe and Planck's law.

Comparing Planck's radiation formula to the classical Rayleigh–Jeans law, we see that the two results agree well at low frequencies but deviate sharply at higher frequencies, with the Rayleigh–Jeans law predicting the UV catastrophe that never occurred (Figure 1.5.2). This new approach, which fit experimentally obtained data, heralded the birth of *quantum mechanics*, which we examine in greater detail in Chapters 2 and 3.

1.6 QUANTUM NATURE OF LIGHT

One of the earliest views of what light is was provided by Isaac Newton early in the eighteenth century. Based on the behavior of light in exhibiting reflection and refraction, he postulated that light was a stream of particles. Although this explanation works well for basic optical phenomena, it fails to explain interference. Later experiments, such as Young's dual-slit experiment, showed that light did indeed have a wavelength and that such behavior could only be explained by using wave mechanics, a concept Newton had argued against a century earlier. By the end of the nineteenth century, wave theory was well accepted but a few glitches remained: namely, the blackbody spectrum of light emitted from heated objects and issues such as the UV catastrophe, upon which classical physics failed. To explain this, Max Planck (the "father" of quantum mechanics) used particle theory once again. He postulated that the atoms in a blackbody acted as tiny harmonic oscillators, each of which had a fundamental quantized energy that obeyed the relationship

$$E = h\nu \tag{1.6.1}$$

where h is Planck's constant and ν is the frequency of the radiation emitted. This important equation may also be expressed in terms of wavelength as

$$E = \frac{hc}{\lambda} \tag{1.6.2}$$

where λ is the wavelength in meters. The ramifications of *quantization*—the fact that the energy of each atom was in integral multiples of this quantity—has far-reaching ramifications for the entire field of physics (and chemistry), and as we shall see in subsequent chapters, affects our entire view of the atom! Einstein also endorsed this concept of quantization and used it to explain the photoelectric effect, which definitely showed light to exhibit particle properties. These particles of light came to be known as photons, and according to the relationship above, were shown to have a discrete value of energy and frequency (and hence, wavelength). Light can be thought of as a wave that has particlelike qualities (or, if you prefer, a particle with wavelike qualities). It was evident from numerous experiments that both wave and particle properties are required to fully explain the behavior of light.

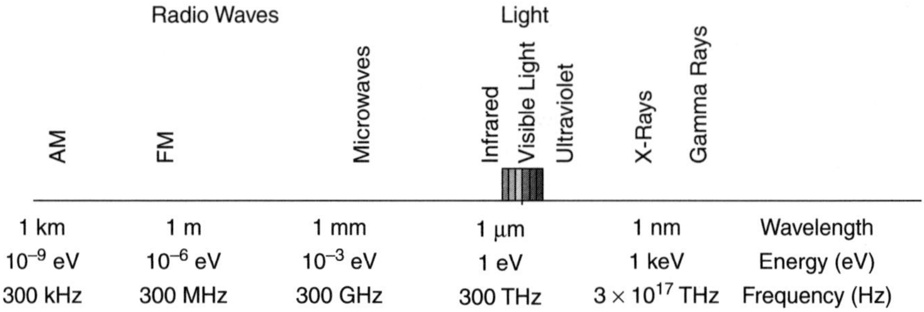

10 LIGHT AND BLACKBODY EMISSION

	Radio Waves			Light			
AM	FM	Microwaves	Infrared	Visible Light	Ultraviolet	X-Rays	Gamma Rays
1 km	1 m	1 mm	1 μm			1 nm	Wavelength
10^{-9} eV	10^{-6} eV	10^{-3} eV	1 eV			1 keV	Energy (eV)
300 kHz	300 MHz	300 GHz	300 THz			3×10^{17} THz	Frequency (Hz)

Figure 1.7.1 Electromagnetic spectrum with energies.

1.7 ELECTROMAGNETIC SPECTRUM REVISITED

Applying the photon model to the electromagnetic spectrum from Section 1.2, we may now find that a photon at a particular wavelength has a distinct energy. For example, photons of green light at 500 nm have an energy of

$$E = h\nu = \frac{hc}{\lambda}$$

or 3.98×10^{-19} J. More commonly, we express this energy in electron volts (eV), defined as the energy that an electron has accumulated after accelerating through a potential of 1 V. It is a convenient measure since the 500-nm photon can now be expressed as having an energy of 2.48 eV (with $1 \text{ eV} = 1.602 \times 10^{-19}$ J). Radio waves, microwaves, and infrared radiation have low energies, ranging up to about 1.7 eV. The visible spectrum consists of red through violet light, or 1.7 through 3.1 eV. UV, x-rays, and gamma rays have increasing photon energies to beyond 1 GeV. The spectrum is shown in Figure 1.7.1 with energies as well as wavelengths shown.

1.8 ABSORPTION AND EMISSION PROCESSES

Light is a product of quantum processes occurring when an electron in an atom is excited to a high-energy state and later loses that energy. Imagine an atom into which energy is injected (the method may be direct electrical excitation or simply thermal energy provided by raising the temperature of the atom). The electron acquires the energy and in doing so enters an excited state. From that excited state the electron can lose energy and fall to a lower-energy state, but energy must be conserved during this process, so the difference in energy between the initial high-energy state and the final low-energy state cannot be destroyed; it appears either as a photon of emitted light or as energy transferred to another state or atom. This is simply the principle of *conservation* of energy. The fact that atoms and molecules have such energy levels and transitions can occur between these

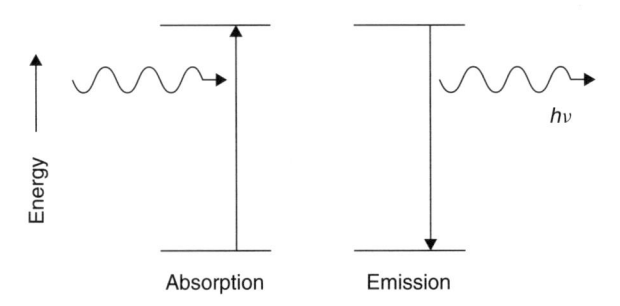

Figure 1.8.1 Absorption and emission of energy.

levels must be accepted for now: compelling experimental evidence is given in Chapter 2.[2]

An atom at a low-energy state can absorb energy and in so doing will be elevated to a higher-energy state. The energy absorbed can be in almost any form, including electrical, thermal, optical, chemical, or nuclear. The difference in energy between the original (lower) energy state and the final (upper) energy state will be exactly the energy that was absorbed by the atom. This process of *absorption* serves to excite atoms into high-energy states. Regardless of the excitation method, an atom in a high-energy state will certainly fall to a lower-energy state since nature always favors a lower-energy state (i.e., the law of entropy from thermodynamics). In jumping from a high- to a low-energy state, photons will be produced, with the photon energy being the difference in energy between the two atomic energy states. This is the process of *emission* (Figure 1.8.1). By knowing the energies of the two atomic states involved, we may predict the wavelength of the emitted photon using Planck's relationship according to

$$E_{photon} = E_{initial} - E_{final} = h\nu_{photon}$$

Examining an incandescent light bulb, we have a thin filament of tungsten metal glowing white-hot and emitting light. The energy is supplied in the form of an electrical current passing through the filament. In doing so, the filament is raised in temperature to 3000 K or more (the atmosphere inside a light bulb must be void of oxygen or the filament would quickly burn). This thermal energy excites tungsten atoms in the filament to high energy levels, but nature favors low-energy states, so the atoms will not stay in these high-energy states indefinitely. Soon, excited atoms will lose their energy by emitting light and falling to lower-energy states. The amount of energy lost by the atom appears as light. The low-energy atoms are now free to absorb thermal energy and to begin the process again. As long as energy is supplied to the system, tungsten atoms will continue this process of absorption of energy (in this case, thermal energy) and emission of light. Not all energy emitted is

---

[2] As we shall see in Chapter 2, gases such as hydrogen exhibit well-defined energy levels. Demonstrations such as the Franck–Hertz experiment prove that such energy levels exist at discrete, well-defined levels.

in the form of visible light. The largest portion of the electrical energy injected into the system appears as heat and infrared light.

Various forms of light use various means of excitation. In a neon sign, for example, an electrical discharge (usually at high voltages) pumps atoms to high-energy states. Gases usually have well-defined energy levels (as for the example of hydrogen in Figure 1.8.2), so their emissions often appear as *line spectra*: a series of discrete emissions (lines) of well-defined wavelengths. In the case of a fluorescent tube, a gas discharge emits radiation that excites phosphor on the wall of the tube. The phosphor then emits visible light from the tube. We examine these sources in detail, in Chapter 2, as they also provide insight into quantization. Other common forms of light include chemical glow sticks (the type you bend to break an inner tube to start them glowing). A reaction between two chemicals in the tube (previously separated but now mixed) excites atoms to high-energy states. In this chapter we examine primarily *thermal light* in which excitation is brought about solely by temperature.

To reiterate, all atoms and molecules have energy levels such as those depicted in Figure 1.8.2, which shows the levels for a hydrogen atom. Light is produced when an atom at a higher energy level loses energy and falls to a lower level in what is called a *radiative transition* (so called because radiation is emitted in the transition). Two such transitions are shown on the figure, one emitting red light and the other emitting blue light. Energy in such a transition must be conserved, so the difference in energy between the upper and lower levels is released as a photon of light. The larger the transition, the more energy the photon will have (e.g., a blue photon has more energy than a red photon). As well as emitting photons, there are also *nonradiative transitions* in which energy is lost without emission of radiation (e.g., as heat). These do not contribute to radiation emission, so are omitted from this discussion.

Evident in Figure 1.8.2 is a level labeled the *ground state* of the atom. This is the lowest energy state that an atom or molecule can have. All atoms and molecules

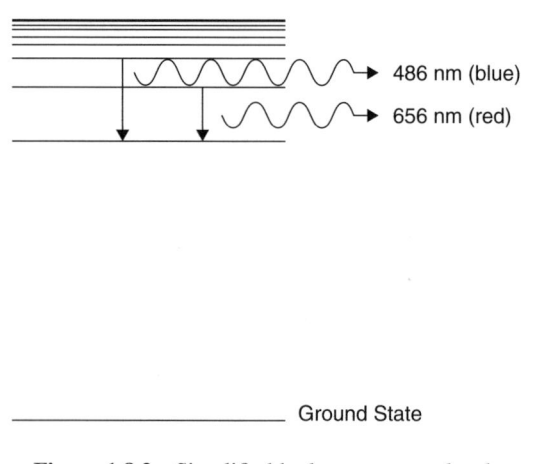

Figure 1.8.2 Simplified hydrogen energy levels.

assume this state at absolute zero (0 K), but frequently, the addition of energy (from electrical or thermal sources) causes the energy of the atom to rise to an upper level. The radiative transitions in Figure 1.8.2 are between two upper energy levels in the atom but could also be between an upper energy level and the ground state. In this case the transition would result in a higher-energy photon than those shown in the figure, and ultraviolet radiation would result.

Knowing where the energy levels are in a particular atom (and these are well known for essentially all atomic and molecular species), the problem now is to determine how many atoms are at a particular energy state. To do this, we use a thermodynamic concept called the *Boltzmann energy distribution*.

1.9 BOLTZMANN DISTRIBUTION AND THERMAL EQUILIBRIUM

Every system has thermal energy. In the case of a gas, consider gas atoms confined to a cylinder. Thermal energy manifests itself as atoms colliding with each other and bouncing off the cylinder walls. It is these collisions of gas atoms with the cylinder walls that are manifested as gas pressure. In a solid, the thermal energy manifests itself as vibrations of the atoms making up the solid; when the vibrations are too much for the interatomic forces keeping the atoms in place, the atoms are no longer tied to a particular spot and start flowing past one another; this is what happens when a solid melts.

Thermal energy also excites atoms (no matter what form they are in: gas, solid, or liquid), raising them to higher energy levels: The more thermal energy that is injected into a system (i.e., the higher the temperature), the more that higher energy levels will be populated. The resulting distribution of energy, describing the population of atoms at each energy level, is governed by Boltzmann's law, one of the fundamental laws of thermodynamics. Boltzmann's law predicts the population of atoms at a given energy level as follows:

$$N = N_0 \exp\left(-\frac{E}{kT}\right) \quad (1.9.1)$$

where N is the population of atoms at the given energy level, N_0 the population of atoms at ground state, E the energy above ground level in Joules, k is Boltzmann's constant (1.38×10^{-23} J/K), and T is the absolute temperature in Kelvin. This is to say that the law predicts an exponential drop in populations of atoms at higher energies. If we assume that a substance has an infinite number of energy levels (a continuum), the population at any given energy can be calculated using this law. The resulting distribution of energy for various energy levels is graphed in a generalized form in Figure 1.9.1.

We can rewrite Boltzmann's law as a ratio of N/N_0, allowing us to predict the distribution of atoms at any given energy level. Consider a generic substance at room temperature (293 K). In order to emit red light, we would need to excite atoms in the substance to 1.7 eV above ground (where an electron volt is equal to

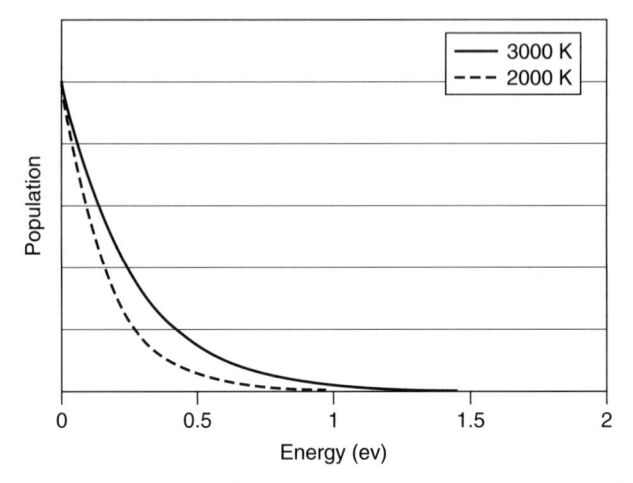

Figure 1.9.1 Generalized Boltzmann distribution of atomic energies.

1.602×10^{-19} J of energy). The number of atoms at this higher-energy state is on the order of 10^{-30}! It is fair to say that there are essentially no atoms with energy this high at room temperature (and this energy level isn't really very high at all; ions have much higher energies). Now, if the temperature is increased to, say, 5000 K, the population of atoms at this energy level increases to 2×10^{-2} or 2%, a sizable concentration at this level which will lead to the emission of light when these atoms lose their energy and return to ground state. The experience of observing red-hot and white-hot objects such as a stove element or glowing metal shows us that anything heated to beyond 1000 K will emit light in the form of a glow.

In a system such as we have described, it is assumed that there are no external sources of energy, so the population of atoms at any given energy level is governed solely by temperature. It is then said to be in *thermal equilibrium*. Examining the formula, we can see that if the temperature of the entire system were raised, the distribution would shift and more atoms would reach higher energies; however, the population of a lower level will always exceed that of a higher level. Higher temperatures result in increased populations of atoms at higher energy levels, and this results in larger energies. Light-producing transitions in such a substance will also have more energy, so more emitted photons from hotter objects will have more energy (and hence be shifted toward the blue region of the spectrum) than the photons emitted by cooler objects. It must also be noted that only at 0 K (absolute zero) will all atoms be at ground state (the lowest energy level).

1.10 QUANTUM VIEW OF BLACKBODY RADIATION

A blackbody absorbs all radiation incident upon it, and emission from such an object depends solely on the temperature of the body. At hotter temperatures it tends to

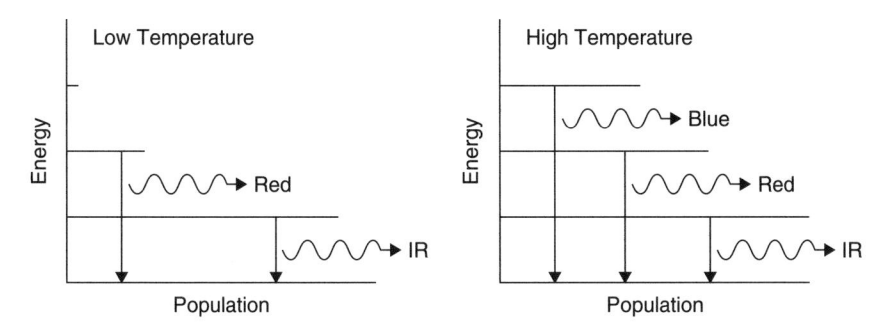

Figure 1.10.1 Energy-level populations at various temperatures.

have a higher blue content, and at lower temperatures the emission is seen as red. Examining the energy levels in this situation (as in Figure 1.10.1), we see that at low temperatures only low energy levels will be excited. The atoms in this case will have enough energy to cause transitions creating a reasonable amount of infrared light and some red light, but very few atoms will have enough energy to allow the production of blue light. As the temperature of the object is increased, higher atomic levels will be excited and blue emissions will be seen. This shift is due to the fact that blue light is literally more energetic than red light, as evident from the Planck relationship $E = h\nu$.

It may also occur to the reader that blackbody radiation is *broadband*. Unlike line spectra produced by most gas discharges, thermal light tends to occur in the form of a continuum of wavelengths spanning a large range. The answer is in the spacing of energy levels in the emitting substance. Gases usually have well-defined discrete wavelengths, due to the fact that each atom in the gas is ultimately identical and completely independent (i.e., no interactions between the atoms in the gas lead to changes in the discrete atomic energy levels). In solids, atoms are closely packed and interact with each other, which serves either to widen energy levels to the point where they overlap or create new electron energy levels. This leads us to speak instead of the energy levels in solids as *energy bands*. In some cases these bands are simply a discrete level that has been broadened (i.e., spans a range of possible energies) through various mechanisms. If the discrete energy levels in Figure 1.10.1 were replaced by wide bands of possible energies, it is easy to see that the average light emitted will be broadband.

1.11 BLACKBODIES AT VARIOUS TEMPERATURES

To get an idea of how blackbody radiation depends on temperature, consider blackbody radiation curves for an object at various temperatures, as shown in Figure 1.11.1. Consider first an extremely hot object at 6000 K. Although a great deal of radiation is emitted in the infrared region of the spectrum, the vast majority

16 LIGHT AND BLACKBODY EMISSION

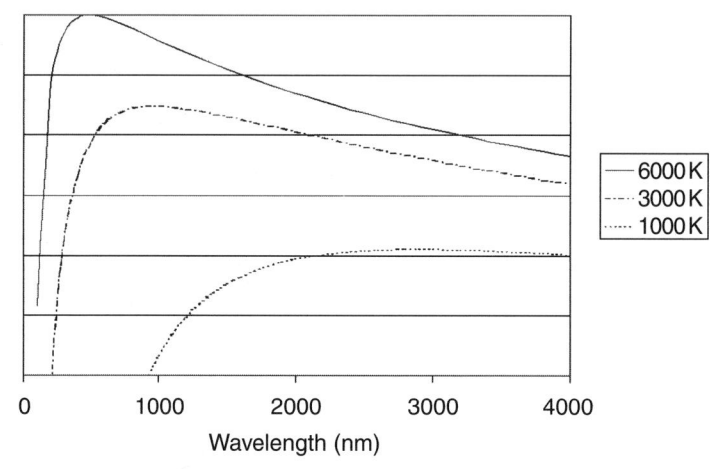

Figure 1.11.1 Blackbody radiation spectra.

lies in the visible range and more specifically, in the blue region of the visible spectrum. Such an object is said to have a *color temperature* of 6000 K and approximates that produced by sunlight. Many commercial lamps, especially those used by photographers, have a color temperature rating like this; it is simply a comparison to an equivalent output from a blackbody radiator.

Considering a somewhat cooler object at 3000 K, the object appears red when viewed since there is a great deal more red light emitted than blue light. Incandescent lamps often have a color temperature around 3500 K, and photographs taken under such light appear reddish yellow and require a blue correcting filter to have accurate color reproduction. This point is also illustrated in the differences between light from a regular incandescent lamp and a halogen lamp. In halogen lamps the filament burns in an atmosphere of halogen gas (e.g., iodine) which allows the filament to burn much hotter than in an ordinary incandescent light bulb without burning out prematurely. The increased temperature results in a shift from the reddish-yellow (warmer) light of an ordinary incandescent lamp to a much whiter light (i.e., containing more blue). Halogen lighting is therefore often used for illumination of objects where truer color rendition is required (e.g., display of artwork). As this object cools to, say, 1000 K, the emission of light shifts even further into the infrared, and only a small amount of red light is emitted (with almost no blue). Such an object is said to be "red hot" and the vast majority of its emission is in the infrared region between 2- and 6-μm wavelengths. The red we see is only a tiny fraction of the total radiation emitted.

Objects at room temperature (300 K) also emit radiation in the infrared centered at about 10 μm. In fact, even objects at a cold 3 K (the background temperature of the universe found in deep outer space) emit radiation at 3-mm wavelengths. Such long wavelengths are in the microwave region of the spectrum and can be detected using sensitive microwave receivers. Cosmologists often call this microwave background the "leftovers" from the "big bang" that created the universe.

1.12 APPLICATIONS

The fact that every object emits radiation can be exploited for a number of uses. For example, the human body at a temperature of about 312 K emits a large amount of infrared radiation centered about a wavelength of 10 μm. This is used for security and convenience lighting purposes by passive infrared (PIR) motion detectors such as that in Figure 1.12.1. Lenses focus radiation from the area under surveillance onto a sensitive pyroelectric detector that detects changes in the incident radiation. Pyroelectric detectors are made from ferroelectric crystals and work by allowing incident radiation to change the temperature of the crystal, which affects the charge on the electrodes. They have a wide response range of 1 to 100 μm and are quite rapid at detecting change. As a human (or any other warm object) walks across the area in front of the detector, the amount of radiation received changes suddenly, triggering a light or alarm.

It may be worth mentioning that clothing and other items normally opaque to visible light are often quite transparent to far-infrared light. To this end a commercially viable security camera is available which, quite literally, has the ability to see through clothing. In this age of tight airport security, this would make it quite impossible to conceal a weapon of any type, including those that do not affect metal detectors. Of course, the privacy ramifications have not yet been dealt with fully. This technology has already been used by border patrols to identify illegal aliens hiding in concealed compartments of trucks. The IR radiation emitted from the humans inside is easily detected by an IR camera, which clearly

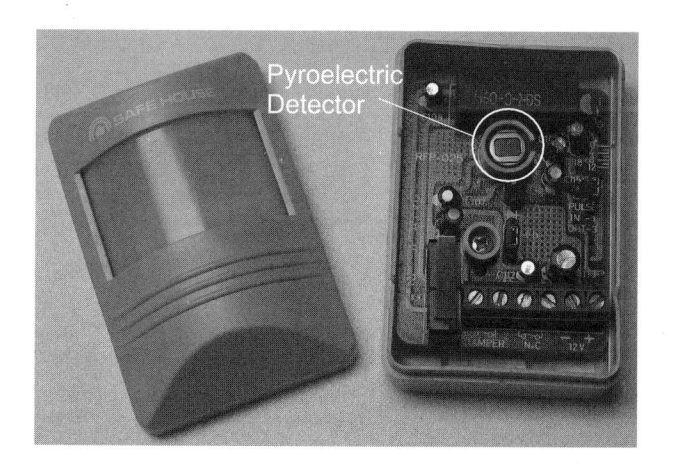

Figure 1.12.1 Passive infrared detector.

shows the outlines of the (warm) figures concealed inside. More mundane applications of blackbody radiation include noncontact thermometers used worldwide by the medical community. These in-the-ear thermometers measure the IR radiation emitted from the eardrum (which is inside the body and hence accurately reflects the body's core temperature) and calculate body temperature from this.

1.13 ABSORPTION AND COLOR

We have seen that atoms at low-energy states can absorb energy and that this energy can be light. If an atom at a low-energy state absorbs a photon with a particular amount of energy, that atom will reach a high-energy state with a final energy equal to its original energy plus the energy of the photon absorbed. This is why substances appear in certain colors. For example, a red liquid appears red when you look through it because red photons are not absorbed by the liquid as are other colors of light (such as blue, green, and yellow). The same is true of a blue liquid, in which blue photons are not absorbed but greens, yellows, and reds are. So why does the blue liquid absorb the lower-energy red photon yet allow the higher-energy blue photon through? The answer lies in the nature of absorption itself, in which the energy of an absorbed photon must match the energy of a transition: If energy levels in a particular atom exist 2.1 eV apart, a 2.1-eV photon is required to excite the atom from the lower level to the upper level. A photon with 1.7 eV of energy falls short of the energy needed to make the jump, whereas a 2.3-eV photon has too much energy.

Another way of looking at absorption is as the opposite of emission. This is expected in nature: If atoms can emit light to lose energy, they must be able to absorb it to gain energy. Because of the quantized nature of energy levels, it is possible to find atomic and molecular species that have energy levels allowing the absorption of photons of essentially any energy. In the case of an atom such as hydrogen, energy levels are specific and sharply defined; however, molecules have broad energy bands that allow absorption (or emission) over a wide spectrum of wavelengths. This is why liquids absorb a range of wavelengths (such as all red and orange light) instead of a specific wavelength such as a low-pressure gas would. This entire concept of atomic emission is dealt with in detail in Chapter 2.

1.14 EFFICIENCY OF LIGHT SOURCES

Since the majority of light sources (natural or artificial) are blackbody radiators, this discussion would not be complete without a discussion of efficiency of various sources. Most practical sources of light involve a heated medium, and as such, most of the radiation emitted is in the infrared region of the spectrum at wavelengths too long to be usable by the eye as light. Some sources are more efficient than others at emitting light at visible wavelengths. To compare light output by various sources, we use a unit called a *lumen*, which represents the power of light emitted by a source.

In terms of a practical light source, we measure efficiency as a function of light output to power input using the unit *lumens per watt*, which measures the number of lumens a light source emits for 1 W of input power. In theory, a perfect light source emitting at the peak sensitivity of the human eye could deliver 622 lm/W. Practical light sources fall far below this figure.

Consider an oil lamp in which kerosene is drawn up through a wick and burned. This is an incandescent system in which heat from the flame causes particles of carbon to burn brightly in the flame. Although kerosene contains a large amount of energy in a given volume, it burns inefficiently, with an efficiency of about 0.1 lm/W. The vast majority of energy in the kerosene is converted to heat rather than light. Electric incandescent lamps with a tungsten filament fare better, with an efficiency of up to 20 lm/W.[3] This still represents an enormous waste of energy, though, since over 90% of the emission from such a lamp is wasted as heat. Other artificial lamp technologies such as fluorescent lamps (discussed in Chapter 2), provide higher efficiencies of about 90 lm/W, and discharge lamps such as the sodium lamp (often used for street lights and easily identified by the yellow color of the light emitted) have the highest efficiencies, at about 150 lm/W.[3] Bear in mind that these last two technologies (fluorescent and sodium lamps) are not thermal light sources at all but atomic emission sources in which electrical energy is converted directly to light, not heat, as in the incandescent lamp.

PROBLEMS

1.1 Halogen lamps burn a tungsten filament in an atmosphere consisting of a halogen gas such as iodine or bromine. Explain why these lamps are (**a**) more efficient and (**b**) produce a whiter light than that of regular incandescent tungsten lamps. Specifically, obtain temperature figures (manufacturers' Web sites often list this) of these lamps and calculate the output wavelength and power for an incandescent and a halogen lamp.

1.2 Cool sources have a distinct red shift, whereas hotter sources appear yellow. Is there a temperature for an emission source at which all visible colors are equally balanced? If not, what is the optimal temperature at which output in the red and output in the blue are at the same level. At this optimal temperature, how much brighter would the yellow output be?

1.3 Passive infrared detectors (PIRs) use the blackbody emission from people to detect their presence. One problem with the use of such detectors in a security system is that pets such as cats can trigger these detectors. Knowing that a cat

---

[3] The values quoted for efficiencies of light sources are recent (2000) and represent midpoints for a given class of lamp. Efficiency of artificial light sources has improved drastically since Edison's first electric incandescent lamp of 1880, which boasted an efficiency of 1.6 lm/W! Today, the average modern incandescent lamp yields an efficiency of about 10 times that value. Other lamp technologies, including the fluorescent lamp, discovered in 1935, have also improved efficiency over the years.

has a warmer body temperature of 38.6°C, as opposed to a human being at 37.0°C, devise a "smart" sensor that can tell the difference. Would it be possible to measure the IR radiation at two discrete wavelengths to accomplish this?

CHAPTER 2

Atomic Emission

When considering thermal light in Chapter 1 we have seen that the spectra produced is a continuum peaking at a predictable wavelength. In contrast to this is the line spectrum exhibited by excited gases. The output in this case is not a continuum but a series of discrete, well-defined spectral components. The analysis of these components provides insight into the inner workings of the atom itself. Going further, we discover an intricate world of quantum mechanics where we see even more complexities in the simple atom involving interactions between electrons.

2.1 LINE SPECTRA

When a gas such as hydrogen is put under low pressure and excited electrically, it emits light. In analyzing the light emitted using a simple diffraction grating, one notices that this light is not a continuum but is actually composed of a series of discrete lines. In the case of hydrogen (chosen because it is the simplest atom, with only one electron), the visible spectrum (between 400 and 700 nm) is actually composed of five discrete lines at 656.3 nm in the red, 486.1 nm in the cyan, 434.1 nm in the blue, 410.2 nm in the violet, and 397.0 nm in the deep violet. This series of lines is called the *Balmer series* after the discoverer, J. J. Balmer, a Swiss secondary schoolteacher, who found the relationship between the wavelength of these lines and the square of an integer n as

$$\lambda = \frac{364.6 n^2}{n^2 - 4} \qquad (2.1.1)$$

where n is an integer with values 3, 4, 5, 6, ... The fact that this formula works for values of $n = 3$ and greater is of great importance, as will be evident as we progress through the chapter.

Further investigations showed that the reciprocal of the wavelength ($1/\lambda$) was a function of a constant (the Rydberg constant) and an integer n. Neither of these relations (Balmer or Rydberg) detail the mechanism of light emission in the atom

---

Fundamentals of Light Sources and Lasers, by Mark Csele
ISBN 0-471-47660-9   Copyright © 2004 John Wiley & Sons, Inc.

22 ATOMIC EMISSION

but rather, attempt to predict its behavior. To fully understand the mechanism of light emission, one must understand the basic structure of the atom as well as the nature of light itself.

2.2 SPECTROSCOPE

Spectroscopes are tools allowing the analysis of emissions such as those from gas discharge tubes. These instruments work by passing incident light through a prism or, more commonly, a diffraction grating to split it into its constituent spectral components. The angle at which the resulting components are diffracted is measured, allowing determination of the exact wavelength. The angle at which the component exits the grating is determined by its wavelength according to

$$\theta = \sin^{-1}\frac{m\lambda}{d} \qquad (2.2.1)$$

where θ is the angle at which the light is diffracted, m the order of the emission (assume that $m = 1$ here), and d the spacing between lines on the grating in meters.

Consider a simple spectroscope as in Figure 2.2.1, used to analyze spectra emitted by a gas discharge tube. Incident light passes through a small slit and is focused by a small telescope (on the left side of the unit), falling onto the diffraction grating, which separates the components. An eyepiece on another small telescope is rotated around the grating until a particular spectral component is found. The angle at which it was diffracted may then be read from the unit and the wavelength, in nan-

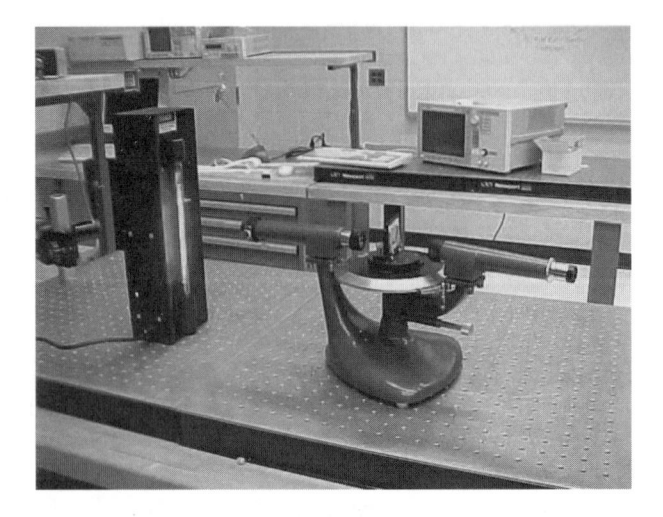

Figure 2.2.1 Grating spectroscope.

Figure 2.2.2 Spectroscope components.

ometers, calculated. Figure 2.2.2 shows a top view of the spectroscope, in which the diffraction grating and vernier scale (from where the angle is read) are seen.

Increasingly complex spectroscopes employ optical arrangements using concave mirrors to increase the dispersion of light diffracted by the grating to improve the resolution of the device (the ability to separate closely spaced lines). Some instruments have optical paths over 1 m long and can discern spectral lines as close as 0.005 nm or better! As well, spectrographs may use photographic film or sensitive photomultiplier tubes (which use the photoelectric effect discussed later in this chapter) to detect very weak emissions.

As mentioned earlier, the spectral discharge from a gas excited by a high voltage consists of a series of discrete lines. Figure 2.2.3 shows the spectrum for several common gases as viewed through a direct-reading spectroscope. By analyzing these lines and their origins, atomic structure of atoms and molecules can be determined. Figure 2.2.3 shows the major emission lines for hydrogen, mercury, and helium, and Table 2.2.1 gives exact wavelengths for each line. The spectroscope used for this analysis is direct reading, in which the lines are superimposed on an illuminated wavelength scale. The accuracy of the unit is about 5 nm.

Example 2.2.1 Identification of an Unknown Gas Using a Grating Spectroscope Consider a gas discharge in which three lines are visible in a spectroscope at 20.55, 23.70, and 16.45 degrees. In this particular spectroscope the angle can

TABLE 2.2.1 Atomic Spectra Wavelengths

Hydrogen	Mercury	Helium
410.2 nm (violet)	404.7 nm (violet)	447.1 nm (blue)
434.1 nm (blue)	407.8 nm (violet)	471.3 nm (blue)
486.1 nm (cyan)	435.8 nm (blue)	492.2 nm (cyan)
656.3 nm (red)	546.0 nm (green)	501.6 nm (green)
	577.0 nm (yellow)	587.6 nm (yellow)
	579.0 nm (yellow)	667.8 nm (red)

24 ATOMIC EMISSION

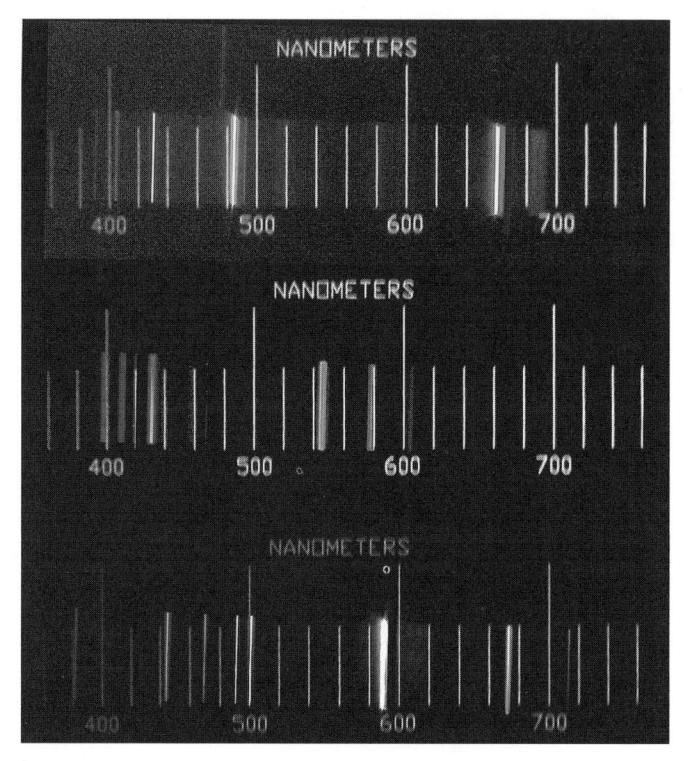

Figure 2.2.3 Gas emission spectra for hydrogen, mercury, and helium.

be read accurately using a vernier scale, shown in Figure 2.2.4. In the example illustrated in the figure, the angle is determined to be 19.8 degrees; 19 is read from the arrow and the eighth is read from the line on the vernier scale that matches that on the scale onto which the grating is mounted. Knowing that the grating is ruled at 600 lines/mm, determine the gas in the discharge tube.

SOLUTION To begin, determine the range of wavelengths possible for each line based on the accuracy with which the angle can be determined. Examining the figure, it is evident that the angle can be read to an estimated accuracy of 0.1 degree. For the line at 20.55 degrees, the actual angle could be ± 0.1 degree and so could be anywhere in the range 20.45 to 20.65 degrees. Knowing that the line spacing is $d = 1/600{,}000$ m^{-1} or 1.67×10^{-6} m, we can substitute each angle into a rearranged equation (2.2) to solve for wavelength. To determine the lower range of possible wavelengths for the unknown line:

$$\lambda = d \sin \theta$$
$$= 1.67 \times 10^{-6} \sin(20.55 - 0.1)$$
$$= 582.3 \times 10^{-9} \text{ m}$$

Figure 2.2.4 Vernier scale.

To determine the upper range of possible wavelengths for the unknown line:

$$\lambda = d \sin \theta$$
$$= 1.67 \times 10^{-6} \sin(20.55 + 0.1)$$
$$= 587.8 \times 10^{-9} \text{ m}$$

so that the possible wavelength range for this unknown line is 587.8 to 582.3 nm.

Similarly, we determine the possible wavelengths of the other two lines to be 672.6 to 667.2 nm and 474.7 to 469.2 nm, respectively. Matching these lines to the known spectra of several gas discharges, we determine that the gas is helium, since a known helium line falls within the range calculated for each line seen in the spectroscope. Of all the gases listed, only helium has a line between 587.8 and 582.3 nm (at 587.6 nm), a line between 474.7 and 469.2 nm (at 471.3 nm), and a line between 672.6 and 667.2 nm (at 667.8 nm). The wavelength of each of these three helium lines falls within the possible range determined for each line.

Emission lines from a gaseous discharge feature a relatively narrow spectral width, meaning that the line as viewed through a spectroscope is narrow. More formally, linewidth refers to the spread of wavelengths covered by the emission line. A spectrally narrow line spans few wavelengths, while a broad source (such as blackbody emission) spans a large range of wavelengths. *Spectral width* is defined numerically as the difference between the highest and lowest wavelength emitted,

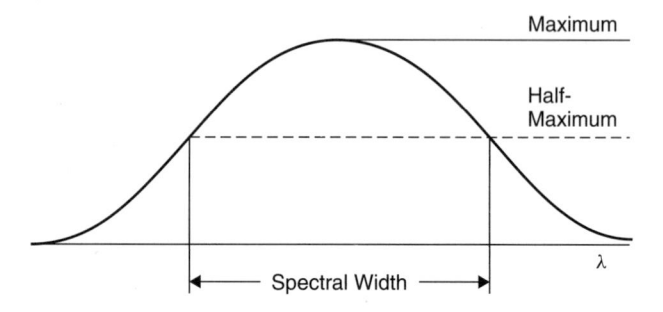

Figure 2.2.5 Definition of spectral width.

as located at the half-maximum intensity point of the output. This is called the *full-width half-maximum* (FWHM) of the output and is the standard way of measuring spectral width, as depicted in Figure 2.2.5.

2.3 EINSTEIN AND PLANCK: $E = h\nu$

Once again we reiterate the importance of Planck's quantization approach. As we have seen in Chapter 1, Planck's constant was originally devised to explain the spectra produced by blackbody radiators. It was actually Einstein who introduced the concept of the photon, based primarily on theory (he was unarguably the greatest theoretical physicist of all time). Photons can be thought of as literally a little packet of light and have energy proportional to their frequency and hence inversely proportional to their wavelength. The mathematical expression of this energy is

$$E = h\nu \qquad (2.3.1)$$

where ν is the frequency in hertz and h is Planck's constant. Substituting wavelength for frequency, we find that

$$E = \frac{hc}{\lambda}$$

where c is the speed of light and λ the wavelength in meters.

Returning to the line spectra of an atom, if one observes the wavelength of a line, the corresponding energy of the emitted photon in joules, or more conveniently in eV, with 1 eV $= 1.602 \times 10^{-19}$ J, may be computed using this relationship. If we now calculate the energy of 400-nm (violet) photons, we find an energy of 3.0 eV. Red photons at 700 nm have an energy of 1.8 eV. This concept of the photon has been confirmed by many experiments, including observations of the photoelectric effect, whereby incident light can cause electrons to be emitted from a metal surface in a vacuum.

2.4 PHOTOELECTRIC EFFECT

The photoelectric effect provides proof of the relationship between energy and frequency as brought to light in the Planck relationship. It also demonstrates the particle nature of light. The effect is observed when photons of light strike a metal surface in a vacuum and electrons are ejected in response to bombardment by these incident photons. This is the principle by which phototubes work: Incident light causes ejection of electrons, which manifests itself as current flow in the tube.

As light of a given wavelength (or frequency) strikes the metal, photoelectrons are ejected with various energies; however, it was found that the maximum kinetic energy of the photoelectrons ejected was sharply defined and was dependent solely on wavelength, not on the intensity of the incident beam of light. More important for our purposes of illustrating the photon concept was the discovery that there is a minimum frequency for incident photons below which no photoelectrons are ejected at all; this minimum frequency depends on the metal used. For potassium, a common metal used for photocathodes, incident photons must have energy of at least 2.0 eV to cause photoelectron ejection (and hence current flow). Photons of red light have energies below 2.0 eV and hence will not cause current flow, regardless of intensity, whereas photons of blue light (which have energies of about 3 eV) cause current to flow in the tube, as illustrated in Figure 2.4.1.

Classical wave theory does not allow an explanation of the photoelectric effect. One might argue that incident waves of light perturb the electrons in the metal, causing ejection. It would be expected, then, that the maximum kinetic energy of the photoelectrons ejected would depend on the intensity of the incoming beam, which it does not. Classical wave theory also cannot account for the minimum frequency required for photoelectron emission. On the other hand, Einstein's photon concept explains this effect as the simple absorption of a photon by an individual electron in an atom of metal. Some electrons simply absorb enough energy to escape the surface of the metal. The threshold energy required to observe this is also

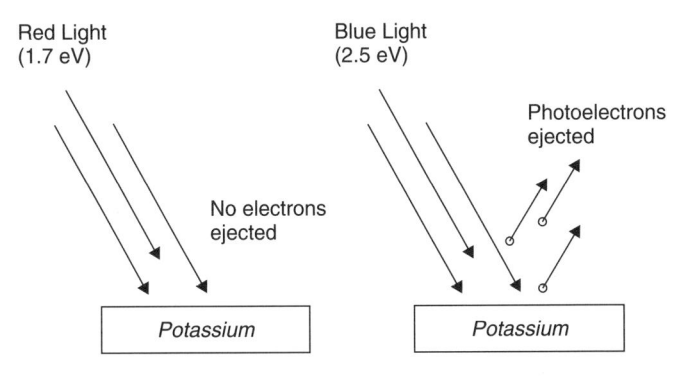

Figure 2.4.1 Photoelectric effect.

28 ATOMIC EMISSION

explained by the fact that electrons will be bound to their atoms of metal by some energy that must be overcome to allow escape. If incident photons of light lack this minimum energy, they can still be absorbed by the atoms of metal but will not give the electrons enough energy to escape their bonds and so will not be ejected.

Similar evidence[1] exists in a host of other experiments, showing the photon concept to be a reasonable explanation of effects observed that cannot be explained through the use of classical wave mechanics. In one such experiment, illustrating the Compton effect, x-ray photons are beamed at a block of paraffin, where they collide with electrons. The resulting collision follows the laws of classical dynamics, with the photon losing energy in the collision, as evidenced by the shift toward a longer wavelength (i.e., it gives energy to the electron during the collision). In this case, the x-ray photons behave exactly like little billiard balls.

2.5 ATOMIC MODELS AND LIGHT EMISSION

The Rutherford–Bohr atomic model was a huge step forward in explaining the mechanism of light emission and was based on experimental evidence. Rutherford performed experiments in which alpha particles (literally, helium nuclei consisting only of neutrons and protons) were scattered by atomic collisions. This demonstrated that atoms have a dense core (the nucleus) where positive charge is concentrated. On the wings of this discovery, Neils Bohr completed the model (called the *Rutherford–Bohr model*), which accounted for the emission spectra seen in hydrogen and other gases. The model postulates that electrons around an atom orbit in a number of possible energy states. In these discrete (allowed) states, electrons orbit the nucleus of the atom according to the laws of Newtonian mechanics. The angular momentum of these orbiting electrons is quantized and limited to a set of values, integer values called the *quantum number*, which represents which orbit the electron is in. Atoms do not radiate energy as long as they are fixed in that orbit. This application of Newtonian mechanics is not surprising, as it is analogous to the moon in high orbit around Earth. The only unusual part of this application was the fact that these orbits could take on only certain limited values (i.e., the fact that they were indeed quantized). The larger the quantum number n, the larger the orbit.

Bohr postulated that the radius of an orbit was given by

$$r = n^2 a_0 \qquad (2.5.1)$$

---

[1] Another demonstration of the photon effect is in the operation of x-ray tubes, where a metal target in a vacuum is struck by high-energy electrons accelerated by a high voltage. Analysis of the spectrum of x-rays produced shows a sharp cutoff at a maximum photon energy (shortest wavelength), regardless of the metal used. The shortest wavelength emitted is dependent solely on the maximum kinetic energy of the high-energy electrons (controlled through the accelerating voltage). It cannot be explained using classical wave theory; however, photon theory shows that the maximum photon energy (and hence shortest wavelength) would be produced by electrons of maximum kinetic energy.

ATOMIC MODELS AND LIGHT EMISSION   29

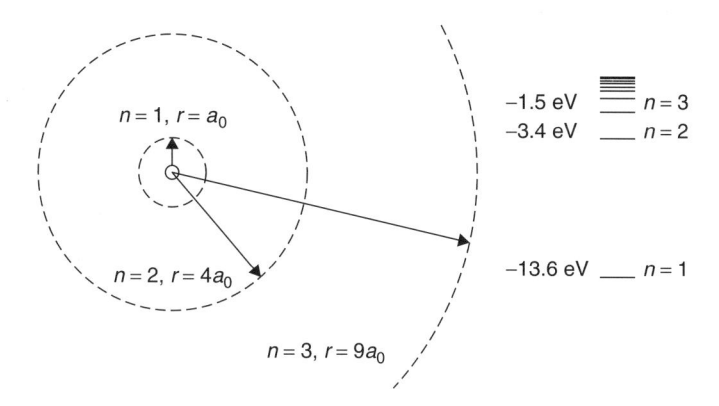

Figure 2.5.1 Hydrogen orbits and corresponding energy levels.

where n is the quantum number and a_0 is the Bohr radius (5.29×10^{-11} m). The Bohr radius (a_0) is a figure that represents the radius of the smallest orbit. In practical terms, it is the highest energy that a bound electron (one orbiting around the nucleus) can have, which occurs at its closest approach to the nucleus. If we now consider an electron of very small mass to be orbiting a proton of much larger mass and use Coulomb's law to determine the energy of the electron (which has a negative charge), classical kinematics allows calculation of the radii of orbits in the hydrogen atom. Newtonian mechanics can be applied to the orbit to render a figure for the energy of an electron in that particular orbit:

$$E = -\frac{13.6}{n^2}$$

where E is the energy in eV. The energy is always negative since the frame of reference is that an unbound electron, one free of the hydrogen nucleus (but with no extra kinetic energy), has zero energy. When the energy levels for various quantum numbers are calculated, we find that electrons in the first Bohr orbit have an energy of -13.6 eV. As the quantum number increases, these levels get crowded closer together, as outlined in Figure 2.5.1, which shows for comparison the radius of the first three orbits of a hydrogen electron, as well as defined energy levels for the hydrogen atom.

When an energy greater that 13.6 eV is supplied to a hydrogen atom in the ground state (i.e., the electron is in the first Bohr orbit, with $n = 1$), the electron will acquire so much energy that it is ejected from the atom altogether. The atom now becomes an ion and has an overall positive charge (the atom has become *ionized*). This is, incidentally, how the energy for the ground state (-13.6 eV) is determined.[2] Having defined energy levels in the atom, Bohr stated further that

---

[2] A more thorough treatment of Bohr's model and the ramifications of this model in terms of reconciliation of classical and quantum theory may be found in *An Introduction to Quantum Physics* by A. P. French and E. F. Taylor (New York: W.W. Norton, 1978).

atoms may jump from one energy state to another (a quantum jump), corresponding to a change in orbit, and in doing so will emit radiation in the form of a photon. The photon will contain the energy difference between the initial, higher-energy state and the final, lower-energy state. This ensures conservation of energy, one of the founding laws of physics. The total energy of the process must remain unchanged, so that

$$E_{\text{photon}} = E_{\text{upper level}} - E_{\text{lower level}}$$

Recall that the emission spectrum of hydrogen consists of discrete lines, each of which can now be explained as corresponding with a jump or transition in the energy states of the atom (a change in orbit, if you will). In the case of the *Balmer series*, each line corresponds to an energy-level change which results in the final energy state being $n = 2$ (the *second orbit*). Higher-energy photons result from larger jumps, such as the violet line, which results from transitions from very high orbital states to the final $n = 2$ state. The red line has relatively little energy; it results from a relatively small jump between the $n = 3$ and $n = 2$ orbital states.

The five visible lines of hydrogen may now be seen as a jump between a higher-energy state with $n = 3$, 4, 5, 6, or 7 and a lower-energy state with $n = 2$. Mathematically, this can be determined by calculating the energy for each Bohr orbit and then the energy lost in a downward quantum jump. For example, consider the transition from $n = 3$ to $n = 2$ as depicted in Figure 2.5.2. The energy associated with each orbit is calculated to be -1.511 and -3.400 eV, respectively. When the transition occurs, the energy of the liberated photon will be equal to the difference in energy between these two levels, or 1.889 eV. This corresponds to a photon at a wavelength of 656.3 nm in the red. Applying the same calculation to the $n = 4$ to $n = 2$ transition yields a photon of 486.1 nm. In the latter case the jump is larger, so the energy lost to the photon is also larger.

Transitions between energy levels do not have to end at $n = 2$ but can end at almost any level, such as $n = 1$ (the *first orbit*). These transitions have very high energies, being so close to the nucleus (from kinematics we know that low orbits close to the

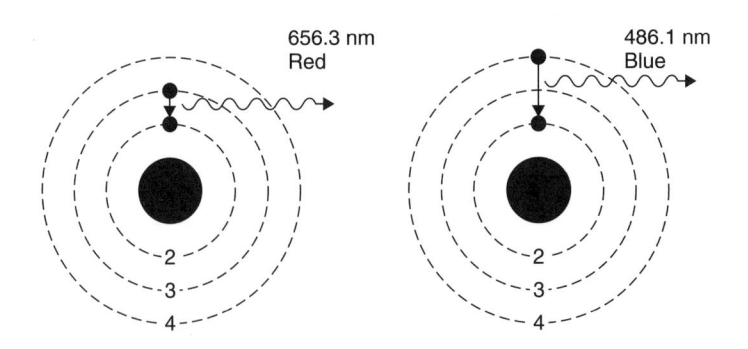

Figure 2.5.2 Photon emission in hydrogen.

center of mass have higher energies), so photons emitted in transitions to this final state of $n = 1$ also have high energies. These lines are observed in the ultraviolet and are called the *Lyman series* of lines. Other transitions end with $n = 3$ and have lower-energy changes, so photons emitted are in the infrared (the *Paschen series*).

2.6 FRANCK–HERTZ EXPERIMENT

The Franck–Hertz experiment demonstrates the fact that energy levels in atoms are indeed quantized into discrete levels. We have assumed this right from the start, beginning with the process of emission of light, and it is a foundation for the quantum approach to explaining light emission. It is considered to be one of the great experiments in physics.

Although the original experiment by Franck and Hertz used mercury vapor,[3] it is more convenient to use neon, since it provides visible evidence of these levels. The experiment, shown in Figure 2.6.1, consists of a gas-filled vacuum tube (i.e., gas at low pressure) with a heated cathode that emits electrons into the gas. These electrons are accelerated toward a grid at a more positive potential than the cathode. This potential is adjustable, allowing the experimenter to give accelerated electrons a specific energy. Electrons then pass through the grid and are collected at the anode, where they show up as current (electron flow is, by definition, electric current).

As the voltage between the cathode and the grid is increased, electrons emitted from the hot cathode are accelerated through the gas and toward the grid. As expected, a rising current is seen at the anode, which collects electrons traveling through the tube. When the accelerating voltage reaches about 18.7 V, the anode current suddenly drops. As the accelerating voltage increases, the current begins to rise again, until the voltage reaches 37.4 V, where another drop occurs. When current through the gas is plotted against accelerating potential, a series of dips in collector current occur at periodic intervals, as shown in Figure 2.6.2. Dips (with corresponding peaks) in collector current are seen at intervals of 18.7 V, as expected. Even more interesting are bands of light and dark that appear in the tube, as evident in Figure 2.6.3. At an accelerating voltage of about 40 V, two bands of light appear as evident in part (*a*), and when the accelerating voltage is about 60 V, three bands of light appear, as shown in part (*b*).

Examining the situation, we find that by definition the accelerated electrons have an energy of 18.7 eV at the point where the first dip occurs, an electron volt being defined as the energy an electron has when accelerated across a potential of 1 V.

---

[3] The original experiment by Franck and Hertz used mercury vapor. The problem with mercury vapor is that emitted photons from the first transition have an energy of 4.9 eV, corresponding to a wavelength of 253.7 nm. This is in the UV region, where normal glass does not transmit. The original experimenters did not have visual evidence of the quantization effect; only the behavior of anode current as grid potential is varied. Later experiments by Gustav Hertz revealed the accompanying photon emission as well as the fact that as potential is increased, spectral emission lines of mercury appear not in numeric order but rather, based on the energy of the upper level of the transition aboveground.

32 ATOMIC EMISSION

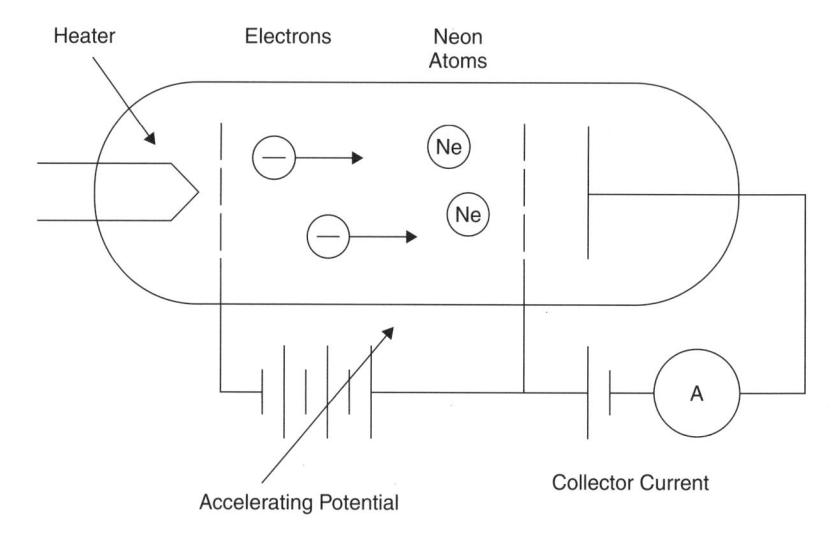

Figure 2.6.1 Franck–Hertz experiment setup.

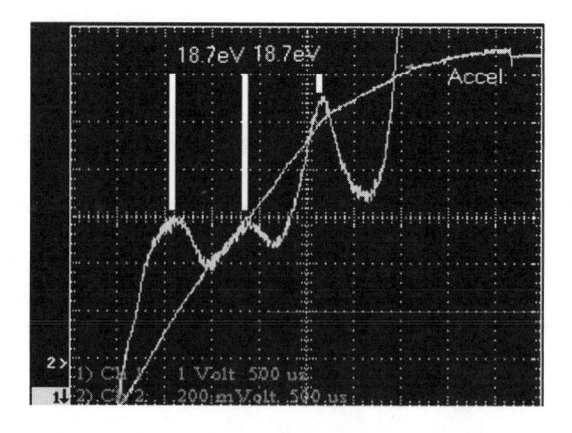

Figure 2.6.2 Franck–Hertz observations: collector current.

This dip corresponds to a known energy level in the neon atom, the first such level above ground.[4] The interpretation of this experiment is that electrons below 18.7 eV do not have enough energy to excite the neon atom to an allowed energy and so pass

---

[4] In reality the situation is more complex than this and involves a number of concepts that we examine in Chapter 3. The actual first level is located around 16.6 eV, but this level has a small cross section (i.e., a low probability of being excited), so the actual first level to absorb energy from the electrons is at about 18.7 eV. Neon atoms at the 18.7-eV level fall to the 16.6-eV level, emitting orange photons. From the 16.6-eV level they fall back to ground by emitting an (unobservable) UV photon. To complicate matters further, these two levels, centered around 18.7 and 16.6 eV, are actually multiple closely spaced energy levels, resulting in photon emission on a number of wavelengths in the red, yellow, orange, and green regions of the spectrum.

(a)   (b)

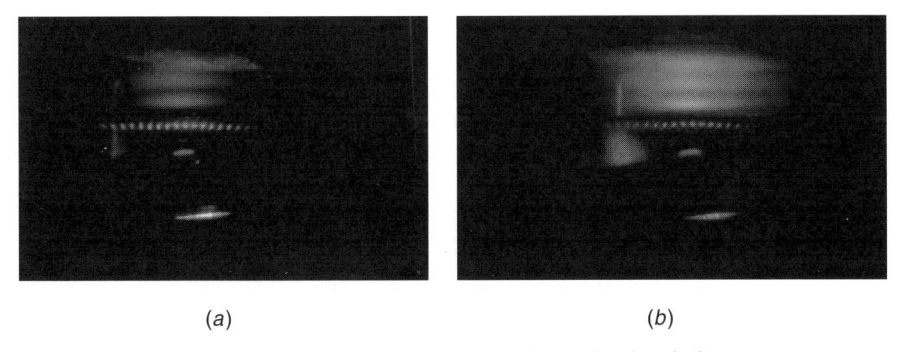

Figure 2.6.3 Franck–Hertz observations: visual emissions.

through the gas unimpeded, being manifested as current through the tube (i.e., electron flow). At 18.7 eV the electrons have enough energy to excite neon atoms to their first excited level upon impact. From basic kinematics we see this as the inelastic collision of the lightweight, excited electron with the more massive neon atom. The result of the collision is that the energy of the electron is totally absorbed by the much heavier atom, pushing the neon atom's energy to the first excited state. The surprising part is that the energy of the neon atom is quantized—it can only take on certain allowed values. Electrons whose energy is below 18.7 eV cannot transfer energy to the neon atom to excite it. The corresponding drop in current occurs because electrons at that energy are no longer flowing though the tube to the anode (i.e., current flow) but rather, are transferring their energy to neon atoms (where they show up as emitted light).

It is evident from Figure 2.6.2 that when the accelerating voltage is ramped from zero to 70 V, three dips in current occur. Electrons emitted from the heater (filament) travel down the tube and acquire energy. Until they reach 18.7 eV, they cannot transfer energy to the abundant neon atoms and so travel unimpeded. When the electrons have 18.7 eV of energy, a good number of them will collide with neon atoms and transfer energy to them. These neon atoms will acquire the energy but will eventually lose it by emitting a photon and returning to ground level (this is actually a two-step process in neon). Electrons, now depleted in energy by the collision, will continue down the tube acquiring more energy until they again reach 18.7 eV, at which point they can again collide with neon atoms and transfer energy to them. The process continues until the electrons reach the anode. With a 70-V accelerating potential, we would expect three dips as well as three corresponding areas in the tube where neon atoms emit light, as shown in Figure 2.6.3(b).

In Figure 2.6.3, electrons in the dark space just above the lower grid are accelerating upward away from that grid toward the top of the tube. They lack sufficient energy to excite neon's energy levels and so retain their energy as they travel down the tube. Eventually, moving upward as they acquired energy, they possess enough energy to excite neon's first energy level, so collisions with neon atoms will result in energy transfer. Excited neon atoms emit light (the lowest band of orange light between

34 ATOMIC EMISSION

the grids in the photo) and in doing so lose energy. Still moving upward, we see a lack of light emission as low-energy electrons travel through, acquiring more energy due to the applied potential. By the time the electrons reach the area of the second band of light, they have again acquired the necessary energy to excite neon's first level, so light emission occurs again as electrons transfer their energy to neon atoms in this region. In Figure 2.6.3(b) the process repeats again, with electrons acquiring energy a third time to transfer to neon atoms by collisions (the third band of light), until they finally reach the top grid and stop accelerating.

The Franck–Hertz experiment demonstrates that atoms do indeed have quantized, discrete energy levels. These levels define the excitation of the atom. An atom that is not excited (and hence is in its lowest-energy configuration) is said to be at *ground state*. As energy is absorbed by the atom, it is elevated to higher-energy states. In the case of neon, the first excited state is observed to be at about 18.7 eV[4]. For hydrogen, the first excited state is observed to be at 3.4 eV above ground state.

2.7 SPONTANEOUS EMISSION AND LEVEL LIFETIME

An atom at an excited state will eventually drop to a lower level and in doing so will emit a photon of radiation in a process called *spontaneous emission*. Excited electrons will not stay at the excited level forever since nature favors a low energy level and so will emit the photon spontaneously after an average time of τ_{sp} called the *spontaneous lifetime* of the level. This time depends on the atomic species involved and can be measured for a given species; some levels have long lifetimes measured in seconds, whereas others are relatively short, on the order of nanoseconds or less. This lifetime determines the ability of the emitting atom to store energy and will affect the efficiency of sources. In lasers, it also factors prominently in determining the probability that laser action can be coaxed from a particular atomic species. Consider Figure 2.7.1, in which two atoms with different spontaneous lifetimes are excited at a start time $t = 0$. The top atom, with a relatively short lifetime, emits a

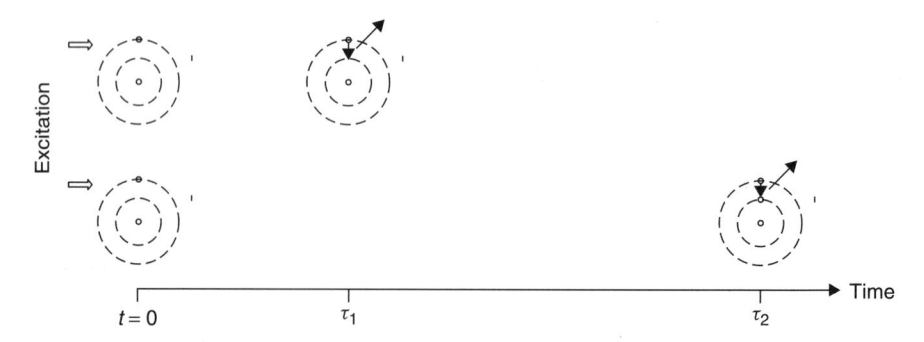

Figure 2.7.1 Spontaneous lifetime.

photon spontaneously at a time $t = \tau_1$, while the second atom, with a longer lifetime, waits until an elapsed time of $t = \tau_2$ before emitting a photon.

As mentioned in Chapter 1, radiation is not *always* emitted from a transition. An electron can lose energy by colliding with tube walls. Such *nonradiative* events are rare, though, in a gas at low pressure (such as that in a discharge tube or neon sign), so we shall not discuss them here. This nonradiative mechanism is important, however, as it is used to depopulate certain energy levels in lasers (including the common helium–neon laser) and so will be dealt with in a subsequent chapter. In addition to losing energy in collisions (primarily in a gas, where atoms are treated essentially as independent entities), electrons that are trapped in a crystal lattice can lose energy by causing vibrations in the crystal. A vibration resulting from a transition, called a *phonon*, produces heat in the crystal and occurs commonly in semiconductor materials. This concept, too, is discussed subsequently, in Section 2.9.

2.8 FLUORESCENCE

Fluorescence is the absorption of radiation at one wavelength and emission at a different wavelength. In many cases it is used to convert otherwise useless ultraviolet emissions into useful visible light, as is the case with a fluorescent tube. Fluorescent tubes contain an inert gas with a small amount of mercury in a glass tube. The inert gas (such as argon) at low pressure allows an electrical discharge (started with the help of a glowing filament, which emits electrons) to be sustained through the tube, which in turn excites mercury atoms in the tube by electron collisions. Mercury then emits light; however, most is emitted in the ultraviolet, where it is invisible. A coating of phosphor on the inside wall of the glass tube absorbs UV radiation and reemits this energy in the visible portion of the spectrum as white light. Without this coating of phosphor, the discharge appears to be the characteristic dull blue color characteristic of a mercury discharge, and the UV radiation is simply absorbed by the glass tube and hence wasted. The entire process is depicted in Figure 2.8.1. By changing the chemical composition of the phosphor,

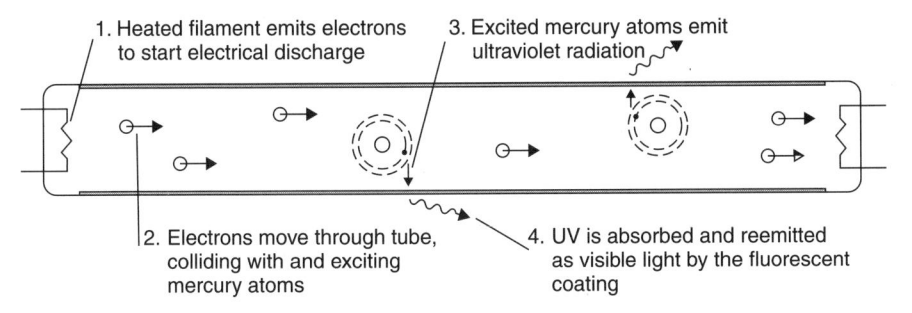

Figure 2.8.1 Fluorescent tube.

36 ATOMIC EMISSION

the emission spectrum of the tube can be controlled. Cool white tubes have a phosphor that emits light primarily in the blue-to-yellow region of the spectrum; warm-white tubes emit light rich in the orange and red region of the spectrum.

The fluorescence process is perhaps most evident in colored neon sign tubes. A pure neon tube containing neon gas glows bright orange, but many modern sign tubes glow blue, pink, green, and many other colors. In most tubes the color is provided by a phosphor coating on the inside of the tube, as evident in Figure 2.8.2(*b*). The actual discharge itself is in mercury vapor in a buffer gas of argon. This is the same as a fluorescent tube used for lighting purposes except that neon sign tubes lack a filament to assist starting—instead, using high voltages (15 kV is standard for many sign transformers). This discharge in low-pressure mercury vapor produces a large quantity of emission in the ultraviolet. A phosphor coating on the inside of the tube absorbs this UV emission and emits light in the visible spectrum. Phosphors can be formulated to emit light of essentially any color.

Fluorescence is also used in vacuum fluorescent displays (VFDs), a popular display technology used on some VCRs and DVD players. In these displays, electrons emitted from a hot filament are accelerated by a relatively low positive voltage of 10 to 15 V through a grid mesh. These electrons then strike anodes which are in the shape of characters or digits. Each anode segment is coated with a phosphor that glows when hit by these electrons. There are numerous anodes, which form segments or dots. When an anode is supplied with a positive voltage relative to the filament (which also serves as a cathode), it will attract the electrons that have been accelerated through the grid. The segment emits light when these electrons strike the phosphor coating. To turn off a segment, anodes are supplied with a negative voltage so that they will repel electrons from their phosphor coating and therefore remain dark. A display of this type is outlined in Figure 2.8.3. In the figure, both (*a*) the elements of a VFD and (*b*) a working display are shown. The filament is visible as well as three lit segments. The filament is overdriven in this case, making it obvious in front of the glowing anodes. In a typical display the filament is heated only to the point where it is barely visible (this provides an adequate source of elec-

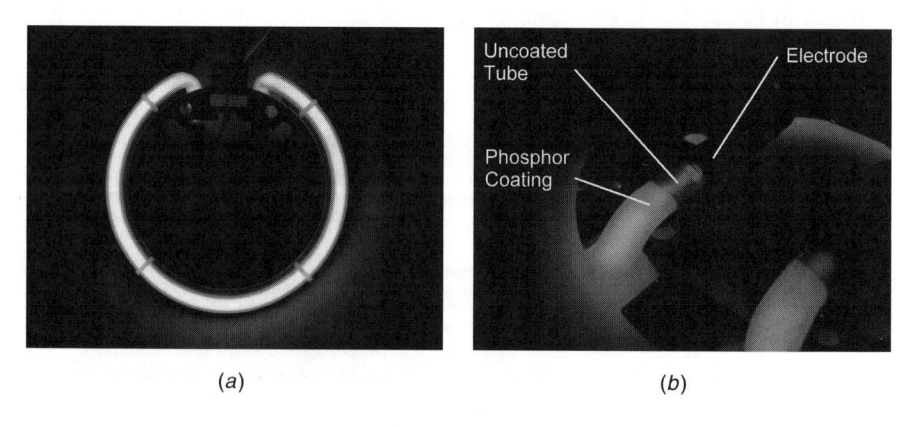

Figure 2.8.2 Neon sign with phosphor coating.

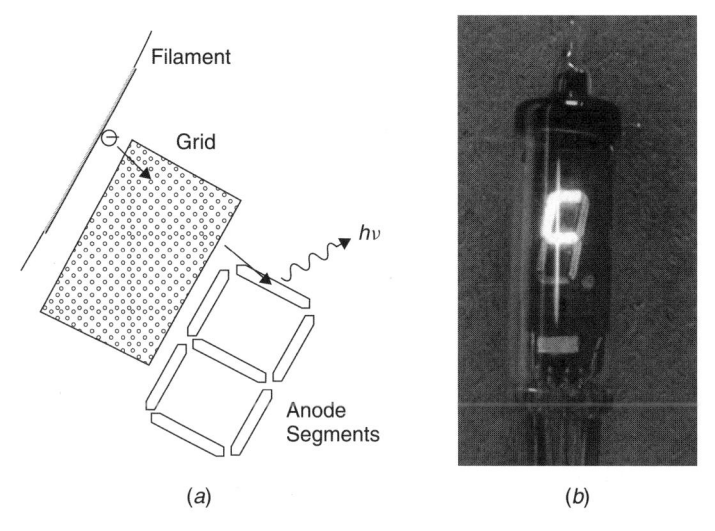

Figure 2.8.3 Elements and photo of a vacuum fluorescent display.

trons to light the anodes) and so is not usually visible, as it is here. Multicolor displays are also possible using this technology; various anodes can be coated with different phosphors to emit light of a desired color.

Aside from the obvious applications in fluorescent lamps (including neon tubes) and VFD displays, fluorescence is also used in television picture tubes, where high-energy electrons are accelerated from an electron gun in the rear of the tube toward the positively charged front of the tube to strike a phosphor coating, producing light emission. In the case of color television, three phosphors are used, to produce red, green, and blue light, which are mixed to produce any color of light desired, including white.

2.9 SEMICONDUCTOR DEVICES

No discussion of light sources would be complete without mentioning semiconductor light-emitting diodes (LEDs). Given their low power consumption and extremely long life, these devices are found in applications ranging from power indicators and alarm clock displays to Christmas tree and traffic lights. We begin by briefly examining the fundamentals of semiconductor devices.[5] Unlike atoms such as hydrogen or mercury, which in a gas form are free and do not interact much with each other, solids do not have discrete, sharply defined energy levels but rather, have energy bands instead. These bands are, in reality, a huge number of closely spaced energy

---

[5] A refreshingly clear explanation of semiconductors and semiconductor devices may be found in *Photonics Essentials* by Thomas Pearsall (New York: McGraw-Hill, 2003).

38 ATOMIC EMISSION

levels, which in many cases overlap to form a continuum of energy. Semiconductors such as silicon or gallium arsenide form crystals in which each atom bonds to neighboring atoms. Interactions between adjacent atoms lead to a splitting of the energy levels of such materials into two distinct bands, the valence band and the conduction band. These two bands are simply energy bands with no energy levels in between (analogous to any other atom that we have examined in this chapter where energies are quantized into distinct levels and cannot assume values between these levels). The *valence band* is the highest-energy band for electrons bound into the atoms of the solid. It is a band fully occupied by the valence (outermost) electrons of the composite atoms. The *conduction band* describes energies for electrons that are free of the atom (as we shall see in Chapter 3, there are constraints on the possible energy states of an electron in any band dictated by the principles of quantum mechanics). A free electron (i.e., one in the conduction band) may move about in the solid and hence may carry current which is, by definition, electron flow.

The ability of a material to conduct electrical current depends on the number of electrons in the conduction band. In a conductor (i.e., a metal such as copper), the valence band is fully populated; in addition, the conduction band is always partially populated by electrons, so these materials conduct current freely. In a perfect insulator (such as glass) the valence band is fully populated and the conduction band void of electrons, with no electrons able to move about in the material current cannot flow. In these materials the energy gap between the valence and conduction band is quite large and cannot easily be jumped by electrons in the valence band, so no free electrons exist in the conduction band to conduct current. A semiconductor resembles an insulator in that normally the valence band is full and the conduction band empty; however, the gap between the bands is relatively small compared to an insulator. With the addition of energy, including thermal energy, electrons may be made to jump the gap between the bands, at which point the free electron is mobile

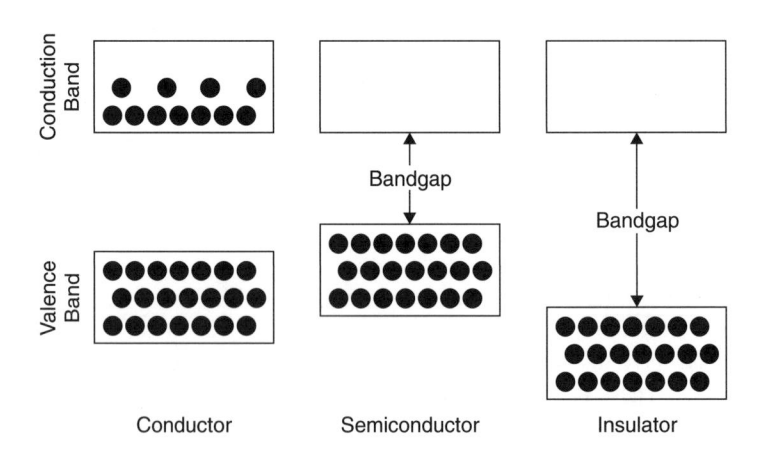

Figure 2.9.1 Electron energy levels in various materials.

and the material can carry current. The energy levels of electrons in each type of material are outlined in Figure 2.9.1.

At its lowest energy state (i.e., with no energy applied and at a temperature of absolute zero) a semiconductor material has all electrons in the valence band, so there are no free electrons. In this state the material does not conduct electrical current (since there are no free electrons to do so) and thus acts like a perfect insulator. To conduct electric current, some electrons must be made to jump from the valence band to the conduction band, where, by definition, electrons are free to travel through the crystal (and hence current flows through the semiconductor). The energy that must be injected into the semiconductor to cause an electron to jump the gap between the valence band (where the electron is bound inside the atom) and the conduction band (where the electron is unbound) is called the *bandgap energy* (denoted E_g). The bandgap literally represents the energy difference between the levels of the two bands.

When enough energy is absorbed by the atom (either by thermal excitation or perhaps by absorption of an incident photon), electrons may move to the next-highest band, the conduction band. As temperature rises, then, we expect a semiconductor material to become more conductive of electric current. The minimum energy to cause this jump is equal to the bandgap energy E_g, so in the case of a photon as the incident excitation source, the energy of the photon ($h\nu$) must be at least equal to E_g. When an electron jumps the gap from the valence band, it creates a vacancy or *hole* in the valence band of the atom (i.e., it is devoid of an electron normally there). A hole, as one might imagine, is a positive charge relatively speaking, since it is the absence of a negative charge that would normally be there to give a net charge of zero. The process results in the production of an electron–hole pair, shown in Figure 2.9.2. The free electron, with energy in the conduction band, may wander through the crystal and eventually fill a hole in a valence band. In doing so, energy of the system must be conserved. This *recombination* process, called *electron–hole*

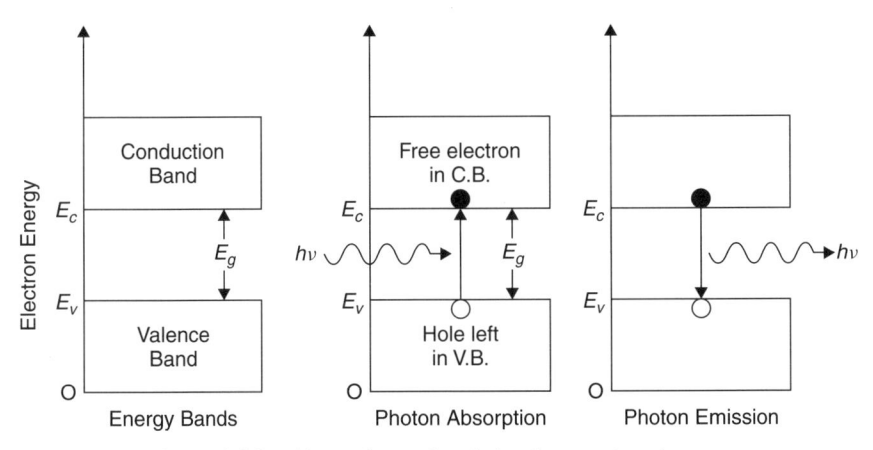

Figure 2.9.2 Absorption and emission in a semiconductor.

40 ATOMIC EMISSION

pair (EHP) *recombination*, results in either emission of a photon or production of heat in the crystal.

The characteristics of the energy bands in a given semiconductor material may be modified by *doping* the material when it is grown. Silicon, in its purest form (called an *intrinsic semiconductor*), is not a good conductor of current. There are just enough electrons in the crystal to fill the valence band entirely, with none left for the conduction band, so it is almost unoccupied. For a given temperature, Boltzmann statistics can be used to predict the population of electrons in a given energy state (e.g., the conduction band), so even at an elevated temperature it will be found that few electrons are promoted to the conduction band. By adding impurities to the silicon (a process called *doping*) it can be made to have an excess of electrons which will then populate the conduction band. Adding dopants such as phosphorus or arsenic creates a type of silicon with excess electrons (in the conduction band) called *n-type* silicon. These dopants feature five valence electrons: Four bond with neighboring silicon atoms, leaving one "spare" electron. The spare electron is easily promoted to the conduction band (usually, by thermal energy), so the addition of an impurity such as this makes the silicon more conductive, since conductivity depends on the availability of charge carriers (in this case, electrons). This material will be unbalanced such that it will have more electrons than holes. In a similar way, the addition of other dopants, such as boron, make the silicon *p-type*, in which case it has a lack of electrons in the valence band and can accept extra electrons; in other words, extra holes are created in the valence band. This dopant has only three electrons, so bonds only to three neighboring silicon atoms, leaving one "hole" in the lattice (an unbonded electron in the silicon host material). Typical doping levels for either p- or n-type materials are small and usually on the order of 0.0001%. Electrons in the bands of each type of semiconductor are depicted in Figure 2.9.3.

When n- and p-type semiconductors are brought into contact, a *p-n junction* (or *diode*) is formed. At the point of contact between the two materials, charge carriers (holes and electrons) diffuse across the junction and combine, forming a layer called the *depletion layer*, which is, literally, depleted of charge carriers, which combine

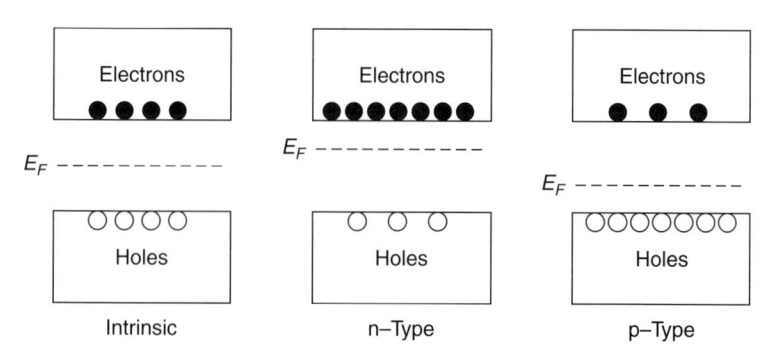

Figure 2.9.3 p- and n-type semiconductors.

rapidly the moment the junction is made. Excess negative charge forms in the p-region of the junction, and excess positive charge forms in the n-region of the junction, a barrier that effectively prevents more carriers from moving across the junction. In a p-n junction the Fermi levels (the energy value that has a 50% probability of being occupied by electrons, designated E_F in Figure 2.9.3) align as required to form equilibrium at the contact point. In turn, the energies of the conduction and valence bands for each of the materials (p and n type) will be moved, with the levels of the n-type material falling. A contact potential develops between the two materials, a voltage that is required to be placed across the junction to make electrons jump the bandgap and allow current to flow. This is the *bias voltage* of the diode and is about 0.7 V for a normal silicon diode to 1.7 V for a red gallium arsenide LED. When the diode is forward biased, meaning that a potential large enough to overcome the contact potential of the junction is applied, electrons are supplied to the n-type material and holes to the p-type material (i.e., current flow through the p-n junction). To maintain electrical equilibrium the excess injected carriers are removed by recombination of electron–hole pairs in which electrons in the conduction band fall to the valence band, the result being the emission of a photon (in the case of an LED) or the production of heat by a nonradiative process (in most diodes used for signal applications). Figure 2.9.4 shows the flow of electrons and holes across the junction in a biased diode.

LEDs can be designed to emit any color desired by changing the dopant concentration and hence the bandgap voltage. Gallium arsenide can be doped to produce LEDs emitting red, orange, yellow, and green light. Blue and violet, on the other hand, require very large bandgap energies, which are not possible using the same technology. An operating LED is shown in Figure 2.9.5. The diameter of the device is 5 mm; however, the device is mostly potting compound. The actual semiconduc-

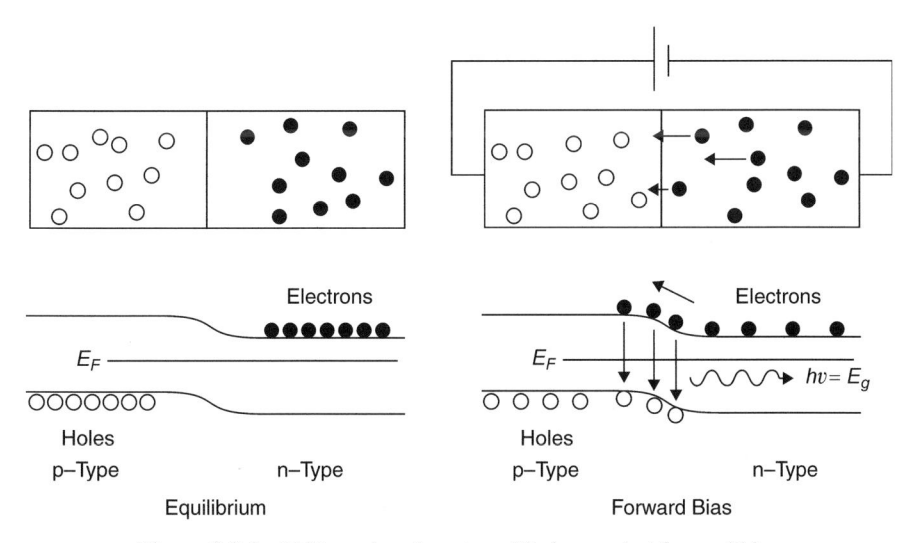

Figure 2.9.4 LED p-n junction at equilibrium and at forward bias.

Figure 2.9.5 Light-emitting diode.

tor crystal is the size of a grain of salt and is held within the cup-shaped terminal on the right. A thin wire connecting the terminal on the left to the crystal is visible in the photo. When operating, an LED requires a series resistance to limit current through the device since a diode connected directly to a voltage source (which exceeds the bandgap voltage) will function as a dead short, drawing as much current as possible from the supply and failing rapidly. When high currents pass through the device, the metals comprising the p-n junction can literally melt, destroying the device. Most common LEDs glow brightly when a current between 10 and 20 mA is passed through them (although today, some high-brightness LEDs specify a maximum current in the hundreds of milliamperes range). If a red LED with a bias voltage of 1.7 V is connected to a 5-V source, the series resistance may be calculated using Ohm's law as follows:

$$R = \frac{V_{\text{resistor}}}{I_{\text{circuit}}} \qquad (2.9.1)$$

where R is the series resistance, V_{resistor} the voltage across the resistor, and I_{circuit} the series current in the circuit. Since this is a series circuit, the current through the resistor is the same as the current in the LED—in this case, 10 mA. With a 5-V supply and 1.7 V across the LED, the expected voltage drop across the series resistor will be 3.3 V. Using formula (2.9.1), the required resistance is found to be 3.3 V/10 mA, or 330 Ω. The circuit is shown in Figure 2.9.6. When an LED of a different color is employed, both the bias voltage and the resistance value will change. A change in supply voltage will also necessitate a change in series resistance.

It is important to note that all semiconductor diodes will not conduct until the forward bias voltage is reached. Figure 2.9.7 shows the experimentally determined

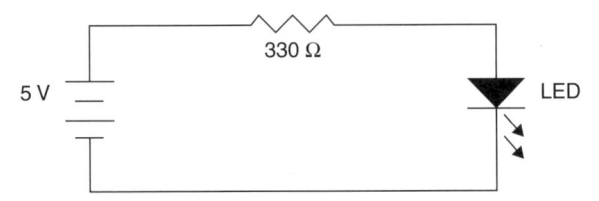

Figure 2.9.6 Use of a series resistance with an LED.

current–voltage (*I–V*) curves for various diodes, including a silicon diode and two gallium arsenide (LED) diodes. No current (except perhaps a small leakage current under 1 µA caused by impurities in the materials) flows until the bias voltage is reached, after which the voltage across the device stays relatively constant and independent of current. In Figure 2.9.7, voltage (on the *x*-axis) is calibrated in units of 200 mV per division. Counting almost 3 divisions before the silicon diode conducts current, the forward bias of this diode is determined to be about 600 mV. Similarly, the red LED is found to have a bias voltage of 1.5 V and the yellow LED a bias voltage of 1.75 V. It is not surprising that the color of emission for an LED depends on the bias voltage: More energy (and hence a larger bandgap) is required to produce shorter wavelengths of light (yellow) than longer wavelengths (red) since the photon energy corresponds to the bandgap.

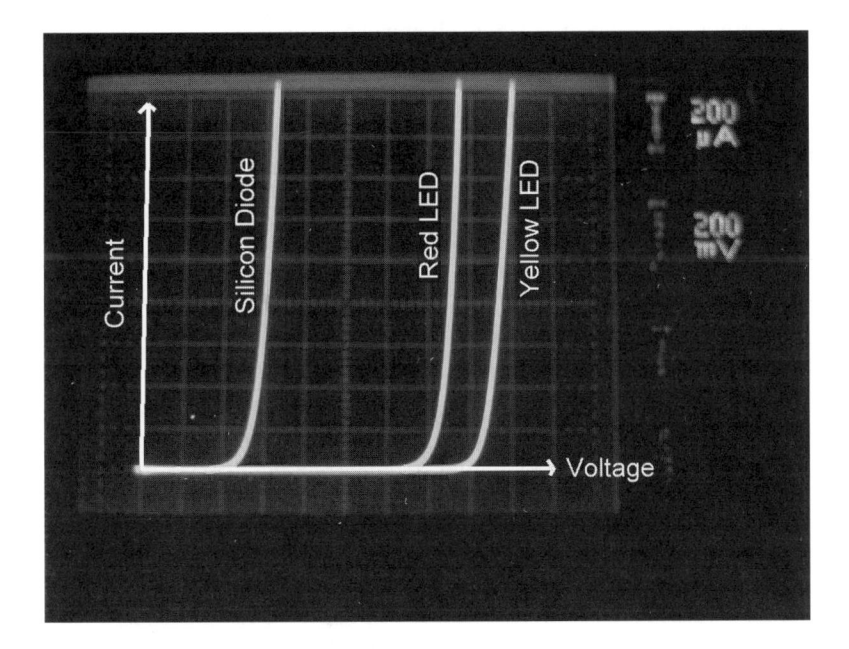

Figure 2.9.7 Current–voltage curves for various diodes.

44   ATOMIC EMISSION

At this point it is necessary to differentiate between two types of semiconductors. As we shall see in the next section, the type of material affects whether or not light will be emitted during conduction. The first type, called *indirect-gap semiconductors*, have energy levels in each band such that an electron making a transition between the valence and conduction bands must experience a change of both energy and momentum. This is dictated by the principle of conservation of both energy and momentum, which occurs during any transition. In the case of a simple indirect-gap diode conducting current, transitions from the valence to the conduction band occur only if energy (sufficient to jump the bandgap) is supplied as well as momentum. The momentum is supplied by physical vibrations in the crystal lattice of the semiconductor material. When a downward transition occurs (from the conduction band to the valence band) the change of energy may result in production of a photon as expected; however, the change in momentum also results in production of a phonon—a physical vibration in the crystal lattice. Most common semiconductors used for electronics purposes, such as silicon and germanium, are indirect-gap semiconductors, and most do not emit photons as a result of recombination but rather, produce phonons, which are eventually manifested as heat in the material.

In *direct-gap semiconductors* the momentum of an electron in either band is almost identical, so no real change in momentum occurs during a transition. Transitions of this type occur in one step, with energy conserved by the emission of a photon and no (or very little) change in momentum required. Gallium arsenide and similar materials used for LEDs and semiconductor lasers are direct-gap semiconductors.[6] In the case of a gallium arsenide diode conducting current, a photon incident on the semiconductor can inject energy, absorbed by an electron in the material, without momentum. A single photon can result in a single electron making the jump across the bandgap. In addition, the recombination of excess electrons and holes in such a device results in production of a photon with energy equal to the jump made across the bandgap—in theory the difference between the top of the valence and the bottom of the conduction bands. An applied forward bias voltage on a gallium arsenide p-n junction generates excess charge carriers near the depletion layer. Excess carriers recombine and photons are emitted.

2.10 LIGHT-EMITTING DIODES

In its simplest form, an LED is simply a diode (or p-n junction) in which flowing current generates electron–hole pairs that recombine and emit excess energy in the form of a photon of light. As current flow through the diode is increased, so is the rate of electron–hole pair production and hence the rate of recombination of

---

[6]Transistors for extremely fast digital logic circuitry and ultrahigh-frequency radio-wave amplification are also fabricated using gallium arsenide, which has a switching time much faster than that of silicon or germanium. The famous Cray supercomputers used logic chips based on this technology. In practice, though, it is a much more expensive technology than silicon, so is used only for applications where the highest possible performance is required.

charge carriers (since they must flow through the device). As such, the output intensity of an LED is expected to be proportional to current flowing through the device.

Consider now a practical device in which radiative and nonradiative processes are competing. Radiative processes, of course, result in production of a photon in order to recombine electrons and holes, but nonradiative processes still occur in all semiconductors. A given semiconductor material is characterized by two lifetimes (refer to Section 2.7), a radiative lifetime and a nonradiative lifetime corresponding to each process. If the radiative lifetime is longer than the nonradiative lifetime (as it is in silicon and other indirect-gap semiconductors), electrons and holes recombine by nonradiative processes long before radiative emissions occur, so no light is produced. This is why silicon diodes do not emit light when conducting (if they did, it would be in the infrared, due to the small bandgap). If, on the other hand, the radiative lifetime is shorter, light is emitted as the primary end product of recombination of electrons and holes. Gallium–arsenide and other direct-gap semiconductors have this favorable situation and so are good choices for light emitters.

Assuming that photons are emitted, their energy should, ideally, be equal to the bandgap energy E_g. The problem here is that electrons in each energy band are not at a single discrete value of energy (in our simplistic model above, we assumed these levels to be the bottom of the conduction band and the top of the valence band) but rather, may have a variety of energy values anywhere inside the bands. We may predict the most probable value as well as the distribution of electrons in a band using a Fermi–Dirac distribution function which describes the probability of occupancy of an energy state. There are a wide variety of possible structures for LEDs, but for simplicity let's consider a common red LED. Statistics show that the maximum population of electrons in the conduction band will occur at $\frac{1}{2}kT$ above the bottom of the band (where k is Boltzmann's constant and T is the temperature in kelvin), where the energy of the bottom of the conduction band is defined by a voltage E_c. Similarly, the maximum population of holes in the valence band will be found $\frac{1}{2}kT$ below the top of the band, defined by a voltage E_v. The concentration of charge carriers (both holes and electrons) per unit energy is plotted in Figure 2.10.1. In this situation we expect, then, to find maximum spectral output at a wavelength corresponding to $E_g + kT$, the point where the concentration is maximum, not at E_g as thought originally. There will also be a spread in photon energies, since there are many possible energies in recombination, also depicted in Figure 2.10.1. The lowest-energy photons at λ_r will be shifted toward the red since these have the least energy. The maximum output of the device is expected at λ_p, where the maximum concentration lies. Photons will then be emitted into the "violet-shifted" side of the peak, as the concentrations of charge carriers taper off slowly throughout the band. The width of each band (both electrons and holes as shown in the figure) is approximately $2kT$ at the halfway point, so a considerable spread in energies is possible (and the spread becomes wider as the device is elevated in temperature).

Using the model described above, built solely on the thermal distribution of electrons in each band, we can predict the emission spectra of LEDs numerically. A typical red LED with a measured bandgap voltage of 1.95 V should have a

46 ATOMIC EMISSION

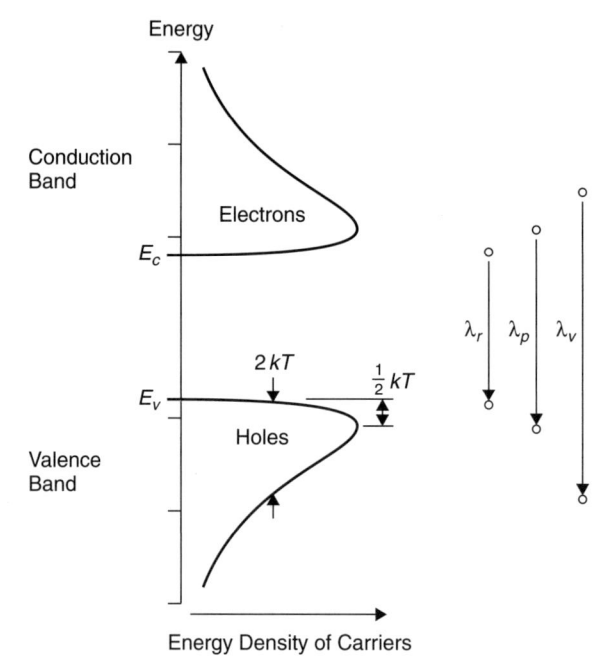

Figure 2.10.1 Energy distributions and possible photon energies.

peak emission of photons with energy equal to

$$E_{\text{peak emission}} = h\nu + kT \qquad (2.10.1)$$

where k is Boltzmann's constant and T is the temperature. Substituting for $h\nu = 1.95$ eV and assuming a temperature of 300 K, the peak emission is expected at 628 nm. Further, we expect a spread of photon energies equal to about $4kT$ since each band has a width of $2kT$. The minimum photon energy will be the bandgap at 1.95 eV (corresponding to 636 nm) and the maximum photon energy will be 1.95 eV + $4kT$ (corresponding to 604 nm). The FWHM expected is then 32 nm.

In addition to the numerical predictions for the spectrum, we expect that the shape of the output spectrum will feature a sharp cutoff at the red (long-wavelength) side of the spectrum, since the minimum photon energy possible is well defined (where a jump between E_c and E_v is the smallest possible energy). This theoretical prediction as well as the output of an actual red LED is plotted in Figure 2.10.2 for comparison. It is evident that the predictions of the output spectrum of the LED, based on energy distributions, are wrong. The spectrum of the actual LED is, as predicted by the model, asymmetrical; however, if you examine each plot carefully you will note that the model predicts a long violet tail on the spectrum and a rather abrupt cutoff on the red side of the emission spectrum. An actual red LED exhibits the opposite behavior! To account for the actual spectrum

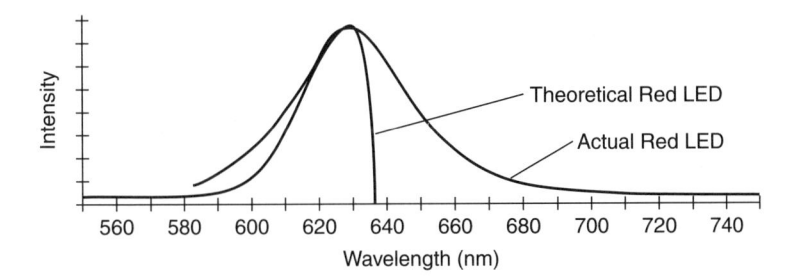

Figure 2.10.2 Predicted and actual output spectrum of a red LED.

of a real device we must factor into the model absorption of emitted light by the material itself (an important concept and one that will be seen again when discussing fast-pulsed lasers such as the nitrogen laser) as well as the effect of energy levels of the impurities added to manufacture p- and n-type semiconductors from which this device is made.

Absorption of light by the semiconductor material itself occurs when photons have an energy at or exceeding the bandgap. For photons with lower energies than the bandgap (i.e., those on the red side of the spectrum), absorption should not occur, since photons lack the energy to excite electrons from the valence band to the conduction band and hence are not absorbed. The effect is seen as a sharper cutoff on the violet side of the spectrum than predicted by the model, as evident in Figure 2.10.2. The effect of impurity energy levels is most obvious when considering the lack of the sharp red cutoff predicted by the model. Transitions with less energy than the bandgap must originate or terminate at energy levels within the gap itself; however, this is logically impossible since this is a forbidden region where no energy levels exist—at least energy levels from the semiconductor itself. The problem with any semiconductor is that impurities such as phosphorus, aluminum, and boron are added to the base semiconductor material to make p- and n-type material for the junction. These atoms have their own energy levels (or should we say *bands*, since they, too, are in the crystal lattice?). Many of these levels exist at energies within the forbidden region (i.e., the bandgap for the semiconductor), so transitions are possible from and to these intermediate levels. In many cases the energy bands from the impurities are at energies just below the conduction band or just above the valence band. The effect is that photon emission is possible at energies lower than the bandgap energy, so the tail on the red side of the spectrum extends into energies that are not possible for the semiconductor material. Of course, this also serves to broaden the spectral width of the emission—practical LEDs typically have a FWHM of 25 to 40 nm. It is possible to design the structure and dopant levels of LEDs to minimize absorption and hence increase efficiency. Efficiencies of up to 30% have been achieved with such optimizations, making the LED about three times as efficient a light source as a tungsten filament lamp.

PROBLEMS

2.1 Determine the order of *visible* emission lines in mercury and the expected accelerating voltage required when observing emissions from a Franck–Hertz tube using mercury. Mercury has the following known energy levels: 4.66, 4.86, 5.43, 6.67, 7.69, 7.89, 8.84, 9.19, 9.52, 9.68, 9.84, and 10.38 eV. Any of these levels (including ground level at 0 eV) may be the final energy level for transitions that emit radiation.

2.2 Many spectroscopes can be equipped with gratings of various lines/mm. What is the advantage of using, say, a 1200-line/mm grating as opposed to a 300-line/mm grating, and vice versa? You must consider the position of the second-order output of each grating, as this will affect the usable operating range of each grating (i.e., the point at which first and second orders overlap).

2.3 A spectroscope is used to observe visible emissions from a gas tube. The grating is rated at 600 lines/mm, and the angle is found to be 30.24 degrees. What is the wavelength of this *visible* emission line (*Hint*: If the answer you obtain is in the infrared, you are probably in the wrong grating order.)

2.4 Using a spectroscope with a grating of 300 lines/mm and an accuracy of ±0.1 degree, the following strong lines are seen from an unknown gas discharge spectrum: 10.10, 9.35, and 11.05 degrees. Identify the gas as mercury, hydrogen, neon, or helium, based on the known lines of each spectrum.

The known lines for neon are (in nm): 540.1, 585.2, 588.2, 603.0, 607.4, 616.4, 621.7, 626.6, 633.4, 638.3, 640.2, 650.6, 659.9, 692.9, 703.2. The known lines for helium are (in nm): 438.793, 443.755, 447.148, 471.314, 492.193, 501.567, 504.774, 587.562, 667.815. Lines for mercury and hydrogen appear in figures in this chapter.

2.5 An infrared LED at 300 K is observed to have a bandgap voltage of 1.412 V. Calculate the wavelength of peak emission. Be sure to factor thermal energy into the picture. What is the expected FWHM of the emission spectrum, ignoring effects from impurities in the semiconductor material?

CHAPTER 3

Quantum Mechanics

We have seen in Chapter 1 many cases in which classical physics fails to describe a process involving light. The ultraviolet catastrophe was the first example of the inability of classical physics to describe such processes adequately. Planck's hypothesis was one of the first involving the quantization of light energy. The concepts developed to explain blackbody radiation later proved useful in describing line spectra emission from excited gases. In Chapter 2 we saw further evidence of the quantized nature of light in the photoelectric effect—and another instance of the failure of classical physics to describe the process adequately. The Bohr theory, one of the first involving quanta, described the origins of the line spectra of hydrogen adequately but failed to describe a number of situations, such as the line spectra of multielectron atoms. Building on Bohr's model, quantum mechanics deviated further away from classical theory by employing wavefunctions to describe the behavior of electrons in atoms.

Quantum mechanics was developed to account for effects seen at the subatomic level which simply cannot be described through classical theory. One such effect is the observation of multiple transitions originating from the *same* energy level. Consider a sodium street lamp viewed through a diffraction grating. At the center of the spectrum a bright yellow line is seen, but under further investigation using a high-dispersion spectrograph, the "line" is actually seen to be two distinct lines. These originate from the same transition between the same energy levels—at least that would be Bohr's view of things. The reality is that although the energy levels for these two lines have the same principal quantum number (i.e., the same n), the levels are actually many levels spaced very closely. In the case of sodium, the energy difference between the two transitions is less than 0.1% of the total energy in the transition.

A second example of this situation occurs in the argon laser. Almost everyone has seen a laser light show where the blue (488 nm) and green (514 nm) beams are featured prominently. The transitions that produce these two beams originate from the same energy levels with the same quantum number n, so clearly there is more to energy levels than simply a principal quantum number.

Fundamentals of Light Sources and Lasers, by Mark Csele
ISBN 0-471-47660-9   Copyright © 2004 John Wiley & Sons, Inc.

3.1 LIMITATIONS OF THE BOHR MODEL

Although the Bohr atomic model was a dazzling advent that explained the atomic structure quite well, it only works with simple atoms such as hydrogen and hydrogenlike atoms, those having only a single valence electron (those in the outer shell of the atom). It does not account for the energy-level structures of complex atoms, which involve two or more valence electrons. Although the model allowed the prediction of transitions in the simple hydrogen atom, it did not work for a complex atom such as neon (which has six electrons in its outer shell) or even for helium (which has two electrons in its outer shell).

One of the major shortcomings of this theory was the angular momentum associated with every Bohr orbit. Orbits in this model are defined by Newtonian mechanics and confined by classical Newtonian laws. Consider the ground state of hydrogen ($n = 1$), which, according to Bohr theory, has orbital angular momentum. This assumption is required since even at ground state the electron is orbiting, and all orbiting objects must have momentum. When quantum states for hydrogen are considered in detail, though, one can prove that the ground state of hydrogen has zero angular momentum. This is not a trivial bit of math and will be spared here, but it does highlight a potential limitation with the Bohr model.

Bohr was half right, and the same results that Bohr obtained—an explanation for energy levels and possible transitions—may be arrived at through a different approach altogether, one that deviates further from classical physics. The new approach, quantum theory, features a wave property for electrons and all other particles. In the world of quantum theory a particle like an electron can act like a wave and be described by wave mechanics.

3.2 WAVE PROPERTIES OF PARTICLES (DUALITY)

Although classical physicists had thought of electromagnetic radiation as a wave for many years, experimental evidence provided by the photoelectric effect had shown that depending on the circumstances, it could exhibit particle qualities. The converse is also true: A particle can exhibit wave behavior. In 1924, Louis deBroglie attempted to calculate the wavelength associated with a particle. By combining the Planck–Einstein relation for photons, $E = h\nu$ (stating that the energy of the photon was related to the frequency or wavelength), with Einstein's famous statement of mass-to-energy equivalency, $E = mc^2$, he speculated that wave behavior was a property of all moving objects and developed a mathematical relationship allowing determination of wavelength. At face value this seems absurd, but under the right circumstances wave behavior is indeed observed, as we shall see in the next section, where an electron, clearly a particle, exhibits wave behavior under the right circumstances.

Central to deBroglie's hypothesis was that any particle with momentum has an associated wavelength. This relation is easily derived from the Einstein–Planck equation combined with the relation for an electromagnetic wave, namely $E = cp$, where c is the speed of light and p is the momentum, as follows:

$$E = h\nu = cp \quad \text{yielding} \quad h\nu = cp \tag{3.2.1}$$

Substituting for $\nu = c/\lambda$, we now get $\lambda = h/p$. Relation (3.2.1) holds true for low-energy particles, where the momentum can be found using a simple relation from kinematics: namely,

$$p = (2mE_k)^{1/2} \tag{3.2.2}$$

where E_k is the kinetic energy of the particle and m is the mass.

Because of relativity, a glitch occurs when considering high-energy particles. As a particle acquires more energy, it approaches the speed of light. As this happens, the apparent mass of these particles changes and must be accounted for. Factoring in relativity, the deBroglie wavelength for a high-energy particle is simply

$$\lambda = \frac{hc}{E_k} \tag{3.2.3}$$

where E_k is the kinetic energy of the particle. The greater the mass and velocity of a particle, the shorter its wavelength. For ordinary macroscopic particles, the wavelength is so short that it can never be measured, but for small subatomic particles with little mass, the wavelengths are well within observable ranges.

Example 3.2.1 ***Wavelength Measurement*** Consider the wavelength of an electron with energy 0.001 eV (or 1.602×10^{-22} J):

$$\lambda = \frac{hc}{1.602 \times 10^{-22}} = 1.24 \text{ mm}$$

This is in the microwave region of the electromagnetic spectrum and is easily measured.

Now consider the wavelength of an electron with energy 2.0 eV (or 3.204×10^{-19} J). The wavelength is found to be 620 nm, which is in the visible spectrum (red) and is easily measured using a spectroscope.

Similarly, an electron with energy 10 keV has a wavelength of 1.24×10^{-10} m. This is in the x-ray region of the electromagnetic spectrum. With wavelengths this short it is not possible to use a diffraction grating (as we use for visible light) since the spacing between the lines needs be so close that such a grating cannot be manufactured. A simple solution is to use a crystal as a diffraction grating. The spacing of atoms in the crystal is just right to allow diffraction to occur when x-rays are beamed at the crystal. In this manner an x-ray spectrometer can be built which works in the same manner as an optical spectroscope, where by measuring the angle of diffraction, the wavelength of the incident radiation may be determined.

52 QUANTUM MECHANICS

Finally, consider a macroscopic object such as a 150-g ball hurled at 20 m/s. The kinetic energy of the ball is

$$E_k = \tfrac{1}{2}mv^2 = 30 \text{ J}$$

so the wavelength is

$$\lambda = \frac{hc}{30} = 6.6 \times 10^{-27} \text{ m}$$

This wavelength is far too short to measure—it is well beyond cosmic rays (the shortest wavelength radiation known). This illustrates why normal everyday (i.e., macroscopic) objects do not exhibit wave behavior (or at least *observable* wave behavior).

3.3 EVIDENCE OF WAVE PROPERTIES IN ELECTRONS

There is much evidence to support the fact that particles such as electrons can and do exhibit wave behavior. Consider the traditional optical principle of diffraction, which conclusively demonstrated the wave property of light. In Young's time-honored experiment, a light beam is incident on a target with two parallel slits (Figure 3.3.1). A pattern forms on a screen after the slits. The pattern clearly shows diffraction occurring when light passing through each slit interferes, to form the fringes. It was deduced that light does this because it is a wave and waves interfere in this manner. If light were a particle, there is no way that particles passing through the slits could produce this pattern.

In 1928, physicist G. P. Thomson attempted the same experiment, with electrons instead of light. Using electrons accelerated by high voltage, he produced a collimated beam of electrons with a deBroglie wavelength of 0.01 nm. This is a much

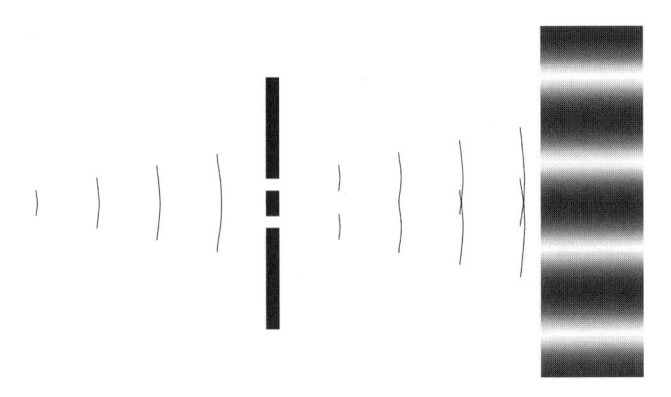

Figure 3.3.1 Young's double-slit experiment.

shorter wavelength than light, so smaller slits would be needed. In his experiment he utilized extremely thin gold foil as a crystal, which would diffract the electron beam as it passed through. The pattern produced was a striking circular diffraction pattern showing without a doubt that electrons were diffracting and hence exhibiting wave behavior. Furthermore, it was shown that magnetic and electrical fields would affect the pattern, proving that the particle exiting the foil was indeed a charged electron.

It is important to understand that in the electron double-slit experiment, that diffraction pattern is still produced even when the beam of electrons is made so weak that only one electron is allowed to pass through the slit at one time. If the electron was "just a particle," one would expect that it passes through one slit *or* the other but never both, giving a simple pattern consisting of two areas on the detection screen. When more than two areas (indeed, when an interference pattern) are seen, we know that even *individual* electrons pass through both slits simultaneously. Only a wave is capable of such behavior.

Since the time of Thomson's original experiment, others have reproduced Young's double-slit experiment using electrons instead of light. The slits are very close together, on the order of 10^{-6} m, which is required given the extremely short deBroglie wavelengths. When electrons pass through the slits and are expanded onto a phosphorescent screen, the resulting pattern looks exactly like that expected with light. Furthermore, electrons have been shown to refract as light does when passing through, say, the interface to a crystal (the air-to-crystal surface, where refractive index changes). Unmistakably, the electrons, normally considered a particle with mass, are exhibiting wave behavior! Quantum mechanics exploits this wave behavior by describing the energies of electrons in the atom (the electrons that create light in making the jump between allowed levels) using wave mechanics. Erwin Schrödinger devised the famous wave equation bearing his name, which describes the states of a bound electron and allowed computation of possible energy levels. The electron is assumed to be bound by forces within the atom, and (more or less) classical wave mechanics are applied to the electron to yield a prediction of how it behaves in this bound state (as if it were a wave). The solution of these complex wave equations yields a prediction as to the probability of finding an electron in a particular area of space around the nucleus of the atom. Unlike the classical Newtonian approach, where an electron is in an exact trajectory around the nucleus, the wave approach simply gives us a probable location for the electron.

3.4 WAVEFUNCTIONS AND THE PARTICLE-IN-A-BOX MODEL

Schrödinger's model pictures electrons in orbit as standing waves in which the orbits are of an exact size so as to allow an electron with a waveform to occupy it in a whole number of wavelengths. As the electron orbits, it comes around and must assume exactly the same position at the end of a complete orbit as it did at the beginning of that orbit (i.e., only an integral number of waves fit inside the orbit). If an electron acquires more energy, its wavelength decreases slightly and it no longer fits inside the orbit. The ground state is now defined as the point where an electron has just enough energy so that a single wave fits inside the orbit. Using this model, one may compute the allowed energy levels of these standing waves and hence the allowed energy levels of the electron.

Let us now model the electron as a particle in a box, as in Figure 3.4.1. The sides of the box represent energy potentials that confine the electron. If we assume Schrödinger's theory that the electron must exist as a standing wave inside this box to be correct, we can describe the behavior of the electron in orbit by the wavefunction ψ which in this case is a simple sine function of the same form as that of a simple harmonic oscillator. At the edges of the box, the wavefunction must be equal to zero since it is a standing wave (i.e., the wave has nodes at the box edges). By substituting integers into the wavefunction, we may identify many possible modes for such a wave—three are shown in Figure 3.4.1. Each mode will have a successively higher energy and in-between modes (those that do not have an integral number of wavelengths—a standing wave—inside the box) cannot exist. This model fits well with observed quantizations of energy levels in atoms.

We may now adapt this model to real particles such as an electron in an orbit by changing the model so that the sides are not infinite but rather are finite in height. This change allows the confined particle (an electron in this case) to leave the box, but only after acquiring significant energy: While in the box the allowed energy levels are well defined. In such a situation, energy levels are lower than in the infinite wall example above but follow the same basic structure. Finally, real particles such as an electron in orbit must be treated in three dimensions. This is quite posssible, although the mathematics becomes somewhat complex.

It is important to note that this wave approach is not limited to predicting the behavior of electrons in orbit but may also be applied to the behavior of a diatomic molecule such as N_2. The nitrogen molecule may be modeled as two atoms

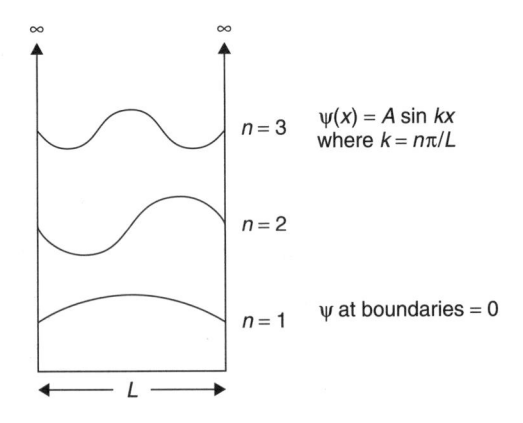

Figure 3.4.1 Model of a particle in a box.

connected by a spring (the bond between the individual atoms). Using an approach similar to that for the particle in a box, we may model this system as a harmonic oscillator (a *quantum* harmonic oscillator) to determine the energy levels allowed for such a system.

3.5 RECONCILING CLASSICAL AND QUANTUM MECHANICS

We now have a confusing situation where a particle like an electron can behave as a wave when it has momentum. The converse is also true, in that photons which normally exhibit wave behavior can exhibit the properties of particles, as demonstrated by the photoelectric effect. Quantum mechanics, using complex wavefunctions, allows predictions of electron behavior, but how do we reconcile this with classical mechanics which works, so well in describing everyday macroscopic events? Let's draw a few conclusions to keep things straight.

- Classical physics such as Newtonian mechanics provides a *very* clear view of how macroscopic particles and everyday objects behave.[1]
- Depending on the circumstances, quantum physics may best describe the properties and behaviors of subatomic particles such as electrons.

Of course, defining those circumstances is the key, but the situation here is identical to that encountered in kinematics. Consider the *correspondence principle*. From kinematics we know that as a particle or object approaches the speed of light, Newtonian mechanics ceases to describe the situation correctly and special relativity must be used. In a similar manner, classical mechanics works well for large, macroscopic objects, but as we delve into the area of the atom, quantum physics best describes situations arising there. Bohr devised the correspondence principle to reconcile classical and quantum mechanics. He stated that at small quantum numbers such as $n = 1, 2, 3, \ldots$, the domain of interest for atomic transitions that emit light, quantum mechanics describes the situation best, but as n becomes larger, up to say 10,000, the predictions of quantum and classical theory will agree. In other words, at large quantum numbers, quantum mechanics simply reduces to classical physics!

Returning to the example of hydrogen, classical physics predicts that the frequency of an emitted photon will be equal to the frequency of the electron's revolution in orbit around the nucleus. Quantum mechanics uses the notion of transitions between energy levels. When considering a quantum number of $n = 2$, quantum mechanics predicts the frequency of emitted photons with great accuracy, whereas classical mechanics fails. When a large quantum number of 10,000 is considered, we find that the classical and quantum mechanical approaches agree within a fraction of a percent!

---

[1] At the other end of the spectrum, namely things that are very large or very fast, classical mechanics also fails to predict the behavior of objects and general relativity must be used.

3.6 ANGULAR MOMENTUM IN QUANTUM STATES

So far in this book we have described atomic energy states by a principal quantum number n, which in the Bohr model represented an orbit. If we are to continue using the analogy of an electron as an orbiting satellite around a planet, we must also account for the angular momentum of the satellite. By knowing the total kinetic energy and the total angular momentum of a planet, we may predict its orbit and energy. Refining the analogy even further, planets have two types of angular momentum: orbital and spin. The spin is an intrinsic angular momentum; in the case of the Earth, it is the spin of the planet that brings about day and night. Furthermore, the orbiting satellite—in our case, a charged electron—generates a magnetic field by virtue of the fact that we have a moving charge. Both effects, magnetic and spin, contribute to the complete description of an energy state, but the largest influence is that of the orbital angular momentum of the electron, so we shall deal with it first.

One major shortcoming of the Bohr model was the failure to account for the hyperfine structure of hydrogen lines. Each line emitted from hydrogen is, in fact, a series of very closely spaced lines. In 1916, the physicist Arnold Sommerfeld suggested that orbiting electrons could follow not only a circular orbit (as in the Bohr model) but also elliptical orbits, and each orbit would have a slightly different energy than that of the circular orbit. His theory was that only certain elliptical orbits would be allowed in the atom fitting the requirements for quantization in the atom. A set of allowed orbits, then, would exist for each principal quantum number n. In the case of the hydrogen emission spectrum, transitions between these closely spaced energy levels give rise to the many closely spaced lines seen.

Every particle in orbit has a constant angular momentum in that orbit equal to its mass times its velocity (where velocity is a vector quantity having the property of direction). The same is true whether an orbit is circular (in which case $l = mv$) or noncircular (in which case $L = mvr \sin \theta$). The later case is depicted in Figure 3.6.1, where angular momentum is the same at every point in the orbit of the particle. When the particle approaches the center (at radius r_2 in the figure), speed increases; when

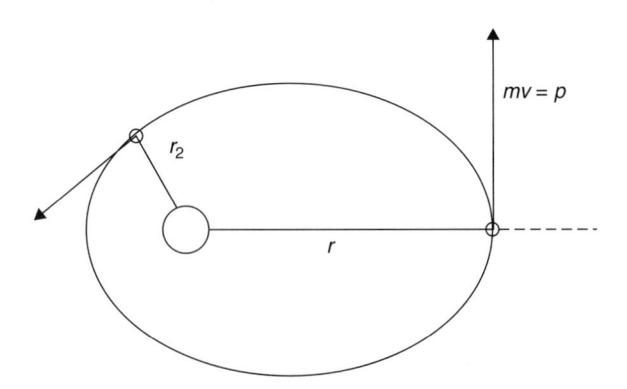

Figure 3.6.1 Angular momentum of a particle.

the radius is larger, the speed is lower. This is simply conservation of momentum, a concept from basic kinematics, which holds true here as well. By analogy, electrons orbiting around a nucleus will also exhibit angular momentum. Having angular momentum implies that the electron has kinetic energy whose amount is limited by the total electron energy (in the case of an atom, the total energy is described by the principal quantum number n). Like other quantities in an atom, angular momentum is quantized into allowed values; this is one result of the solution of wave equations (Schrödinger wave equations), which describe the behavior of the electron.

Angular momentum consists of two key components: orbital angular momentum and spin angular momentum. Spin is an intrinsic property of electrons and is manifested by the magnetic moment created by spinning electrons. Orbital angular momentum is somewhat more straightforward from a classical perspective and is examined first. We now introduce a quantum number, l, to describe orbital angular momentum. This number, the *orbital quantum number*, describes the *magnitude* of the orbital angular momentum. Values for l are constrained by the principal quantum number n and can have integer values of zero to $n - 1$. For a principal quantum number of $n = 1$, l always has a value of zero. This indeed states that the lowest orbit has an angular momentum of zero—at odds with Bohr's original model! For a principal quantum number of $n = 2$, l can be 0 or 1, meaning that an electron in the $n = 2$ state can have zero angular momentum, corresponding to a circular orbit, or some discrete value of angular momentum, corresponding to an elliptical orbit. In either case ($= 0$ or 1), the electrons have *almost* the same energy (i.e., they are still both in the $n = 2$ principal state). Both quantum numbers together, n and l, are required to fully describe the energy of a given state (along with a few others that we'll get to later in the chapter).

3.7 SPECTROSCOPIC NOTATION AND ELECTRON CONFIGURATION

One can use a vector notation (in which each quantum number, including the ones that come later in this chapter, are listed in order) to describe a given quantum state, but the notation commonly used to describe atomic energies was one devised by spectroscopists long before quantum mechanics. Features of the atomic spectra (emission lines) were defined as sharp, principal, diffuse, and fundamental, based on how they appeared when viewed. These descriptions fit those provided by the orbital quantum number, in that sharp lines were found to be associated with the $l = 0$ state and diffuse lines with the $l = 2$ state. For any given value of n, then, there are a number of possible states of l, and each value of l is assigned a letter as follows:

Sharp	s	$l = 0$
Principal	p	$l = 1$
Diffuse	d	$l = 2$

Fundamental	f	$l = 3$
	g	$l = 4$
	h	$l = 5$

For states $l = 0$ to $l = 3$ the system uses the letters s, p, d, and f; higher-order states after $l = 3$ continue with consecutive letters g, h, i, and so on.

So what do these numbers l *really* represent? For given values of n (the principal quantum number from the Bohr model corresponding to radius), one can apply wave mechanics to solve the complex wavefunctions for electrons and to determine the probability of finding an electron in a particular space. Put a different way, Bohr described electrons in circular orbits where the energy depends solely on distance from the nucleus. In quantum mechanics an electron having a particular energy can be described as having a good *probability* of being in a certain defined area. In each area, called a *probability distribution*, the angular momentum (and hence energy) is the same—analogous to a noncircular orbit in classical kinematics. These areas of probability can be various shapes and sizes. We begin by solving the radial part of the wavefunction (i.e., in one axis). Solutions for the first three orbitals—1s, 2s, and 2p—are plotted as probabilities in Figure 3.7.1. The x-axis is in units of Bohr radii and represents the distance of the electron from the nucleus, while the y-axis is the probability of finding the electron at that radius.

For the 1s orbital, the probability is high that the electron will be found at a radius of a_0 from the nucleus: exactly what Bohr would have predicted for the hydrogen atom's $n = 1$ state. This illustrates how quantum mechanics leads us to the same answer as Bohr (at least for a hydrogenlike atom) derived, albeit using a completely different approach. For a 2s orbital the probability is highest that the electron will be found at a radius of about $5a_0$ from the nucleus, but there is also a good probability of finding the electron much closer to the nucleus.

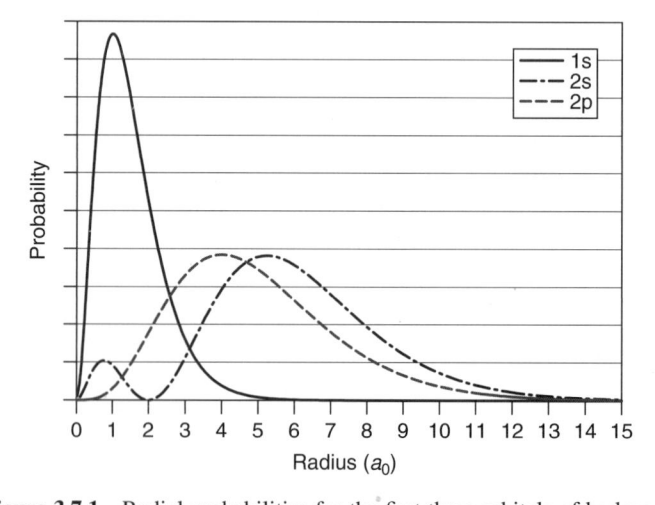

Figure 3.7.1 Radial probabilities for the first three orbitals of hydrogen.

We must remember that the electron occupies three-dimensional space, so although Figure 3.7.1 shows us the most probable radius from the nucleus in one axis, there is an angular component as well in the solution of the wavefunctions that shapes these orbitals. Several are depicted in Figure 3.7.2 as three-dimensional figures. Orbitals with zero angular momentum (i.e., $l = 0$ or s orbitals) have a probability distribution similar to that of a Bohr orbit; the 1s orbital, for example, has the appearance of a sphere around the nucleus. As we move toward electron states with more angular momentum (i.e., p and d orbitals), these probability distributions often take the form of lobes and toruses around the nucleus.

For the $n = 2$ state there are two possibilities for l and hence two possible electron configurations. When $l = 0$, the orbital resembles two concentric three-dimensional spheres in what is termed the 2s configuration, and when $l = 1$, the orbital does not resemble a circle at all but rather three sets of two lobes, like dumbbells about the center, each in a different axis in what is the 2p configuration. For $n = 3$, even more orbitals are possible, with three possible states for orbital angular momentum ($l = 0, 1,$ or 2). It is apparent that the term *orbit* is misleading and why in quantum physics we speak of probability distributions (although the terms *shell* and *orbital* are still used by convention as they are in this book).

Each value of l represents an electron orbital, and each can hold a maximum number of electrons before it is completely filled. Once filled, additional electrons will begin to fill the next-highest orbital based on order of energy. An s orbital can hold a maximum of two electrons, a p orbital a maximum of six electrons, a d orbital a maximum of 10 electrons, and an f orbital a maximum of 14 electrons. Consider the hydrogen atom with a single electron in a 1s orbital at ground state—the orbitals are similar to Bohr's circular orbit. The sodium atom at ground state, however, has a total of 11 electrons. Two electrons will populate the 1s state first, filling this level completely, followed by two more in the 2s state and six electrons in the 2p state, also filling these levels completely. This leaves a single electron in the next outermost orbital, the 3s state. Its electron configuration is described as $1s^2 2s^2 2p^6 3s^1$, where the superscript after each description for l denotes the number of electrons

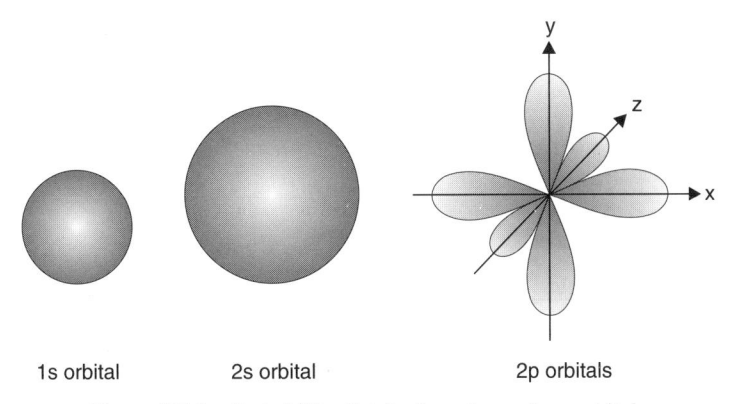

Figure 3.7.2 Probability distributions for various orbitals.

in that state. Electrons in the outermost, unfilled orbitals are termed *valence electrons*. When the outer electron in the sodium atom is excited, it can be elevated to a higher state, so the electron configuration might, as an example, be $1s^2 2s^2 2p^6 4p^1$ or $1s^2 2s^2 2p^6 5d^1$ for an excited atom. The reasons for the particular way in which orbitals are filled and why, for example, an s orbital can hold only two electrons will be seen later in the chapter, after additional quantum numbers have been introduced.

Transitions can occur between any energy state to yield photon emission, and unlike the Bohr model, where electrons could be in only one principal quantum state ($n = 2$, $n = 3$, etc.), many more levels are now available. With angular momentum now in the picture, each level of n will now have n possible values of l. If the outer electron of the sodium atom is elevated in energy from its lowest or *ground-state* energy ($n = 3$) to an $n = 5$ level, there are five possible energies that it can assume there (with l ranging from 0 to 4), so a large number of transitions are possible.

3.8 ENERGY LEVELS DESCRIBED BY ORBITAL ANGULAR MOMENTUM

The question begs: Why do electron energies depend on the orbital quantum number anyway? This is, after all, at the root of explaining why the Bohr model fails to explain the spectra of multielectron atoms (like helium, which has two electrons completely filling the 1s orbital). We would expect that all electrons in an $n = 2$ state have exactly the same energy, regardless of orbital configuration. Some electrons in higher orbitals (e.g., p or d) will have larger angular momentum, others will be closer to the nucleus, but either way, we would expect the same total energy for each electron here. This is not the case, and electrons in various orbital configurations do indeed have different energies.

The answer lies in the shielding effect of electron shells. Consider, as illustrated in Figure 3.8.1, atoms of hydrogen and sodium, each with a single electron in the

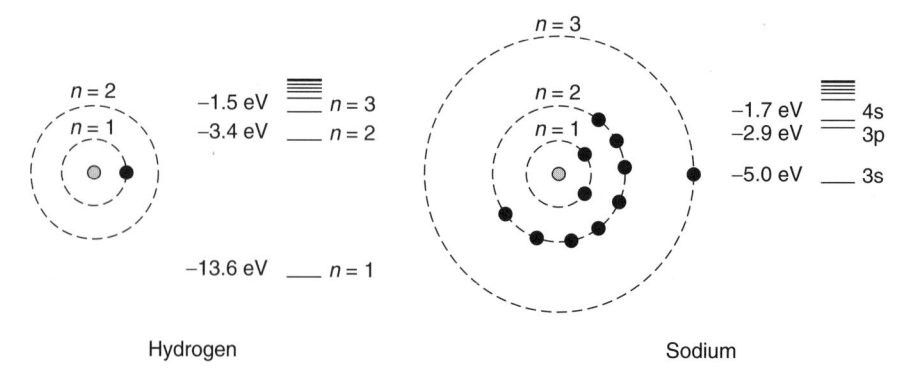

Figure 3.8.1 Energy shells and levels in hydrogen and sodium.

ENERGY LEVELS DESCRIBED BY ORBITAL ANGULAR MOMENTUM   61

outer shell. This outer electron is called the *valence electron* because it lies outside the closed shells below it (e.g., the $n = 1$ shell of sodium, which is full at two electrons and hence called closed, or the $n = 2$ shell with eight electrons, which is also full). At face value the energies of electrons in the outer shell of the sodium atom should be calculable by the same methodology as Bohr applied to the hydrogen atom. Spectroscopic studies, however, reveal that electrons in the $n = 3$ orbit have significantly *lower* energies than electrons of the hydrogen atom in the same $n = 3$ orbit. As an example consider the hydrogen atom, with its $n = 3$ orbit energy of -1.5 eV. The comparable orbit in the sodium atom is the 3s orbit (ground state for sodium), which in this case has an energy of -5.0 eV, which is even lower than hydrogen's $n = 2$ orbit energy!

The reason for the shift in energy levels is attributed to a *shielding effect* of the completed inner shells, which serve to lower the energy of the levels outside these. In effect, the inner shells, complete with all electrons, prevent the solitary electron in the outer shell from "feeling" the full attraction of the positive nucleus of the atom. The farther away these outer-shell electrons are from the nucleus, the more closely their energies match those of the electron in the hydrogen atom at the same state. It is found that an outer electron of a sodium atom in a 3s state has energy much lower than an electron in a hydrogen atom also in the 3s state. An electron with a higher orbital quantum state, such as $l = 2$ (in a 3d state), will have an energy almost identical to that of an $n = 3$ electron in the hydrogen atom at -1.5 eV, as it is farther from the nucleus (using the Bohr analogy of shells). Similarly, 4d and 5d orbitals have energies almost identical to that of hydrogen's $n = 4$ and $n = 5$ state, respectively. Figure 3.8.2 illustrates the dependence of energy levels on the orbital quantum number.

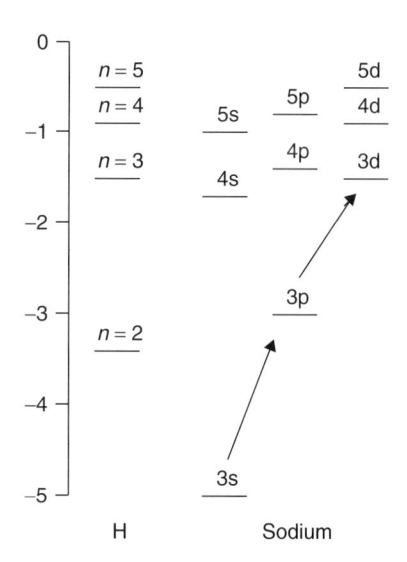

Figure 3.8.2 Dependence of energy on orbital quantum number.

Consider the outer electron of sodium, normally in the 3s state (third orbit, $l = 0$). When energy is injected into the atom, that electron may rise to the 3p state next (remember that in the Bohr model this state does not exist and the next state is simply $n = 4$). As yet higher energies are attained by the electron, it can achieve a 4s state. This state is *still* below that of the hydrogen's $n = 3$ state! This model, then, allows for a host of intermediate states between the Bohr states of $n = 2$ and $n = 3$ and explains many emission lines which are impossible to explain using the Bohr model. In a multielectron atom (such as helium with two valence electrons) the direction of the angular momentum of each electron will also contribute to splitting of energy levels. If two electrons are orbiting in the same direction, their total angular momentum will be large but they will meet less often than if they were orbiting in the opposite direction. The repulsive force between the electrons is hence lower, and the resulting total energy of that level is lower. This effect is called *orbit–orbit interaction*.

3.9 MAGNETIC QUANTUM NUMBERS

We have now described the orbital angular momentum of an electron with the term l, but an electron with angular momentum is analogous to a current loop and will exhibit a magnetic moment. This magnetic moment will (generally speaking) be characterized by the magnetic quantum number m_l. An alternative way to think about this parameter is that it represents the *direction* of angular momentum of an electron. Assuming that we've chosen an axis for the atom, the orientation of this angular momentum (or orbital) with respect to this axis gives rise to m_l. It can be thought of as the three-dimensional tilt of an elliptical orbit, as depicted in Figure 3.9.1. This number is confined by the orbital quantum number l and may assume integer values ranging from $-l$ to $+l$. For example, an s orbital with $l = 0$ always assumes a magnetic quantum number of $m_l = 0$, while a p orbital with $l = 1$ can have numbers $m_l = -1, 0,$ or $+1$. The range of numbers increases as the orbital does, with five values possible for a d orbital, and so on. In all, there are $2l+1$ possible values for m_l for a given value of l.

It is somewhat puzzling perhaps why this is called the magnetic quantum number as opposed to tilt or some other description. The answer lies in the conditions

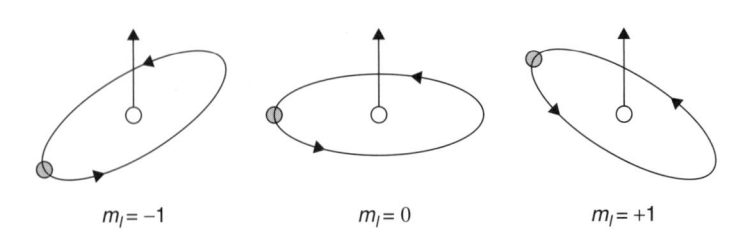

Figure 3.9.1 Representation of the magnetic quantum number.

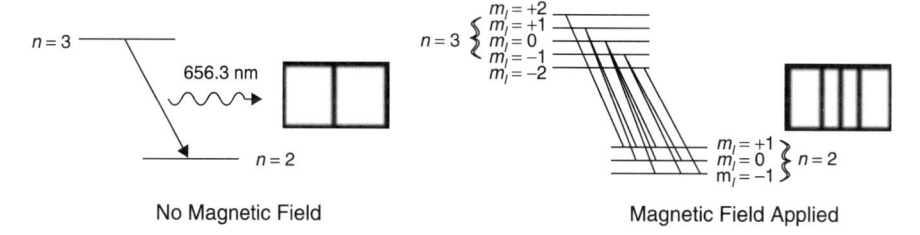

Figure 3.9.2 Zeeman effect on the red hydrogen line (energy levels and resulting spectra).

required to produce an energy difference between levels with various values for m_l. Under the influence of a magnetic field, certain spectral emission lines can be split into a number of hyperfine lines in what is called the *Zeeman effect*. Figure 3.9.2 depicts the effect of an external magnetic field on the red emission line of hydrogen.

With no external magnetic field, the transition looks like a simple one between the $n = 3$ and $n = 2$ levels, resulting in the emission of a single wavelength. When a magnetic field is applied, the alignment of various magnetic moments becomes apparent as a number of spectra lines appear. Although it appears from Figure 3.9.2 that many hyperfine transitions (perhaps 15) are possible, there are selection rules that determine which transitions are unlikely (called *forbidden*) and which are *allowed*. The simple rule is that a transition is allowed (i.e., is quite probable) if the change in orbital angular momentum (l) is -1 or $+1$ (but not zero). A transition is also allowed if the change in the magnetic quantum number (m_l) is -1, 0, or $+1$. Referring back to the figure, it is apparent that the transition from $n = 3$, $m_l = +2$ to $n = 2$, $m_l = +1$ obeys these rules but that no other transition originating from the $n = 3$, $m_l = +2$ level will. Furthermore, only three spectral lines are visible in the split because there are only three *unique* hyperfine transitions (i.e., changes in quantum numbers). Further application of selection rules can be seen in the spectrum of sodium observed (shown in Figure 3.13.1), in which observed transitions correspond to a jump between an s and a p orbital (a change in l of 1) but never from a p to another p orbital. Selection rules will be expanded in Section 3.12.

3.10 DIRECT EVIDENCE OF MOMENTUM: THE STERN–GERLACH EXPERIMENT

One of the great experiments in quantum physics, the *Stern–Gerlach experiment*, showed that angular momentum was indeed quantized and could be described by an integer quantity (i.e., it could only assume certain, allowed, values). We have seen in Chapter 2 that energy levels are quantized (in the Franck–Hertz experiment), but the properties of angular momentum are also important to understand. This is certainly an important feature since our application of angular momentum to electron energy levels requires this to be a quantized quantity.

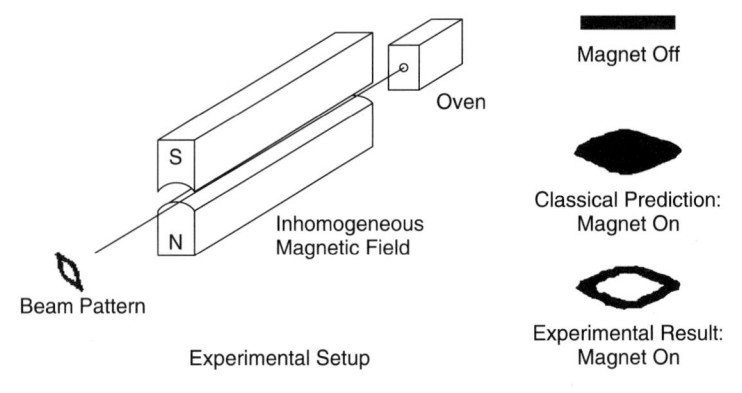

Figure 3.10.1 Stern–Gerlach experiment.

The original experiment, depicted in Figure 3.10.1, involved the deflection of a beam of neutral silver atoms emerging from a hot oven via a magnetic field and onto the target of a photographic plate. Silver atoms were used since the outer electron, like that of a hydrogen atom, has no angular momentum (an S orbital with $l = 0$), so no interaction with an external magnetic field would be expected. The beam of atoms was directed through an inhomogeneous magnetic field whose field could be varied and directed toward a photographic plate, where it could be detected. When no magnetic field is applied, the beam appears as a line on the plate. Thinking classically, one would expect that the spinning outer electron (spinning like an orbiting satellite) acts as a dipole that has a magnetic moment.

When a magnetic field is applied to the beam, the expectation is that this dipole can have any orientation, and hence the deflection of these atoms by the applied magnetic field will be in all orientations, producing a continuous range of deflections. The results of the actual experiment show that the beam was deflected into two discrete areas with no regard to the intensity of the magnetic field applied. If the magnetic field was above a certain threshold, it caused the deflection to occur in these two discrete areas; magnetic fields below this threshold value caused no deflection to occur.[2]

At the time it was a startling fact that the electron interacted with the magnetic field at all. This outer electron has no angular momentum and so does not possess a magnetic moment (caused by a current loop in classical analogy). The implication is that the electron has an intrinsic momentum (as opposed to an orbital momentum). This new type of angular momentum, called *spin*, is so named because an electron spinning on its axis would create a magnetic moment for which there would be two natural values for clockwise and counterclockwise rotation. Further evidence of spin is provided by the fact that the deflection occurred in only two areas. Orbital angular

---

[2] In reality, application of a low magnetic field causes the beam pattern to become spread out, due to the wide range of speeds of atoms emerging from the oven. Actual splitting of the beam into two distinct areas, indicating quantization, occurs only when the magnetic field has reached a threshold value.

momentum occurs in integrals of $2l + 1$ states (since there are that many states of m_l for each value of l). One would then expect to see $2l + 1$ discrete areas in the pattern, not two. It appeared that this new type of angular momentum, spin, occurred in quantized units of $\frac{1}{2}$.

3.11 ELECTRON SPIN

Continuing with the Bohr analogy, if we consider the electron to be orbiting the nucleus like a miniature planet, that electron will also have a spin associated with it. In a classical view, a spinning charge produces a magnetic moment that interacts with other magnetic fields, but this model fails to account for the quantized nature of this spin (for which experimental evidence is provided in the form of the Stern–Gerlach experiment), so it is a purely quantum concept, despite the fact that the name suggests a classical foundation.[3]

Spin is not, unto itself, important (spectroscopically speaking). What affects energy levels is the way in which spin interacts with angular orbital momentum in what is called $l-s$ *coupling* (where the s represents spin). The spin of an electron can assume two possible values, $-\frac{1}{2}$ and $+\frac{1}{2}$. When spin is added and subtracted from orbital angular momentum (l), the effects of spin on energy levels can be seen. Consider Figure 3.11.1, which shows (in a Bohr model type of approach) an orbiting electron. Depending on the direction of the spin of the electron, the energy level of that electron will change. When the orientation of the spin momentum is in the same direction as the orbital angular momentum, the resulting energy level is slightly higher than when the orientation of the two momentums is in the opposite

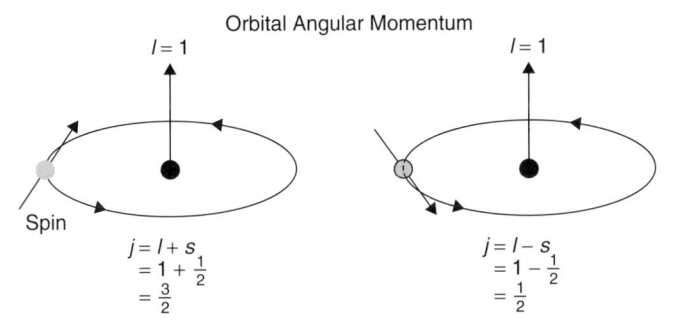

Figure 3.11.1 Electron spin and $l-s$ coupling.

---

[3] Indeed, quantum mechanics is a field unto itself. In this chapter we illustrate the basic parameters that affect atomic levels and transitions but do not, by any means, provide a complete treatment of the topic. A much more detailed analysis of quantum mechanics may be found in *Introduction to Quantum Physics* by A. P. French and E. F. Taylor (New York: W. W. Norton, 1978).

direction. This effect is designated by a subscript j, which is a *combination* of l and s. In this case, $l = 1$, so j can assume values of $\frac{1}{2}$ or $\frac{3}{2}$, depending on the orientation of spin relative to orbital angular momentum.

The effect of spin on energy levels can be quite small but is evident in the hydrogen spectrum. When a hydrogen line such as the red line at 656.3 nm is examined using high-resolution spectroscopy, the "single" line is actually found to be a doublet of two very closely spaced lines, separated by only about 0.02 nm (most spectrographs lack enough resolution to discern the individual lines). The transition is commonly referred to as the $n = 3$ to $n = 2$ transition in hydrogen. The actual lower state for the transition ($n = 2$) is found to be two energy states very close together, with electron spins in opposite directions. The slightly higher energy state results from the electron spinning in the same direction as the orbital angular momentum, the lower state where spin is opposite, as shown in Figure 3.11.2.

In Figure 3.8.2 the designation of the 2P levels is determined by adding $l + s$ and $l - s$. The subscripts are determined in the same manner as in Figure 3.11.1. The resulting levels are then designated as $2P_{3/2}$ and $2P_{1/2}$. Called a *doublet*, this results in the production of two closely spaced spectral lines (doublets occur in P orbitals in an atom where there is a single valence electron, such as hydrogen or sodium). Note that j is an absolute value, and should a negative number result, the negative sign is simply ignored.

The effect of the spin of electrons on the energy levels is most dramatic in a multi-electron atom such as helium, which has two electrons in the outer shell. If the spins of each electron are parallel, the resulting energy levels will be lower than if the spins are opposing. The reason for the behavior is that electrons with parallel spins have a greater chance of being closer together than if their spins are opposite. The closer together the electrons are, the higher the resulting energies. (When two negative charges oppose each other, a higher energy will result.) Transitions between otherwise identical atoms would show higher energies in an atom having electrons with parallel spins than in an atom with opposing spins. Called the *spin-spin effect*, this serves to split energy levels. It is responsible for splitting of the yellow sodium D lines, which unlike the hydrogen lines, which are 0.02 nm apart, are 0.6 nm apart and easily discerned with an inexpensive diffraction grating.

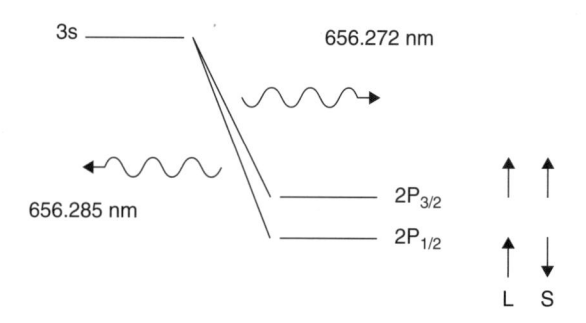

Figure 3.11.2 Hydrogen fine structure.

3.12 SUMMARY OF QUANTUM NUMBERS

Before applying quantum theory to real atoms, let's summarize what we've got so far:

- The principal quantum number n is analogous to the Bohr orbit number and gives the major energy state for the atom.
- The orbital quantum number l describes the *magnitude* of orbital angular momentum of the electron. Values for l are constrained by the principal quantum number n and can have values of zero to $n - 1$.
- The magnetic quantum number m_l describes the *direction* of orbital angular momentum. It is confined by the orbital quantum number l and may assume integer values ranging from $-l$ to $+l$.
- Finally, spin is the intrinsic angular momentum of an electron and may assume values of $-\frac{1}{2}$ or $+\frac{1}{2}$.

The final piece to the quantum puzzle is the *Pauli exclusion principle*, which simply states that no two electrons in an atom can have *exactly* the same set of four quantum numbers. For any value of n, we have n possible values of l. For each value of l, we have $2l + 1$ possible states for m_l. Finally, we have two possible states for spin. Multiplying these possibilities together, we find that there are $2n^2$ possible energy states for each principal quantum number n. Figure 3.12.1 outlines all of the possible states for an electron in an $n = 3$ state.

For an $n = 3$ state, three states for angular orbital momentum (l) are possible, corresponding to an s, p, and d orbital. For each of these values, various magnetic quantum numbers (m_l) are possible—in the case of a 3s orbital, this value can only be zero, while for a 3d orbital, five values are possible. Finally, the electron can assume a spin of $-\frac{1}{2}$ or $+\frac{1}{2}$. It is now apparent that for any given principal quantum number n, there are indeed $2n^2$ possible energy states. Following the Pauli principle we can also determine that there are also $2n^2$ available places in a given shell. For example, an $n = 1$ shell can hold only two electrons, while an $n = 2$ shell can hold a total of eight electrons: two in the 2s orbital and six in the 2p orbital.

The Pauli principle factors in the building of the periodic table of elements as well, as depicted in Figure 3.12.2, which outlines the electron configuration of the first 18 elements. Beginning with hydrogen, the $n = 1$ shell fills until it is full with helium (it can hold only two electrons, so there are only two elements on this row). The $n = 2$ shell begins to fill with lithium and is complete with neon, which has eight electrons in the $n = 2$ shell along with the two in the $n = 1$ shell. Similarly, the $n = 3$ shell begins to fill with sodium. Since no two electrons can have exactly the same quantum numbers, each shell is limited in the number of electrons that can populate it.

Elements with completely filled s and p shells appear in the rightmost column. These are inert or noble gases and include helium, neon, argon, krypton, and xenon. These elements are characterized by a lack of reactivity—they do not normally combine with any other atomic species to form molecules, as, for example,

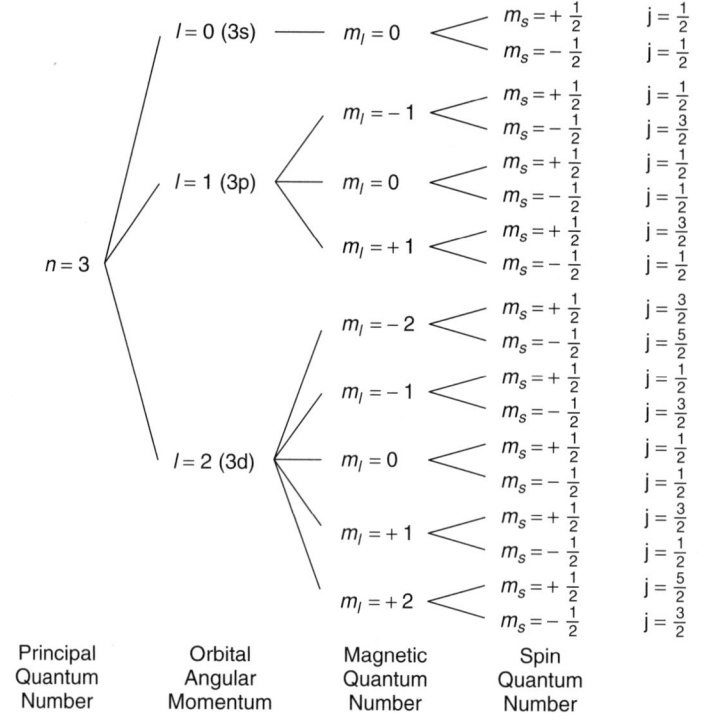

Figure 3.12.1 Summary of quantum numbers.

oxygen and nitrogen do.[4] Next to these elements are the halogens, such as fluorine and chlorine, which need only one electron to complete the p shell. These are very chemically reactive and form compounds readily. Elements appearing in the same column in the periodic table usually exhibit similar chemical characteristics.

Finally, we summarize the basic selection rules for "allowed" transitions involving a single electron, taking into account new quantum numbers introduced in this chapter:

- The change in orbital quantum number ℓ for an allowed transition must be -1 or $+1$.
- The change in the magnetic quantum number m_ℓ must be -1, 0 or $+1$.
- The change in spin s for an allowed transition must be 0.
- The change in j $(= \ell + s)$ must be 0, -1 or $+1$ but a transition from $j = 0$ to $j = 0$ is not allowed.

---

[4] There are a few situations in which inert gases may be made to react and form compounds. First, some inert gases may react with fluorine to form, for example, argon hexafluoride (ArF$_6$), which is somewhat unstable. Second, in some lasers (excimers) an excited diatomic molecule called a *dimer*, such as argon fluoride (ArF) or krypton fluoride (KrF), may be produced. These lasers are examined in detail in Chapter 10.

EXAMPLE OF QUANTUM NUMBERS: THE SODIUM SPECTRUM   69

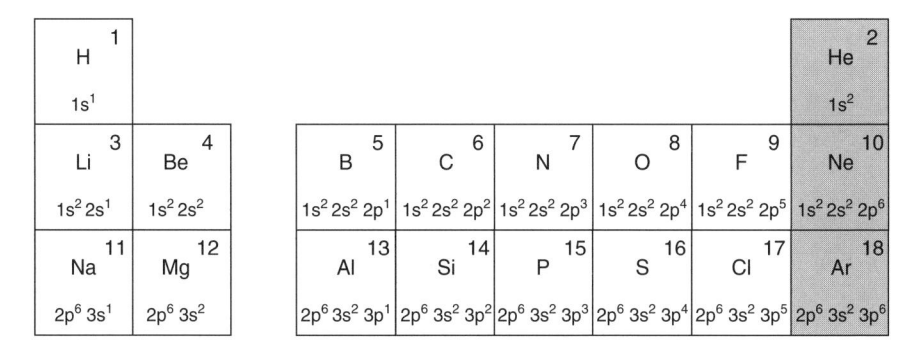

1 H $1s^1$							2 He $1s^2$
3 Li $1s^2 2s^1$	4 Be $1s^2 2s^2$	5 B $1s^2 2s^2 2p^1$	6 C $1s^2 2s^2 2p^2$	7 N $1s^2 2s^2 2p^3$	8 O $1s^2 2s^2 2p^4$	9 F $1s^2 2s^2 2p^5$	10 Ne $1s^2 2s^2 2p^6$
11 Na $2p^6 3s^1$	12 Mg $2p^6 3s^2$	13 Al $2p^6 3s^2 3p^1$	14 Si $2p^6 3s^2 3p^2$	15 P $2p^6 3s^2 3p^3$	16 S $2p^6 3s^2 3p^4$	17 Cl $2p^6 3s^2 3p^5$	18 Ar $2p^6 3s^2 3p^6$

Figure 3.12.2 Building the table of elements.

3.13 EXAMPLE OF QUANTUM NUMBERS: THE SODIUM SPECTRUM

So boiling down all of the possible quantum numbers used to describe a state, we conclude that there are a total of $2n^2$ states of essentially the same energy level for each principal quantum number n in the hydrogen atom. Each state has a slightly different energy, though giving rise to multiple transitions. Spectroscopic notation may also be enhanced to include the contributions from spin and spin-orbit coupling as a fractional subscript.

Consider the line spectrum of sodium, which is dominated by a pair (doublet) of bright yellow lines (the D lines) at 589.0 and 598.6 nm, observable with most spectroscopes and shown in Figure 3.13.1. Sodium is a hydrogenlike atom and has only one electron in its outer shell, a $3s^1$ configuration at ground state. The energy transition that creates this yellow emission is from the 3p to the 3s level in the sodium

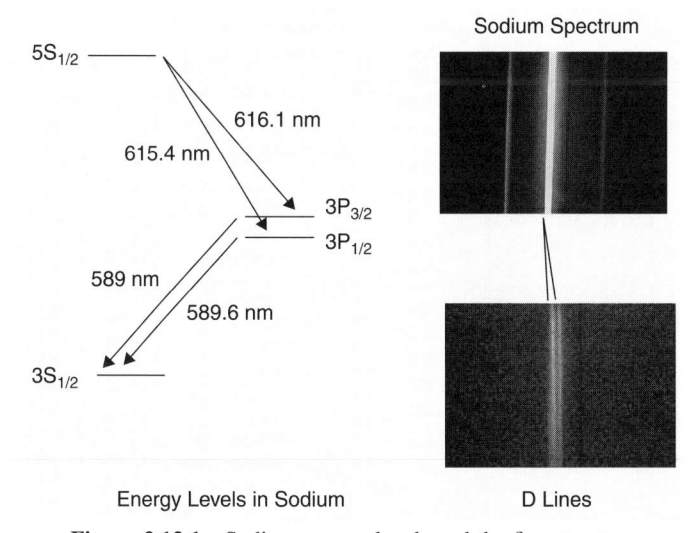

Figure 3.13.1 Sodium energy levels and the fine structure.

atom—levels that would be considered the same level (i.e., $n = 3$) in the Bohr model. Energy must be supplied to the sodium atom to raise its electron to the higher 3p level, perhaps by collision with electrons. There is no guarantee that an electron will enter this level, as it may well enter any other level, such as an $n = 4$ state (or higher), in some cases producing emission lines other than those shown. The energy of the 3s level (which is the ground state for sodium) is lower than that of the 3p level, as expected since electrons in a 3p orbital are more shielded from the attractive force of the nucleus (or, put a different way, an electron in the 3s orbital is more tightly bound to the nucleus). It is also noteworthy that the energy level of any $n = 3$ level for sodium is below that for hydrogen.

The 3S state shown in Figure 3.13.1 is actually designated as $3S_{1/2}$. The subscript $\frac{1}{2}$ is a number referring to a state that is derived from $l - s$ coupling: It represents a combination of orbital angular and spin angular momentums. This number is found by adding and subtracting numbers for l and spin together. In this case the angular momentum is zero and spin has a value of $\frac{1}{2}$, so the only possible number is $\frac{1}{2}$ (remember that a negative result for j, in this case $-\frac{1}{2}$, is changed to a positive number).

The actual upper level for the yellow transition, the 3p level, is found to be two closely spaced levels, $3p_{1/2}$ and $3p_{3/2}$. These two levels have the same principal quantum number and the same orbital quantum number. Again, the subscripts are found by adding and subtracting a spin of $\frac{1}{2}$ from $l = 1$ to yield $\frac{1}{2}$ and $\frac{3}{2}$. The $\frac{3}{2}$ number represents a condition where the electron is spinning in the same direction as orbital angular momentum, while the $\frac{1}{2}$ number represents opposing spins. The difference between the two levels is brought about by electron spin, and the energy difference between these two 3p levels is only 0.0021 eV. Spinning electrons create a magnetic field that interacts with the magnetic field created by orbital motion. Further splitting is possible by applying an external magnetic field[5] in the Zeeman effect (outlined in Section 3.9).

Often, another number is added to denote the multiplicity of a level of given angular momentum l. In this case the level becomes $3^2p_{3/2}$ where the superscript 2 shows that the fine structure turns each level (except $l = 0$) into a *doublet*: two levels with slightly different energies. Be forewarned that there are various spectroscopic notations in use, and they may vary from the one described here. One popular variation is to number the principal quantum number starting from the last filled shell in the atom.

In addition to the yellow D lines, a red line and a green line are visible in the spectrum. Like the D lines, both lines are actually two closely spaced lines. In the case of the red lines, the transitions are shown to originate from the 5S level and terminate in the two 3P levels. The green lines, similarly, originate from a 5D level (not shown on the energy-level diagram) and terminate with the two 3P levels. Overall, there are four doublets resembling the D lines in the visible spectrum of sodium.

---

[5] This is the Zeeman effect, which splits energy levels by application of an external magnetic field. The applied field will produce one hyperfine level for each (quantized) component of the z-component of angular momentum. In the case of sodium, each of the three levels involved (two upper, one lower) will be split into multiple levels, resulting in many fine lines in the spectrum.

3.14 MULTIPLE ELECTRONS: THE MERCURY SPECTRUM

Unlike sodium, mercury has two valence electrons. These two electrons may have their spins in opposite directions, in which case they effectively cancel each other out, resulting in a spin of $s = 0$. The subscript j, then, is simply the orbital angular momentum l. If the spins are in the same direction, the contribution from spin is $s = 1$. Consider the 6S state. $l = 0$, but spin can contribute a value of 0 or 1, so the 3S state becomes two states: $6S_0$ and $6S_1$. The 6P state, with $l = 0$ or 1, can now yield values for j of 0, 1, and 2. These levels are designated as 6^3P_0, 6^3P_1, and 6^3P_2, where the superscript 3 denotes that the multiplicity of the level is three.

Where two electrons are considered, there are two basic types of spin states, triplets and singlets. A *triplet* results from a group of three symmetric states for spin in which the resulting total spin is -1, 0, or $+1$, yielding j values of 0, 1, and 2, as stated above. A *singlet* results when the spin of an excited electron is antiparallel to that of the remainder of the atom or molecule: an asymmetric state where the total combined spin is 0. Consider the simplest two-electron example in helium with two electrons: If the excited electron has a spin in the direction opposite to that of the remaining ground-state electron (which stays in a 1s state), a singlet results (with the excited species called *parahelium*); if the spin of the excited electron is in the same direction parallel to that of the remaining ground-state electron, a triplet results (with the excited species called *orthohelium*).

Transitions resulting from this splitting of energy levels in an atom such as mercury show a triplet of three spectral lines (Figure 3.14.1). Unlike the closely spaced lines of the doublet in the previous example, these lines will be separated by a much larger spacing, since the energy difference between these levels is much larger. In the case of the $6P_2$ level, both electrons are spinning in the same direction as that of angular orbital momentum, so the energy is the lowest in this case. Another feature of multielectron atoms is the fact that no two electrons can have exactly the

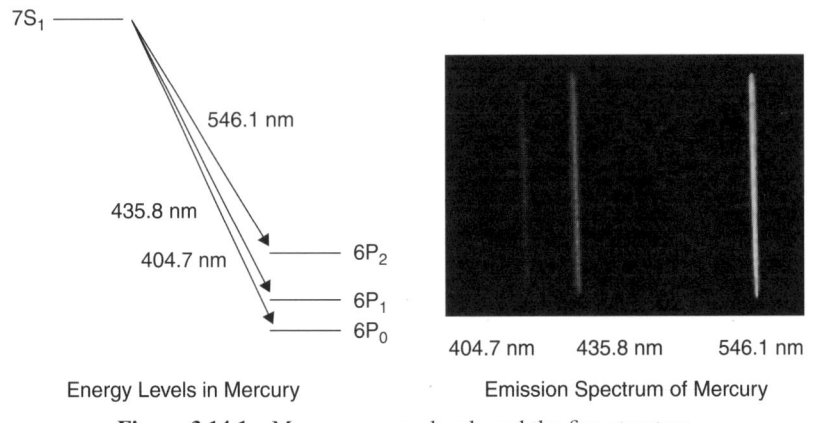

Figure 3.14.1 Mercury energy levels and the fine structure.

3.15 ENERGY LEVELS AND TRANSITIONS IN GAS LASERS

The transitions that bring about light emissions in a spectrum tube also give rise to laser action. In a helium–neon (HeNe) laser, for example, the active lasing species is the neon atom, which has a red transition between two upper-energy levels at 632.8 nm, the red light emitted by a common HeNe laser (Figure 3.15.1). A neon atom at ground state has an electron configuration of $1s^2 2s^2 2p^6$. It is an inert gas in that all outer orbitals are filled, so it is not reactive (it does not react to form compounds). If neon is excited sufficiently, it can achieve a level of $1s^2 2s^2 2p^5 5s^1$ (the excitation mechanism, discussed in Chapter 4, involves helium). From that level it can fall to the $1s^2 2s^2 2p^5 3p^1$ level and in doing so emit a photon of red light at 632.8 nm. This is the transition used for the red HeNe laser.

As one might imagine, the two levels involved are not singular levels but rather a group of closely spaced energy levels. The upper level ($2p^5 5s^1$) is actually four hyperfine levels, and the lower level ($2p^5 3p^1$), 10 separate levels! Unlike sodium or hydrogen, there is more than one electron in the outer shell; in the case of neon, a total of eight electrons exist in the outer ($n = 2$) shells. These electrons interact to produce many possible energy levels. As a result, there are numerous transitions at which the helium–neon laser can operate. Some transitions are favored

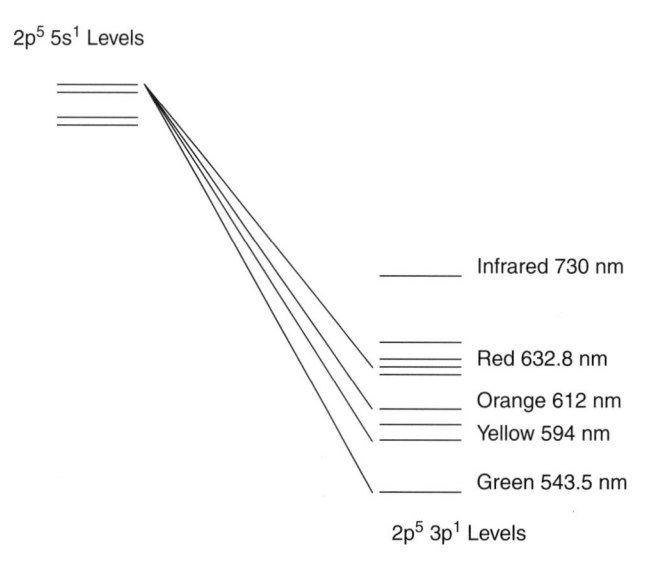

Figure 3.15.1 Energy levels for neon transitions.

over others, and not all will produce laser light.[6] Although the red transition at 632.8 nm is the strongest of the visible transitions (and the most commonly available), a green transition also exists at 543.5 nm, allowing a green HeNe laser. In all, there are 10 possible wavelengths between red and green—at least four are commercially available as off-the-shelf HeNe lasers (the red and green mentioned above plus a yellow and orange transition). These energy levels are shown in Figure 3.15.1 as well as the corresponding emission wavelengths.

3.16 MOLECULAR ENERGY LEVELS

Most transitions from gas discharges (e.g., those in a HeNe or argon-ion gas laser) have very well defined wavelengths of the type illustrated throughout most of this chapter. These are *electronic transitions*, which involve an energy change in only electrons of the species. Precise and well-defined energy states characterize these transitions, which are responsible for most visible gas laser transitions as well as those in the ultraviolet. Other energy levels possible are *vibrational* and *rotational levels*, brought on, in molecules, by various supported modes of movements of individual atoms relative to each other. In the simplest case, a diatomic molecule such as nitrogen (N_2) or hydrogen (H_2) is composed of two atoms which are free to vibrate only in certain allowed ways. When such a molecule is formed, electrons in the outer orbitals of the individual atoms are shared with neighboring atoms to form complete electron shells. In the case of a hydrogen molecule, the single electrons from each hydrogen atom combine to form a shared orbital for both atoms, this orbital belonging to the molecule as a whole rather than to an individual atom. The resulting molecule has the property that the electrons are divided symmetrically between the two hydrogen nuclei and there is no net displacement of electrical charge. As a consequence, each atom then "sees" an apparently filled $n = 1$ shell with two electrons in it, which leads to great stability (in other words, it will not react with other atoms to form yet other molecules), and the total energy of the molecular system is decreased when the molecule is formed (and the lowest energy state is always the most stable one).[7] Such shared bonds, called *covalent bonds*, are formed, in the case of the hydrogen atom, with electrons of opposing spin, as dictated by the Pauli exclusion principle. You will recall that this principle states that no two electrons may have exactly the same set of quantum numbers, so two electrons can occupy the same molecular orbital only if they have opposite spin.

This molecular bond is not rigid (as you may have been led to believe while using molecular model sets) but rather, is flexible and may be stretched in various ways as

---

[6] In the case of transitions in the helium–neon laser it is necessary to suppress stronger transitions in order to allow weaker transitions to be manifested. A green HeNe laser, for example, has special optics to ensure that the red transition is not amplified in the laser. This is illustrated in Chapter 6 when wavelength-selective optics are discussed.

[7] The covalent bond might also be thought of as overlapping atomic orbitals; in the case of hydrogen, two half-filled 1s orbitals overlap to produce a completed shell. This is not an exact view of the situation, but it does provide a simple conceptual model.

the atoms move. Different motions require different energies, which correspond to photon energies in the infrared region of the spectrum. Modeling a diatomic molecule as two weights attached by a spring (the covalent chemical bond between the two atoms), we may utilize kinematics to predict the behavior of this system, which resembles a simple harmonic oscillator. Allowed modes of vibration (and hence corresponding energy levels) are depicted in Figure 3.16.1.

As the nuclei of the two hydrogen atoms deviate from the normal separation for a hydrogen molecule, the energy increases and the vibrational mode (denoted by v in the figure) increases. Not surprisingly, a molecule with more energy tends to vibrate more. Transitions can take place between two of these vibrational levels, resulting in a purely vibrational transition with energies corresponding to transitions in the infrared region. It is also possible to combine vibrational states such as those shown with electronic states to form hybrid *vibronic levels* and transitions. It is these type of transitions (from one vibronic energy level to another) that are responsible, for example, for the molecular nitrogen laser output at 337.1 nm in the ultraviolet.

A more complex molecule, such as carbon dioxide, is depicted in the model of Figure 3.16.2. This molecule features a single carbon atom chemically bonded to two oxygen atoms with double covalent bonds acting as springs leading to vibrational modes each having unique levels. Such double bonds are simply the sharing of two electrons in a manner similar to that discussed previously. Of course, one would expect a double bond to be stiffer than a single bond and so have correspondingly higher vibrational energies. Several modes are shown in the figure,

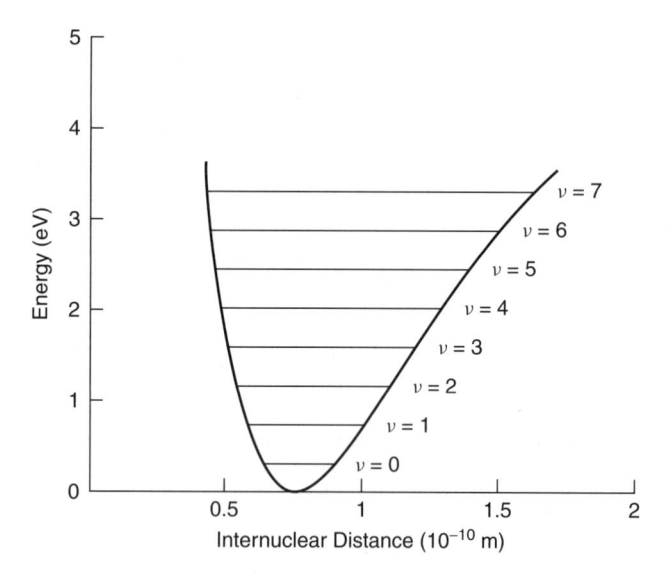

Figure 3.16.1 Vibrational energies in the hydrogen molecule.

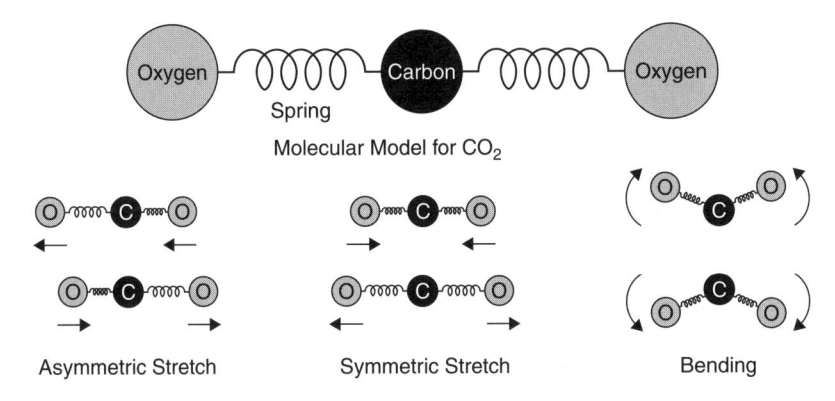

Figure 3.16.2 Model of the CO_2 molecule and vibrational modes.

including an asymmetric stretch (where one bonds stretches while the other compresses), symmetric stretching (where both bonds stretch or compress simultaneously), and a bending motion. Energies for such vibrational levels are again quite low (i.e., the difference between allowed energy states is quite minute), and hence transitions between allowed vibrational modes correspond to the infrared and far infrared regions of the spectrum [e.g., the carbon dioxide laser output, generated primarily by transitions between vibrational modes, is in the infrared at 10.6 μm, which corresponds to an energy of 0.117 eV; unlike the nitrogen laser, electronic levels are not involved here].

Picturing each molecular bond as a spring between two weights, it is easy to imagine the oscillator (in a classical sense) as being confined to oscillating on certain discrete frequencies. It could oscillate at its natural resonant frequency v_0 or at twice this frequency, but not at a noninteger multiple such as 1.5 times this frequency, since this would not allow a standing wave in the spring (and would hence quickly be dampened). This is the quantized nature of atomic energy levels and applies to these types of levels as well. Oscillating at $2v_0$, we expect the energy level for this vibration to be twice as large as that of v_0. In Figure 3.16.3, the energy levels for carbon dioxide as well as the corresponding vibrational modes are outlined. Lasing transitions (those responsible for the actual laser output) are shown in the figure between a higher-energy allowed vibrational state and two lower-energy allowed vibrational states.

The energy levels for the CO_2 molecule are assigned a three-number designation that describes the vibrational mode as well as the frequency. In the case of the bending mode, two modes (010 and 020) appear in the figure. Mode 010 corresponds to a frequency of v_0, while 020 corresponds to a frequency of $2v_0$, which has twice the energy of mode 010. Energy is provided to the CO_2 molecule from nitrogen in what is termed a *four-level system* (which we examine in detail in Chapter 5). This *pump energy level* in nitrogen is also a vibrational energy level (logical since electronic levels would correspond to a much higher energy). Nitrogen (N_2) is composed of

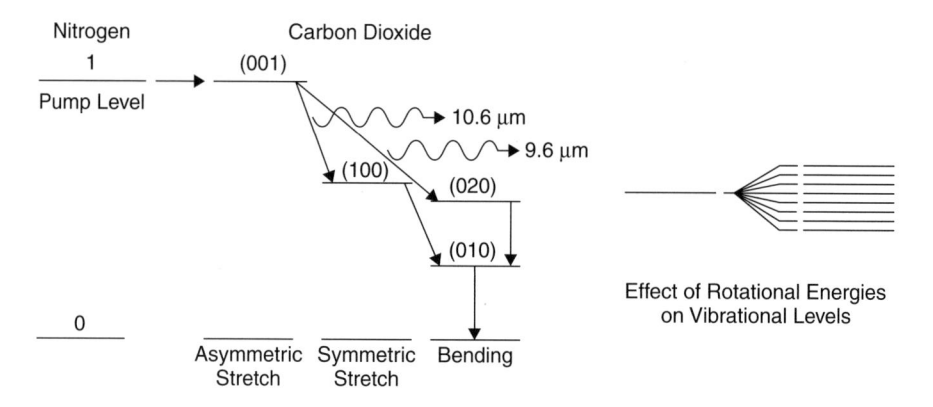

Figure 3.16.3 Energy levels in the CO_2 laser.

two nitrogen atoms held together by a bond and vibrates only in certain allowed modes. The energy of one of these purely vibrational modes (not to be confused with the vibronic modes mentioned earlier in this section) corresponds closely with the upper lasing level for the CO_2 laser, so nitrogen acts as a pump.

As well as vibrational energies (which usually correspond to transitions in the infrared), there are *rotational energies* involved in this system. These energies originate from various modes in which an outer atom rotates about a central atom—in the case of CO_2, an oxygen atom literally rotates around the axis of the carbon and other oxygen. As one might imagine, there is a great deal of freedom to rotate in a given molecule, so there are many more possibilities for rotational modes in a large molecule than there are vibrational (although it, too, is constrained and even rotational levels are quantized). With many possibilities, rotational energy levels appear in bands of tightly clumped energy levels. Purely rotational energy levels are possible as well; these correspond to transitions in the far infrared and are seen in many organic molecules. In the CO_2 laser, vibrational energies combine with rotational energies to turn each vibrational level into a tight group of energy levels, as depicted in Figure 3.16.3. The result is that the two transitions of the CO_2 laser centered at 9.6 and 10.6 μm are actually a series of closely spaced spectral lines spanning the region from about 9.2 μm to almost 11 μm. In Chapter 6 we examine methods for tuning one specific line for a single-wavelength laser.

Aside from the preceding examples of vibrational, rotational, and hybrids of these two types of levels, an energy level in a species might well involve all three types of levels. This leads to energy levels that are quite broad (we speak of energy *bands* at this point), so the resulting laser line is also quite broad spectrally. Dye lasers usually involve all three types of energy-level mechanisms, which leads to a continuum output over a given range. By tuning the feedback mechanism (covered in Chapter 6), one can selectively amplify a given wavelength: These lasers are truly tunable.

3.17 INFRARED SPECTROSCOPY APPLICATIONS

The most dramatic example of the effect of vibrational energy levels (and evidence of various bond energies) is the application of infrared spectroscopy. This technique allows analysis and identification of complex molecules by their absorption "signature" of wavelengths in the infrared. In this technique, incident infrared energy causes covalent bonds between atoms in molecules to vibrate. Because bonds between atoms of different masses have different spring constants, they vibrate at wavelengths characteristic of the particular chemical bond involved. A single carbon-to-carbon bond, for example, will vibrate at a different wavelength than a carbon–oxygen bond, so the features of a particular molecule (indeed, the molecule itself) can be identified by the IR absorption spectrum. The carbon–oxygen bond, for example, will always vibrate at the same wavelength, regardless of where the bond is in a given molecule, so the appearance of an absorption peak at a specific wavelength indicates the presence of that bond in a sample molecule.[8]

A typical instrument used for such analysis is shown in Figure 3.17.1, in this case an infrared gas analyzer used to identify an unknown gas (and the concentration of the gas) by its infrared absorption spectrum. Broadband infrared radiation from an

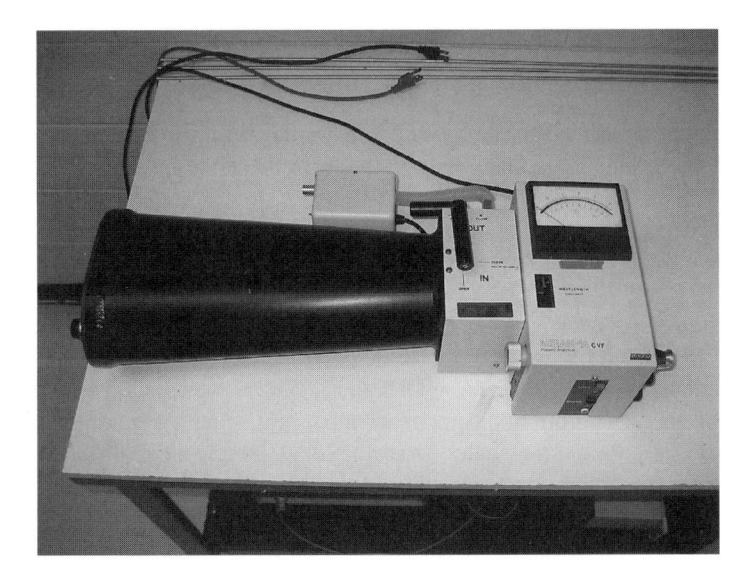

Figure 3.17.1 Infrared gas analyzer.

---

[8] This, too, is a bit naive since the vibrations of a bond will certainly be affected by the rest of the molecule, but features of a particular molecule such as a carbon–oxygen bond are always found in a certain narrow wavelength region. In the case of a carbon–hydrogen bond (the most common in organic molecules), the exact wavelength at which the absorption appears is determined by the hybrid orbital formed between the two atoms and indicates the chemical family to which this bond is attached.

78 QUANTUM MECHANICS

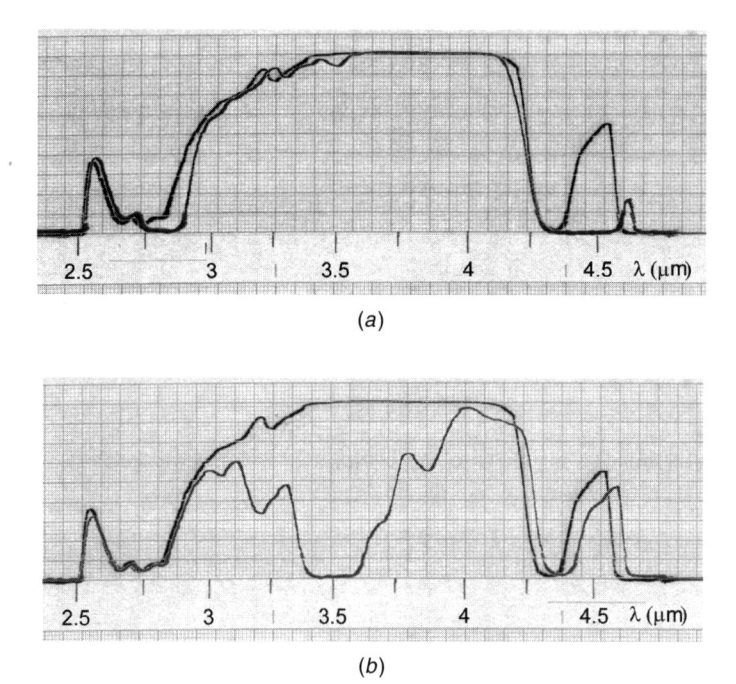

Figure 3.17.2 Infrared absorption spectrum of CO_2 and propane.

internal source is beamed through a cell containing the gas to be analyzed. In this case the gas cell (the large black unit on the left side of the instrument) contains a folded-mirror arrangement yielding a path of 20 m for the radiation. Infrared radiation exiting the cell is then analyzed using an infrared spectrophotometer to determine which wavelengths were absorbed by the molecules of the sample in the cell.

The infrared spectrum between 3 and 7 μm[9] is called the *functional group region* and is used to identify the type of molecule based on the presence of certain groups of atoms, such as —CH_3, —CH_2, —OH, and —NH, as well as doubly- and triply-bonded carbons. Absorption spectra from this region identify common chemical groups that compose the molecule: for example, all simple alcohols have a similar absorption pattern in this range since they all feature an —OH (oxygen–hydrogen or *hydroxyl*) group as well as a —CH_3 (*methyl*) group which terminates the molecular chain. Other regions of the spectrum are used to identify the individual molecule based on its individual "fingerprint."

Consider the infrared absorption spectrum of carbon dioxide (CO_2) in the near-IR region of 2.5 to 4.5 μm. Comparing a dilute sample of this gas to the reference spec-

---

[9] It is common practice to express infrared wavelengths as a wavenumber with units of cm^{-1}. To convert to wavelengths, the inverse is multiplied by 0.01, so it now has units of meters. λ (m) = 0.01/wavenumber (cm^{-1}).

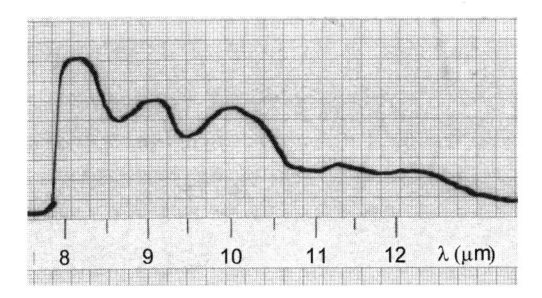

Figure 3.17.3 Infrared "fingerprint" of propane.

trum of air in Figure 3.17.2(*a*), we find that there are no discernible absorption peaks caused by the sample; in other words, CO_2 does not absorb infrared radiation in this band. The reason is that there is only one type of bond in this molecule, a carbon-to-oxygen double covalent bond, which has an allowed vibrational mode with energies corresponding to wavelengths outside the range we are examining. Contrast this to the absorption spectrum of propane shown in Figure 3.17.2(*b*), which reveals a complex structure of absorption wavelengths.

Propane, which has the chemical structure CH_3—CH_2—CH_3 contains numerous features which are evident in this region, including absorption in the area of 3.2 to 3.4 μm, corresponding to carbon–hydrogen bonds in the molecule. Most interesting, perhaps, are sets of absorption "peaks" (which, I suppose, really should be called "valleys"), which correspond to specific functional groups such as the —CH_3 and —CH_3 groups of which this molecule consists. At this stage we are able to identify component groups; however, for the exact chemical composition, the region to examine is the fingerprint region. In this case the absorption spectrum for propane between 8 and 12 μm (Figure 3.17.3) reveals a pattern of peaks unique to propane alone.

Simple gas analysis is performed by matching peaks in the sample spectrum to known reference spectra for various gases. Advanced infrared spectroscopy instruments utilize interferometry techniques, combined with Fourier transform mathematics, to yield high resolutions in which closely spaced peaks can be resolved. Aside from an interesting demonstration of molecular energy levels, infrared spectroscopy is an extremely important analytical tool, with uses ranging from forensics to environmental science.

PROBLEMS

3.1 Calculate the deBroglie wavelength for electrons at 0.01 eV, 1 eV, and 10 keV.

80 QUANTUM MECHANICS

3.2 Show that quantum mechanics and classical mechanics converge by considering the energy of an electron in hydrogen at $n = 2$, $n = 200$, and $n = 2000$ states. In each case, calculate the energy using both classical and quantum means and compare these.

3.3 Electron microscopes use high-energy electrons instead of photons of light since these have shorter wavelengths and hence offer better resolution of small objects. To study viruses, a wavelength of 0.1 nm is required (as opposed to 400 nm for violet light). Calculate the accelerating voltage required to produce electrons at this wavelength.

3.4 Describe all possible states (in spectroscopic notation) for electrons as follows:

(a) A hydrogen atom with an electron in the P orbital
(b) A mercury atom with an electron in the D orbital
(c) A helium atom at ground state

3.5 For an argon-ion laser, there are commonly 10 transitions which lead to visible light emission. Ten are shown on the simplified energy-level diagram shown in Figure P3.5, but there are at least five more possible transitions.

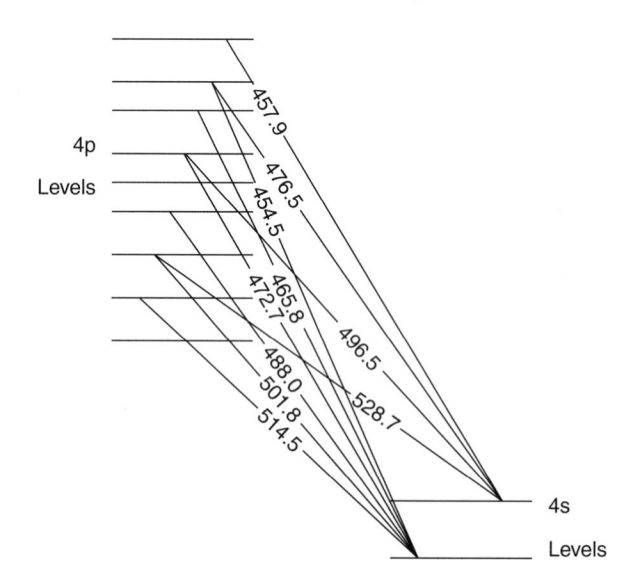

Figure P3.5

(a) Identify the other five possible transitions on the diagram.
(b) By calculating the energies of the known transitions on the diagram, deduce the wavelengths of these transitions.

3.6 Electron shells usually fill as 1S, 2S, 2P, 3S, and so on; however, after 3P it is found that the 4S level fills *before* the 3D level since its energy is actually lower (see Figure 3.8.2). Knowing this information, show the electronic configuration for:

(a) Neon (atomic number 10)
(b) Argon (atomic number 18)
(c) Potassium (atomic number 19)
(d) Titanium (atomic number 22)

(These configurations can be verified in the *CRC Handbook of Chemistry and Physics*) (Boca Raton, FL: CRC Press.)

CHAPTER 4

Lasing Processes

In this chapter we examine the basic principles and processes that make the laser possible. Throughout the chapter we use gas lasers as an example, the most common being the helium–neon (HeNe) laser. This was one of the first lasers, discovered in the early 1960s. Its red output has found uses in everything from holography to bar code scanning. A few years ago it was the most common laser, but its dominance has been replaced by cheaper semiconductor lasers. Still, it makes an excellent example of how a typical laser operates.

The basic laser itself consists of a glass tube filled with helium and neon gases in a ratio of about 10 parts helium to 1 part neon. The internal pressure is low (about 1.8 torr, where 760 torr is 1 atmosphere), allowing a sustained electrical discharge. The electrical energy required is supplied by a high-voltage power supply, which is often encapsulated in a small block of epoxy material. When energized, the tube glows a bright pink color. Most common HeNe lasers have a red output at 632.8 nm; however, the quantum mechanics of the neon atom (which is the active lasing species) also allows lasing transitions in the orange, yellow, and green. The discharge itself takes place between a small anode and a much larger cathode through a tiny capillary tube called a *plasma tube*. At either end of the laser are cavity mirrors. One mirror is fully reflecting, the other partially reflecting. The tiny portion of light (typically around 1%) that is transmitted through the front mirror is the actual laser beam itself. The structure of a typical HeNe laser is depicted in Figure 4.0.1.

Essentially all modern HeNe lasers have a coaxial design similar to that of Figure 4.0.1, in which the plasma tube is in the center of the laser with the cathode surrounding it. The large cathode and surrounding glass envelope act as a reservoir for the gas mixture. This reservoir is required since some gases, such as helium, have a tendency to diffuse out of the tube given their low atomic mass. The operating conditions of many lasers also result in consumption of the laser gas, so a large reservoir extends their useful life. Many large gas lasers have an actual gas reservoir attached to replenish used gas as the laser operates. (We examine the argon laser in detail later; argon lasers usually have elaborate systems to ensure that the gas pressure stays constant.)

Fundamentals of Light Sources and Lasers, by Mark Csele
ISBN 0-471-47660-9   Copyright © 2004 John Wiley & Sons, Inc.

84 LASING PROCESSES

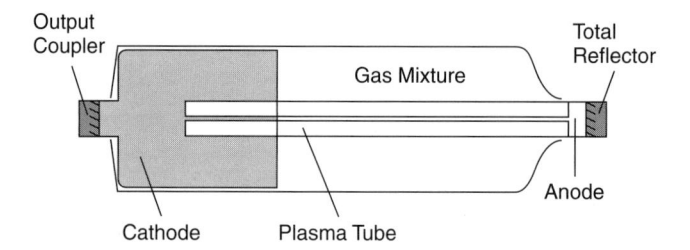

Figure 4.0.1 Elements of a helium–neon laser tube.

As we progress through the chapter we examine the key processes involved in this and other lasers. In subsequent chapters we further refine our understanding of lasing processes by examining individual elements in detail.

4.1 CHARACTERISTICS OF COHERENT LIGHT

In the first few chapters we have examined *incoherent light*, the light normally produced by gas discharges, blackbody sources, and the like. We've also examined the basic quantum mechanics behind these sources. Lasers, like any other light sources, are governed by the same rules and principles, but laser light is unlike any other source found in nature. It has three special properties that lead to its usefulness in many applications: coherence, monochromaticity, and collimation (directionality) as shown in Figure 4.1.1.

Coherence is the most interesting property of laser light. This property states that all photons emitted from a laser are at exactly the same phase; as waves they all "crest" and "valley" at the same time. Coherence is brought about by the mechanism of the laser itself (stimulated emission), in which photons are essen-

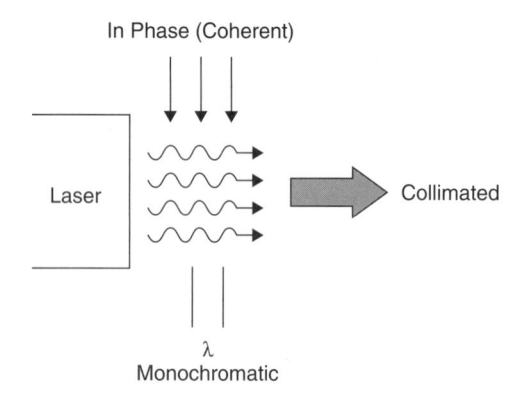

Figure 4.1.1 Properties of laser light.

tially copied. To stay in phase it is required that all emitted photons are at exactly the same wavelength (or very, very close). If some photons are at a different wavelength, the phase of those photons relative to others will be different and the light will not be coherent. They must also be highly directional, all moving in the same direction. The property of coherence, then, gives rise to the other two properties of laser light.

Coherence is not trivial and is brought about by the amplification mechanism of the laser. Think of a beam of incoherent light as having a large number of photons, all acting as waves, with random phases relative to each other. As with any other wave, constructive and destructive interference will occur. The beam is a jumble of all sorts of photons at various wavelengths and phases. In a laser beam all photons travel in lockstep with each other, cresting at the same time. The photons complement each other, rendering high amplitudes (high intensities in our case). Indeed, although a 1-mW laser beam does not sound like much power, consider that this intensity is packed into a spot 1 mm in diameter. As a density this represents 1.2 kW/m^2, a level surpassing the intensity of the sun on the surface of the Earth!

Monochromaticity is the ability of the laser to produce light that is at one well-defined wavelength. It is a requirement for coherence since photons of varying wavelengths cannot be coherent (i.e., in phase). When white light is dispersed through a prism, you note that it is composed of an infinite number of wavelengths of light covering the entire visible spectrum as well as into the UV and IR regions. Now consider an emission line from a gas discharge. These lines are much narrower when viewed on a spectroscope. The range of wavelengths spanned by such a line depends on many factors, such as gas pressure, but suffice it to say that it is finite and could well be 0.1 nm in width. As we shall see in this chapter, the light produced by most lasers is much narrower, spectrally. A standard HeNe laser will produce a red light centered at 632.8 nm with a linewidth of only 0.002 nm! That is much narrower than any incoherent source can produce. Some lasers can produce multiple output lines; the argon-ion laser produces up to 10 lines in the blue–green region of the visible spectrum. Each of those lines is also (spectrally) very, very narrow. The argon laser has many allowable lasing transitions. (Quantum mechanics shows that argon has intricate and complex energy-level structures.) Each of those transitions lases, and many can lase simultaneously, but each produces a spectrally narrow output.

Collimation is the property of laser light that allows it to stay as a tight, confined beam for large distances. It can be thought of as the spread in a beam of light (called *divergence*). This property of laser light makes it possible to use the laser as a level in construction or to pinpoint speeders on a highway. The simplest explanation for the highly directional output of the laser is in the mechanism of the laser itself. In our example of the HeNe laser, photons of light inside the tube must be reflected by the mirrors many times (hundreds of times) before exiting through the front of the laser tube. To do this, photons must be very well aligned to the axis of the tube: Photons emitted at even a slight angle to the tube axis will be bounced into the walls of the tube and will not contribute to the output beam. All laser beams have some amount of divergence. Some lasers with very high gains such as nitrogen lasers have wider divergences, whereas others, such as the HeNe laser, have low divergences. The

4.2 BOLTZMANN DISTRIBUTION AND THERMAL EQUILIBRIUM

Recall from Chapter 1 that atomic populations of a system without an external source of energy are governed solely by temperature according to a Boltzmann distribution (Figure 4.2.1). This distribution predicts an exponential drop in atomic populations of higher energies. Recall that a central concept of quantum mechanics is the condition that atomic energies are confined to certain allowed, discrete states. Applying Boltzmann's law, we may now predict the population of atoms at each discrete energy state.

Consider mercury atoms, discussed in Section 3.14. The first energy level is at 4.66 eV above ground, or (multiplying by 1.602×10^{-19} J/eV) 7.46×10^{-19} J. If we have 1 mole of atoms (6.02×10^{23} atoms, a standard quantity in chemistry and physics) at ground state, then at room temperature (293 K) the number of atoms at this high-energy state is on the order of 10^{-57}! It is fair to say that there are essentially no atoms with an energy this high at room temperature, so no light will be emitted from the mercury, as no transitions are possible. Now, if the temperature is increased to, say, 2000 K, the population of atoms at this particular energy level increases to 10^{12}. A sizable concentration at this energy level will lead to emission of light when these atoms lose their energy by emitting photons and falling

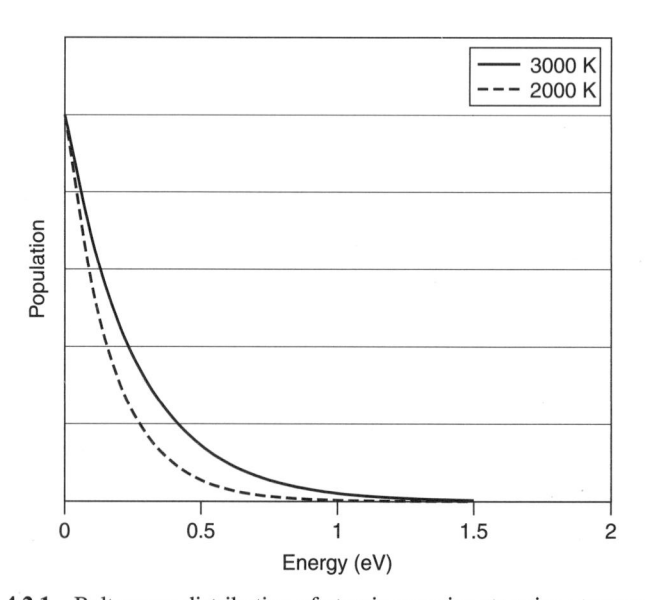

Figure 4.2.1 Boltzmann distribution of atomic energies at various temperatures.

back to ground state. Experience tells us that anything heated to 2000 K (e.g., an electric stove element) will emit light in the form of a glow.

Such a system is said to be at *thermal equilibrium*. The population of atoms at any given energy level is governed solely by temperature, since there is no external source of energy. As we have seen from the calculations, if the temperature of the entire system were raised, the distribution would shift and more atoms would reach higher energies. However, the population of a lower level will always exceed that of a higher level. Had we chosen an incandescent lamp (a "blackbody radiator"), the situation would be the same. Electrical energy heats a tungsten filament. Many atoms at high energies drop to a lower-energy state and in doing so, emit a photon of light. The difference in energy between the upper and lower energy states for the transition is manifested as the photon energy and hence the wavelength of the emitted light. In such a system, there are a large number of upper and lower levels, and these levels may span a range of energies so that the output from such a source is *broadband*. It is not at a single discrete wavelength like the line spectra emission from our gas, but rather, spans a range of wavelengths. No matter how the example above is considered, the situation still arises where there is a higher population of atoms at lower than at higher energy states.

4.3 CREATING AN INVERSION

Consider now a situation where energy is injected into a system at thermal equilibrium to cause a population of atoms at a higher energy level to be greater than that of a lower level. Such a nonequilibrium condition is indeed required for lasing action in almost all lasers.[1] Energy may be injected selectively to pump an upper energy level from which transitions occur to a lower level (in doing so, photons of light are emitted). It is not necessary to excite all upper energy levels in the gas: The requirement is that a higher population of atoms exists at the upper level of a particular transition than at the lower level of that same transition. The reasons for this will become clear when considering rates of stimulated emission.

Pumping is the process of supplying energy to the laser medium to excite the upper energy levels. It may be accomplished by any number of means, including electrical, thermal, optical, chemical, or nuclear. Most common lasers utilize electrical pumping (as seen in the previous example with the helium–neon laser) or optical pumping, as shown in Figure 4.3.1. In the case of optical pumping, light from a flashlamp or an arc lamp is focused onto a rod containing the lasing atoms. Common laser rods include ruby (chromium ions in an aluminum oxide host glass) and YAG (neodymium ions in a yttrium–aluminum–garnet host glass). The lasing atoms (chromium or neodymium in this case) absorb photons of incident pump light and

---

[1]Essentially, all lasers require population inversion to work. Only one special class of lasers [lasers without inversion (LWIs)] operates without such inversions. In this laser, energy is extracted from a metal vapor such as sodium without an inversion. See the reference in *Physical Review Letters*, Vol. 70, No. 21, pp. 3235 (1993) by E. S. Fry et al.

88 LASING PROCESSES

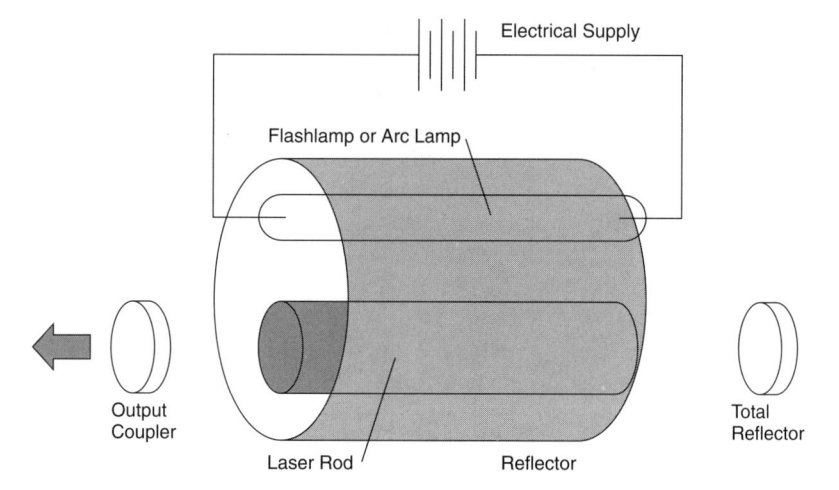

Figure 4.3.1 Optical pumping for lasers.

become excited to upper energy levels. An electrically pumped laser (like most gas lasers) uses an electrical discharge to excite atoms directly to high energy levels. In some lasers (e.g., the HeNe), the actual lasing atom is neon, which is pumped with energy via a multistep process involving the transfer of energy between two atomic species. In this case, helium serves to pump the neon atoms when a helium atom (excited by the electrical discharge) collides with a neon atom, transferring energy to the neon atom.

In addition to optical and electrical pumping methods (the most common), other ways of injecting energy into a laser include chemical reaction and nuclear pumping. Regardless of the method, the end goal of pumping is to excite high-energy states within the lasing medium so that the population of atoms at a high-energy state is greater than the population of atoms at a low-energy state for the lasing transition.

One might now wonder how pumping can be effected to ensure that the upper level is filled without filling the lower level (i.e., creating an inversion). The answer lies in the design of the pumping source or method to ensure that the upper level is pumped selectively while the lower level is not. Consider the HeNe gas laser, which consists of a glass tube containing both helium and neon gases in which an electrical discharge occurs. The discharge, similar to that in the tubes of a neon sign, excites helium atoms by electron collision. The vast majority of energy is wasted in the process, some of it in the pink glow given off by the tube, still more in the form of heat. Some excited helium atoms (with a specific energy of 20.61 eV) can collide with neon atoms and in doing so transfer energy to them at an almost identical energy level (the target is the neon level at 20.66 eV). The energy levels involved are depicted in Figure 4.3.2. This energy level for neon, which accepts energy from the helium, is the upper level for the transitions that produce the visible HeNe laser output. As expected for a multielectron atom such as neon, quantum mechanics predicts a splitting of energy levels into a series of closely spaced levels. In the case of neon, the actual upper level at 20.66 eV splits into four discrete levels.

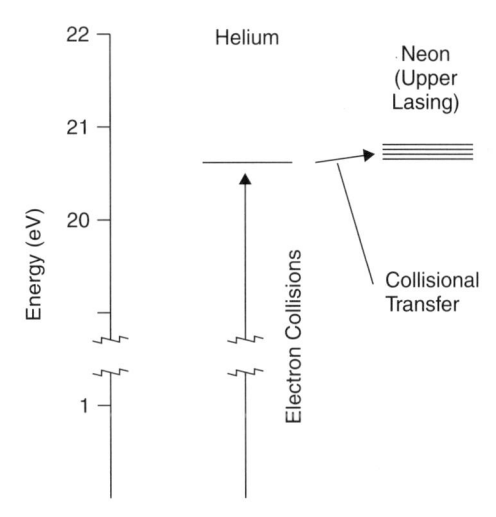

Figure 4.3.2 Pumping mechanism in the helium–neon laser.

Through this process neon atoms are then pumped, quite selectively, to an upper energy level from which lasing transitions can occur (helium lacks a level close to the lower lasing level of neon, so it does not receive energy from collisions with helium). This scheme ensures that the upper energy level for the transition, which produces laser light, has a higher population of neon atoms than the lower level. We can contrast the situation in the HeNe laser to the Boltzmann distribution of neon at thermal equilibrium in Figure 4.3.3 to see the inversion of the levels.

In Figure 4.3.3 the populations of neon atoms are compared both at thermal equilibrium and in an operating laser. The two levels of interest are the $2p^5 5s^1$ and $2p^5 3p^1$ levels. (This is the configuration of the excited neon atom from Chapter 3, which describes the state of electrons in the atom. Recall that neon at ground state has a configuration $2p^6$ or, more fully expanded, $1s^2 2s^2 2p^6$.) These are the upper and lower lasing levels for the 632.8-nm red transition, the common red HeNe laser. As we shall see in Chapter 5, numerous other lasing transitions are poss-

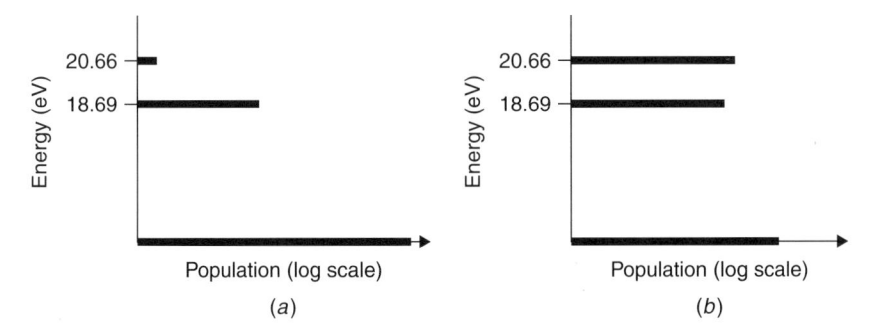

Figure 4.3.3 Population energies of neon atoms (*a*) at equilibrium and (*b*) in a HeNe laser.

ible given neon's many energy levels (e.g., electron spin leads to a hyperfine structure where the $2p^5\,3p^1$ level is actually 10 closely spaced levels, giving us 10 possible laser transitions!). Using Boltzmann's equation, it is possible to calculate the populations of atoms at these two levels (or at any other level). The energy of the upper level is 20.66 eV (or 3.31×10^{-18} J). Knowing that the photon is at 632.8 nm, we calculate the lower level to be at 2.99×10^{-18} J, and the rest is trivial. As expected, the population of atoms at the 20.66-eV level for neon in thermal equilibrium is extremely low, with $N/N_0 < 10^{-100}$ for room temperature neon. Even if there were a few atoms at these two levels involved in the lasing transition, there would still be many more at the lower level than at the upper level.

In an operating laser the upper level is selectively pumped with energy so that (1) there are many more atoms at the upper level than when at equilibrium, and (2) there is a higher population of atoms at the upper level than at the lower level. Because excited helium atoms (Figure 4.3.2) transfer energy to the 20.66-eV neon level and not the 18.69-eV level, the lower level will not fill from collisions the way the upper level does (the level for helium is not at all close to the 18.69-eV level). This situation produces the inversion we are seeking.

It is no accident that helium is used in a HeNe laser: Helium has an excited energy level very close to the level of neon used as the upper level for the lasing transition. Direct electrical discharge in pure neon will not excite the upper level enough to cause an inversion (the lower levels would fill as well in that case).[2] It must also be noted that the neon never ionizes in the HeNe laser described (i.e., loses an electron). Neon atoms in the HeNe laser remain neutral; they just acquire more energy, and a valence electrode is promoted to a higher orbital. Examination of the electron configuration for the upper energy level ($2p^5\,5s^1$) reveals that all six valence electrons are still intact. In some lasers, the energy levels involved in lasing occur in the ion of the atomic species, but in this particular case it remains a neutral (but excited) atom. Ionized neon will lase under the right conditions, but the neon-ion laser is completely different in mechanism and spectral characteristics from the common HeNe laser.

4.4 STIMULATED EMISSION

If a photon of light at an energy of, say, E is fired into a cell containing atoms at ground state, it is possible that it will be absorbed and, in the process, pump the atom to a higher-energy state. The energy of the incident photon must be at least equal to that of the upward transition. This, of course, is absorption. If the atom is already in a high-energy state, though (the upper-energy level we refer to continu-

---

[2] Pure neon will, in fact, emit visible laser light under the right conditions. Such a laser is similar to that of a nitrogen laser, in which a fast transverse discharge current of thousands of amperes excites upper levels preferentially, producing an inversion. Neutral neon atoms will then emit green light at 540.1 nm. This laser is, however, pulsed (about 1-ns pulses) and cannot operate CW since the lower level has a much longer lifetime than the upper level, and once full, violates the population inversion criteria required for lasing action.

ally), and a photon of the correct wavelength comes along (the wavelength corresponding to a transition from the upper state to a lower state in the atom), it can stimulate the excited atom to emit a photon of exactly the same wavelength and phase as the incident photon, leaving two photons exiting this process going in exactly the same direction. The fact that these two photons are identical makes the emitted radiation coherent and monochromatic, two key properties of laser light. The fact that they are emitted in the same direction will play a role—along with a well-aligned cavity—in making the light collimated, the third key property of laser light. In essence, the original photon is *amplified* by this process, which is called *stimulated emission*. The word *laser* is an acronym for "light amplification by stimulated emission of radiation"; stimulated emission is the *key* process here. The process is diagrammed in Figure 4.4.1.

Of course, the atom that emits the photon loses its energy in the process and must be pumped to an excited state again or it will reabsorb another photon. This is why population inversion is (generally speaking) required: If inversion is not maintained, atoms will absorb rather that emit photons of light. This is a not trivial outcome. When the population of atoms at the lower state exceeds that of the upper state, emitted photons of laser light are actually absorbed and laser action is not possible. This effect can actually be demonstrated by passing light from a gas laser (e.g., a HeNe laser) through a cell containing the same—but unexcited—gas as the laser. In the process of passing through the cell, the power of the laser will actually be measurably reduced.

At this point, the analysis of the situation becomes statistical. Statistically speaking, we require a higher rate of stimulated emission than absorption. Each of these rates can be computed. Another consideration, though, is spontaneous emission. This creates an alternative, unwanted, pathway, allowing atoms in the upper energy state to lose their energy by emitting a photon of light spontaneously. In the case of

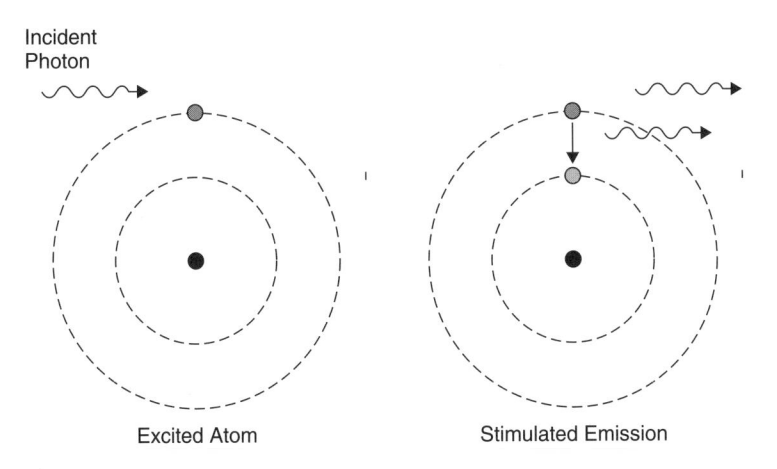

Figure 4.4.1 Stimulated emission from an excited atom.

92 LASING PROCESSES

the helium–neon laser, this spontaneous emission shows up as the pink glow from the tube (which is, of course, composed of discrete lines like any other low-pressure gas discharge). Spontaneous emission robs the upper level of atoms that would otherwise be available for stimulated emission. Finally, we must consider attenuation of emitted laser light by the gain medium as well as the mechanisms of the laser itself (windows, mirrors, etc.). Generally, all mechanisms causing a reduction in laser gain (absorption in the lasing medium and optics, etc.) are referred to as *losses*, which we consider later.

4.5 RATE EQUATIONS AND CRITERIA FOR LASING

The general criteria for a net photon gain to occur (i.e., a *laser gain*) is that the rate of stimulated emission must exceed that of spontaneous emission plus that of all the losses. We begin by determining, mathematically, the parameters involved. The rate of absorption of photons depends on the number of atoms in the lower state (i.e., the number of atoms available to absorb photons) as well as the energy density of incident photons. The second parameter should be obvious: More photons to absorb leads to a higher rate of absorption. Mathematically, the rate of absorption may be stated as

$$r_{\text{absorption}} = B_{12} N_1 \rho \tag{4.5.1}$$

where B_{12} is a proportionality constant called *Einstein's coefficient*, N_1 the number of atoms at the lower-energy state, and ρ the energy density. The energy density in this case is specific: It represents the number of photons that have the exact energy for the transition between energy levels E_1 and E_2. Similarly, the rate of stimulated emission depends on the number of atoms at the upper state and can be written as

$$r_{\text{stimulated}} = B_{21} N_2 \rho \tag{4.5.2}$$

where B_{21} is Einstein's coefficient, N_2 the number of atoms at the upper energy state, and ρ the energy density.

Finally, we must calculate the rate of spontaneous emission. This rate does not depend on incident energy density—atoms emit photons spontaneously regardless of external conditions—but solely on the number of atoms at the upper energy state available to emit a photon:

$$r_{\text{spontaneous}} = A_{21} N_2 \tag{4.5.3}$$

where A_{21} is Einstein's coefficient for spontaneous emission and N_2 is the number of atoms at the upper energy state. It might be noted that the A coefficient for absorption is related to the spontaneous lifetime as

$$A_{21} = \frac{1}{\tau} \tag{4.5.4}$$

where τ is the spontaneous radiative lifetime of the upper state. Simply put, equation (4.5.3) states that the rate of spontaneous emission is equal to the number of atoms available in the upper state divided by the spontaneous lifetime (in seconds). It is calculated in units of s^{-1} (the rates are in terms of number of atoms per second). Level lifetime is an important consideration in determining laser action and is dealt with in detail in Chapter 5.

Example 4.5.1 The Role of Level Lifetimes The effect of level lifetimes on operation of a laser can be illustrated in a TEA (transverse electrical discharge at atmospheric pressure) nitrogen laser. Nitrogen has an exceptionally short upper lasing level lifetime of about 1.5 ns (at atmospheric pressure). Assume that nitrogen gas is placed in a long tube (like most other gas lasers) as in Figure 4.5.1 and is quickly pumped with energy to cause a population inversion. Assume now that a single photon is emitted spontaneously at the end of the tube (labeled A in the figure) and travels down the length of the tube being amplified along the way to produce an intense laser pulse. After 1.5 ns the laser pulse has traveled 0.45 m down the tube to the point labeled B. At this point, remaining excited nitrogen molecules in the tube (in front of the optical laser pulse now racing down the tube between points labeled B and C in Figure 4.5.1) have reached their spontaneous radiative lifetime and begin to emit light spontaneously, dropping to the lower level of the transition. Molecules in this state are strongly absorbing and absorb photons readily in the pulse traveling down the tube: The laser absorbs its own pulse and has no output.

This situation seems to indicate that the tube cannot exceed 0.45 m in length; in reality, nitrogen lasers are not usually constructed using a longitudinal arrangement but rather a transverse arrangement in which the electrodes are parallel to the discharge tube. An electrical pulse is launched down the tube preceding the laser pulse, ensuring that the nitrogen gas is excited just before the laser pulse arrives to be amplified. This scheme ensures that the excited molecules do not stay in that state long enough to radiate radiation spontaneously before they get a chance to amplify the pulse traveling down the tube.

As an aside, consider that in the nitrogen molecule the lower level has a much longer lifetime than the upper level. In such a system, continuous-wave (CW) laser action may not be possible, as is the case here. The mechanism of the nitrogen laser itself must ensure that the upper level is quickly pumped with energy (in much less than 1.5 ns) to generate a population inversion. In the case of a transverse-

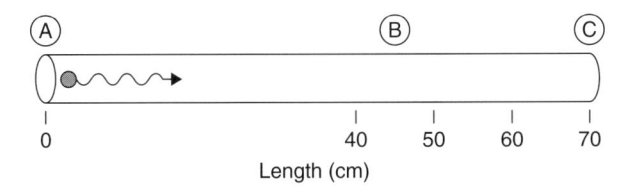

Figure 4.5.1 Hypothetical nitrogen laser tube.

94 LASING PROCESSES

discharge laser such as a nitrogen laser, a discharge of 10,000 A or more occurs. This mechanism is covered in detail in Chapter 10.

Let us proceed by equating rates at thermal equilibrium. It is assumed, then, that there is no external energy input to the system. We equate the rate of upward transitions (absorption) with the total rate of downward transitions (spontaneous and stimulated emission):

$$r_{absorption} = r_{stimulated} + r_{spontaneous}$$

or

$$B_{12}N_1\rho = B_{21}N_2\rho + A_{21}N_2 \tag{4.5.5}$$

We now use Planck's blackbody radiation law as well as Boltzmann's law to solve for the energy density ρ. Recall from Chapter 1 that Planck's radiation law for cavity radiation is

$$\rho = \frac{8\pi h\nu^3}{c^3} \frac{1}{\exp(-E/kT) - 1} \tag{4.5.6}$$

We can substitute into (4.5.6) as follows using Boltzmann's law to ratio the populations of atoms at each energy level: $\exp(-E/kT) = N_2/N_1$ provided that E is the energy difference between levels E_2 and E_1, so that

$$\rho = \frac{8\pi h\nu^3}{c^3} \frac{1}{N_2/N_1 - 1} \tag{4.5.7}$$

Rearranging to solve for energy density in the original rate equation (4.5.3), we now find that

$$\rho = \frac{A_{21}}{B_{12}(N_1/N_2) - B_{21}} \tag{4.5.8}$$

The two Einstein B coefficients represent the probability of an atom making the upward transition E_1 and E_2 and the downward transition E_2 to E_1, respectively. These are equal, so we shall now use the generalized term B:

$$\frac{8\pi h\nu^3}{c^3} \frac{1}{N_2/N_1 - 1} = \frac{A_{21}}{B} \frac{1}{N_1/N_2 - 1}$$

yielding

$$\frac{8\pi h \nu^3}{c^3} = \frac{A_{21}}{B} \tag{4.5.9}$$

Now that we have a solution for the Einstein rate equations, we can draw some mathematical conclusions.

Consider the ratio of stimulated to spontaneous emission. As we have already discussed, the rate of stimulated emission must exceed that of spontaneous emission for laser action to occur (in order to have amplification occur):

$$\frac{r_{\text{stimulated}}}{r_{\text{spontaneous}}} = \frac{B_{21}N_2\rho}{A_{21}N_2} \tag{4.5.10}$$

Substituting for the Einstein coefficients, this ratio simplifies to

$$\frac{r_{\text{stimulated}}}{r_{\text{spontaneous}}} = \frac{c^3 \rho}{8\pi h \nu^3} \tag{4.5.11}$$

It is further possible to solve the ratio outright by substituting for ρ using Planck's law from equation (4.5.6). We know the energy levels involved, hence we know the frequency of emitted photons ν.

An important conclusion at this stage is that for this ratio to be large (i.e., stimulated emission rate exceeds spontaneous rate), the energy density of incident photons must be high. This is the actual flux of photons within the laser cavity itself. Unless the gain of the medium is extremely high in order to create a huge flux of photons as they pass down the tube, cavity mirrors will be required to contain photons within the cavity to create further amplification. Indeed, the vast majority of lasers require cavity mirrors for oscillation. Furthermore, these mirrors are usually very efficient (>95%) reflectors. We also note that the ratio of these rates also depends on the inverse of the frequency, cubed. In other words, higher frequencies (shorter wavelengths) have lower ratios and require higher energy densities to operate. As an aside, consider that the first laser operated at microwave frequencies, where much lower energy densities are required to favor this ratio over optical frequencies.

The last concern is the rate of absorption of photons. We must ensure that photons emitted are not absorbed within the medium itself. If the medium absorbs more photons than were emitted by stimulated emission, the laser cannot work. This ratio is calculated directly from equations (4.5.1) and (4.5.2) as

$$\frac{r_{\text{stimulated}}}{r_{\text{absorption}}} = \frac{N_2}{N_1} \tag{4.5.12}$$

96 LASING PROCESSES

But in our case we have a population inversion where $N_2 > N_1$, so the rate of stimulated emission will indeed exceed the absorption rate. This relation proves the necessity of population inversion for laser action (for which we have given only a qualitative argument so far). This is a simple introduction to the rate equations, which can become considerably more complex, especially when multiple levels other than just the two lasing levels are involved (which they are in a real laser). Rate equations and the types of energy-level systems involved in real lasers are examined in much greater detail in Chapter 5.

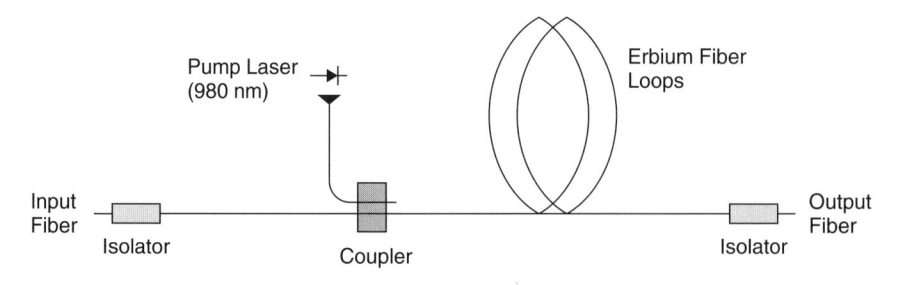

Example 4.5.2 Fiber Amplifiers In telecommunications systems employing laser emitters and long runs of fiber-optic cables, fiber amplifiers are used to boost weak signals optically. A practical amplifier consists of a long (10- to 20-m) section of glass fiber doped with erbium ions (Er^{3+}). A pump laser at 980 nm is coupled to the amplifier fiber via a coupler. This pump radiation is absorbed by the erbium atoms in the fiber, exciting them to an upper level which rapidly decays to a level 0.80 eV above ground state. This level, which has a relatively long spontaneous radiative lifetime (τ_{sp}, about 10 ms long) can amplify incoming signals via stimulated emission, producing a net optical gain at 1549 nm. A diagram of this amplifier is depicted in Figure 4.5.2.

Atoms in the fiber are excited continually by the pump laser. Weak incoming photons from the input fiber stimulate the emission of more photons in the fiber amplifier, producing a cascade of identical photons, which in effect amplifies the original signal. The fiber amplifier is just that: an amplifier. Unlike a laser, it lacks the ability to oscillate and hence requires a "seed" signal to amplify (i.e., the input signal).

Like any other system with excited atoms, spontaneous emission can occur from the upper-energy level of the atoms in the fiber. Spontaneous emission in the band of wavelengths at which the amplifier operates (centered around 1549 nm, the wavelength of the laser used for fiber communications) will be amplified producing *amplified stimulated emission* (ASE), which is seen as noise in the output of the device. When no input signal is present, the output of the amplifier consists of a good deal of ASE, as evident in Figure 4.5.3(*b*), which shows the output as analyzed by an *optical spectrum analyzer* (OSA), showing the amplitude at each wavelength. In a communications system, ASE is noise.

Figure 4.5.2 Fiber amplifier.

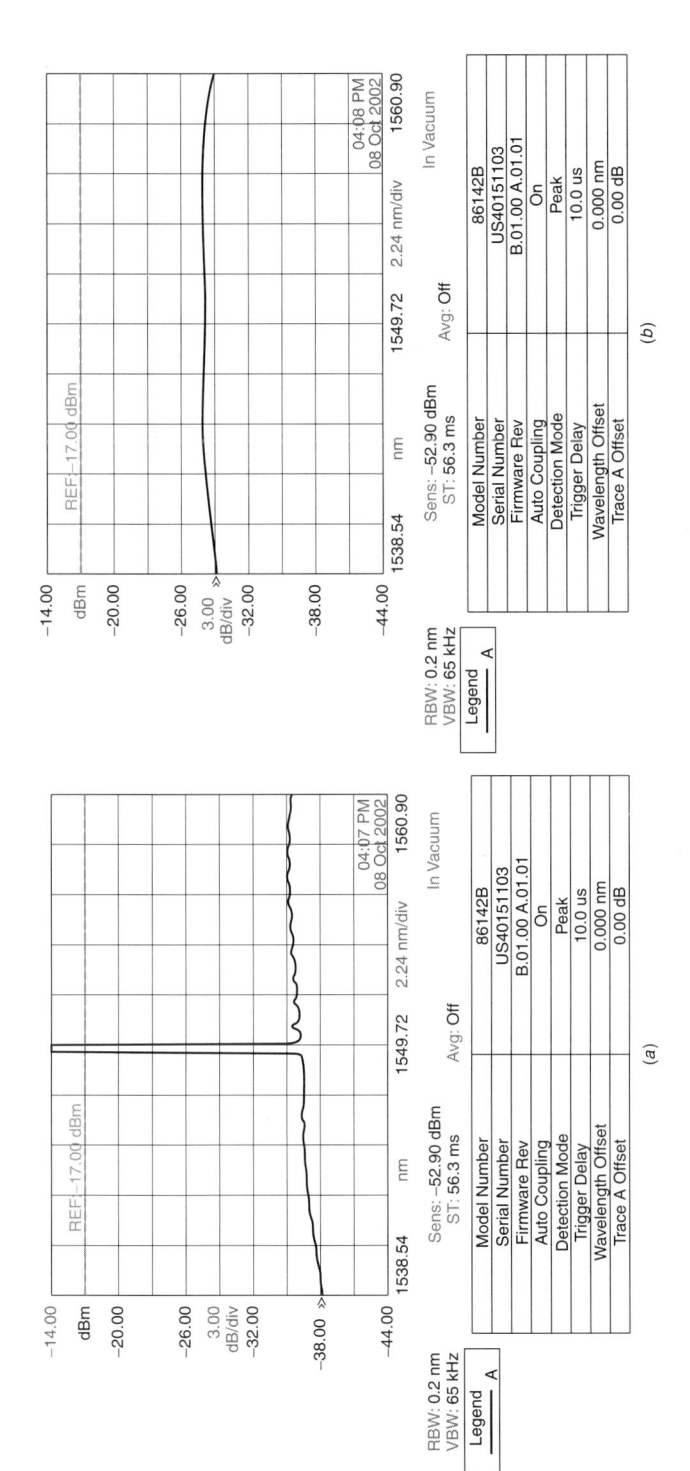

Figure 4.5.3 Fiber amplifier output.

97

When a signal is fed into the fiber amplifier, the output consists of a large amount of signal seen as a spike at 1549 nm (the input signal highly amplified). The surprising part of this output, seen on the OSA in Figure 4.5.3(a), is the low background noise seen in the output around the spike at 1549 nm. When the amplifier is amplifying an input signal spontaneous emission is reduced drastically (in this case over 9 dB of reduction in noise in the output of the amplifier; the amplitude scale is identical in the two outputs shown in the figure, allowing comparison). When amplifying a signal (in this case a large signal which essentially saturates the amplifier), excited atoms of erbium which would, in the absence of a signal, emit spontaneously are coaxed into emitting coherent light; that is, they emit coherent light stimulated emission before they get a chance to emit light spontaneously, hence lowering the rate of spontaneous emission. This stimulated emission process (the key process by which this amplifier works) causes excited atoms to emit coherent light. Erbium, the species responsible for amplification here, has a relatively long spontaneous radiative lifetime in the upper lasing level, giving the process of stimulated emission a good probability of occurring before spontaneous emission occurs.

4.6 LASER GAIN

Assuming that we've generated population inversion (by pumping energy into the lasing medium, in this example a gas, to excite upper lasing levels), laser gain may now occur. Laser gain (or optical gain) is a measure of how well a medium amplifies photons by stimulated emission. Consider a single photon traveling down a laser tube and stimulating an excited atom to emit a photon of the same frequency and phase—this is *laser gain*. Thinking of the stream of photons traveling down the tube as an electromagnetic wave, we can see that the *power* of the wave increases if the rate of stimulated emissions exceeds that of spontaneous emissions. As the wave travels farther down the tube, the power increases as a function of length. The power increase, mathematically, is

$$\exp(gx) \tag{4.6.1}$$

where g is the optical gain coefficient of the laser medium and x is the distance down the tube. No surprises, the power increases exponentially as the wave travels down the tube. Consider where one photon stimulates another, giving us two; these two then stimulate two more, so we get four, and so on. The exponential nature of this gain is evident. The gain coefficient, g, represents the optical power gain per unit length. Mathematically, it can be expressed as $\Delta P/\Delta x$, the change in power for a given length of lasing medium, or more correctly,

$$g = \frac{\Delta P}{P \Delta x} \tag{4.6.2}$$

which yields a quantity in units of length^{-1} (e.g., m^{-1}). The standard model for picturing a gain medium, shown in Figure 4.6.1, is to view a thin slice of a medium.

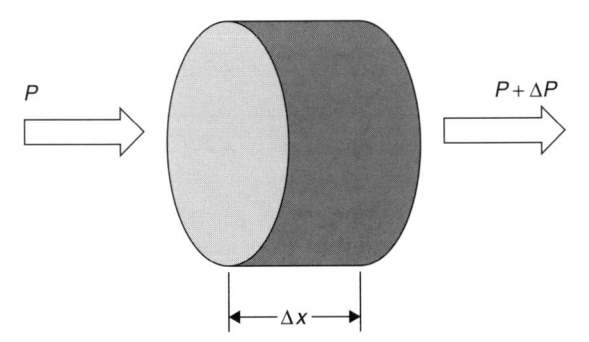

Figure 4.6.1 Gain in a laser medium.

A stream of photons with power P enters the gain medium, which has length Δx, and exits with increased power $P + \Delta P$. Gain is proportional to the net rate of stimulated emission, which is in turn equal to the rate of stimulated emission minus the rate of absorption. The ramifications for practical laser construction are simple: The longer the lasing medium, the more power will increase per pass down the tube. When attempting to lase a weak transition (as many gases have), a longer tube will allow a larger power increase per pass through the tube. This may be important when considering losses in the laser.

The gain of a laser may be expressed in various ways, such as percent per pass, percent per meter, or m^{-1}. As an example, consider an ion laser tube 80 cm long with a gain of 12% per pass. Each time the beam traverses the gain medium (the laser tube) it gains 12% in power. This gain may also be expressed as 15% per meter (12%/0.8 m), or $0.15\ m^{-1}$. Although this figure may seem high, and it is for many lasers, consider losses in the laser that must be overcome. Assuming that we use aluminum mirrors, which reflect about 85% of incident light, this laser cannot work. The loss at one mirror alone (15% loss) is greater than the gain down the entire tube! Total losses in the laser, including the portion we extract as our output beam, losses in the mirrors, losses at the tube windows, and absorption (or reabsorption) within the tube itself, cannot exceed 12% in this case. Clearly, laser cavities must be designed for minimal loss, issues covered in Section 6.2.

Example 4.6.1 Laser Gain Calculation Consider a small pulsed mercury-ion laser (with output in the green) which has a tube 0.1 m long and a gain of 50% per meter. The gain in this particular laser is exceedingly large; few other gas lasers have gains of 50% per meter! Assume that the total reflector has a loss of 0.1% (these are special, dielectric mirrors). Most ion lasers have Brewster windows; in this case, assume that each has a loss of 0.8%. Finally, the tube has an attenuation of $0.1\ m^{-1}$. The parameters are included in Figure 4.6.2. Given these parameters, calculate the maximum transmission, in percent, that can be extracted through one of the mirrors as an output beam.

100   LASING PROCESSES

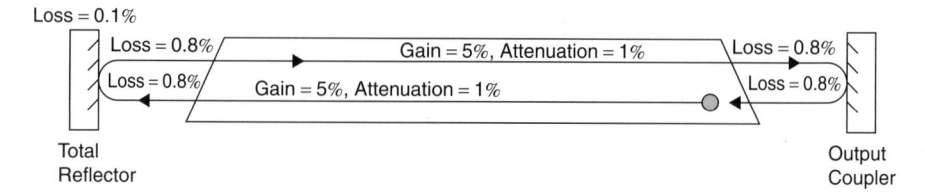

```
Loss = 0.1%
   ┌─Loss = 0.8%─────Gain = 5%, Attenuation = 1%─────Loss = 0.8%─┐
   │ Loss = 0.8%     Gain = 5%, Attenuation = 1%      Loss = 0.8%│
   └─────────────────────────────────────────────────────────────┘
   Total                                                   Output
   Reflector                                               Coupler
```

Figure 4.6.2   Gain and loss in a mercury-ion laser.

SOLUTION   To solve this, sum the total loss in the laser on a round trip through the tube. Radiation in the cavity passes down the tube, being amplified 5% (50% × 0.1 m) in the single pass. At the same time, the pass down the tube also attenuates the power by 1% (this is absorption in the tube and will be discussed shortly). In this case the attenuation is specified as 0.1 m^{-1} (0.1 per meter) or 1.0% per pass for the 0.1-m tube length. Passing through the Brewster window the beam loses 0.8%. Another 0.1% is lost at the total reflector, and another 0.8% as the beam passes once more through the window to reenter the tube. The situation is repeated as the beam is amplified and a loss occurs at the other end of the tube, where the beam reflects off the output coupler.

Total loss is calculated as 1% (attenuation during forward pass) + 0.8% (exiting the tube via the window) + 0.1% (total reflector loss) + 0.8% (entering the tube) + 1% (attenuation) + 0.8% (exiting the tube) + 0.8% (entering the tube) = 5.3%. Total gain is calculated as two passes of 5% each, or 10%. We could then extract 10 − 5.3 = 4.7% as an output beam (the output coupler could impose a loss of 4.7%, which includes losses in the mirror itself plus the portion transmitted). This would not be optimal, however, as it barely allows the laser to oscillate. In reality, the output transmission of the mirror would be designed to be about 2.0%, to ensure that the energy density inside the cavity is more than enough to sustain laser action. It is also obvious why certain features of real lasers are designed the way they are. First, the mirrors in almost all lasers are dielectric types. Ordinary aluminized mirrors reflect only 85% of the incident light; such losses are not tolerable in a system like this. The second feature of many lasers are the angled Brewster windows on the tube ends. From basic optics theory you will recall that at a specific angle a window will have almost zero loss to a certain polarity of light. Ordinary, uncoated glass windows perpendicular to incident light have a reflective loss of about 8%. Again, such losses from a plane window on a plasma tube are not tolerable in most lasers, so windows at Brewster angles or windows with special antireflective coatings are required. We examine practical laser design in Chapter 6 and as we shall see, many practical lasers have very low gains.

So the requirements for laser oscillation are not solely that population inversion occurs (i.e., the population of atoms at the upper lasing level exceeds that at the lower level) but also that the gain be high enough to overcome any losses in the cavity. These losses include the portion we extract as an output beam. The required laser gain also depends on the inverse of the spontaneous lifetime of the lasing

4.7 LINEWIDTH

We often think of a laser transition as a single, sharp, well-defined wavelength. Indeed, compared to the emission characteristics of an incoherent source, it is, but under closer examination we see that a laser line does indeed have a finite bandwidth. The primary reason to broaden a line's spectral width (the range of wavelengths it covers) is the *Doppler effect*, commonly associated with train whistles. Gas molecules in a laser tube are often hot and travel at high speeds. Assuming that the average speed of a gas molecule is v and that gas molecules can move in either direction at this speed, the spread in frequencies seen in the output of the laser will be

$$\nu = \nu_0 \left(1 \pm \frac{v}{c}\right) \tag{4.7.1}$$

where ν is the output frequency of the laser, ν_0 the center frequency ($=\lambda_0/c$), and v the average velocity of gas molecules. This situation is outlined in Figure 4.7.1, where a gas molecule is seen emitting radiation. If the molecule moves toward the observer when emitting radiation, the frequency of the radiation appears to increase (as does the sound of an approaching train blowing a whistle, in which case the frequency of the whistle is heard to increase in pitch as the train approaches). Similarly, the frequency decreases if the molecule speeds away from the observer when emitting a photon of light (in this case, ν_2 in the figure). This will yield two values for the frequency of the laser, one minimum and one maximum, which will define the range of outputs of the laser. The easy way to calculate gas molecule velocity is to use kinematics and equate kinetic energy to thermal energy as follows:

$$\tfrac{1}{2}mv^2 = \tfrac{1}{2}kT \tag{4.7.2}$$

where m is the mass of the molecule, v the velocity of gas molecules (unknown), k is Boltzmann's constant, and T is the temperature in kelvin.

The range of frequencies possible is now found by computing v given molecular mass and temperature, then substituting into the equation above to find the minimum

Figure 4.7.1 Doppler effect on gas molecules.

and maximum frequencies. Although this will yield a reasonable answer, we know from kinematics that the velocities of gas molecules follow a Maxwell distribution.[3] Using such a distribution will give a more precise answer. In this manner we may compute the full width half maximum (FWHM) of the resulting output. This will yield the width (in hertz) of the optical gain profile

$$\Delta \nu = 2\nu_0 \sqrt{\frac{2kT \ln 2}{Mc^2}} \qquad (4.7.3)$$

where $\Delta \nu$ is the FWHM of the output (in hertz), ν_0 the center frequency (at maximum gain), k is Boltzmann's constant, T is the temperature of the gas in kelvin, M the atomic mass of the atom or molecule, and c the speed of light.

Example 4.7.1 Linewidth of a HeNe Laser Consider the linewidth of a HeNe laser. The laser has a gas temperature (in the actual plasma tube itself) of 150°C and an output wavelength of 632.8 nm in the red. In this laser the active lasing atom is neon (helium is present only as an energy-level pump). Neon has an atomic number of 10 and so has an atomic mass of 20.

SOLUTION First, calculate the mass of the atom as

$$\frac{20 \text{ amu}}{6.02 \times 10^{23} \text{ mol}^{-1}} = 3.32 \times 10^{-23} \text{ g}$$

This is then substituted into the Maxwell distribution formula:

$$\Delta \nu = 2\nu_0 \sqrt{\frac{2kT \ln 2}{Mc^2}}$$

$$= 2 \left(\frac{3 \times 10^8}{632.8 \times 10^{-9} \text{ m}} \right) \sqrt{\frac{2(1.38 \times 10^{-23})(423) \ln 2}{3.32 \times 10^{-26} \text{ kg} \, (3 \times 10^8)^2}}$$

$$= 1.56 \text{ GHz}$$

This corresponds to a wavelength FWHM of 0.002 nm. Indeed, this is small compared to the FWHM of any incoherent light source but is still a far cry from a "single line."

You will note that linewidth is a function of temperature, as depicted in Figure 4.7.2. Gas lasers that operate at high temperatures, such as the argon-ion laser, have larger linewidths than that of the helium–neon laser, which operates at a relatively low temperature. Despite the fact that the graph shows temperatures up to 6000 K, HeNe lasers would normally never exceed 450 K. In the case of the argon laser, though, high temperatures are unavoidable: Discharge currents of 10

---

[3]Equating the kinetic energy as done here gives the root-mean-square (RMS) speed of gas molecules. The problem with this value is that it does not recognize the variations in speeds of individual gas molecules as the Maxwellian distribution does. A good explanation of this is provided in *Physics* by D. Halliday et al. (New York: Wiley, 1992).

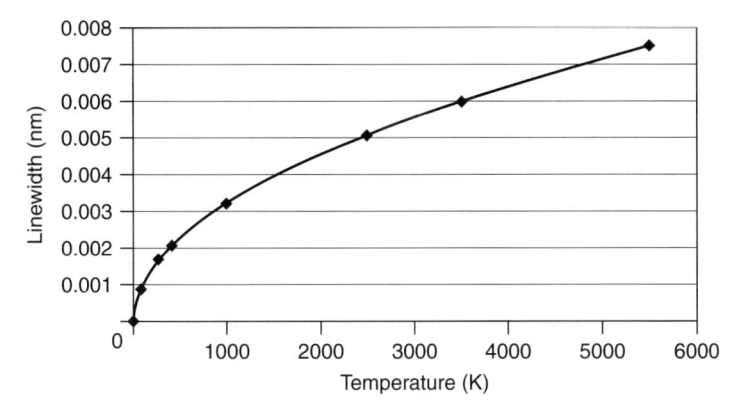

Figure 4.7.2 Linewidth versus temperature for a HeNe laser.

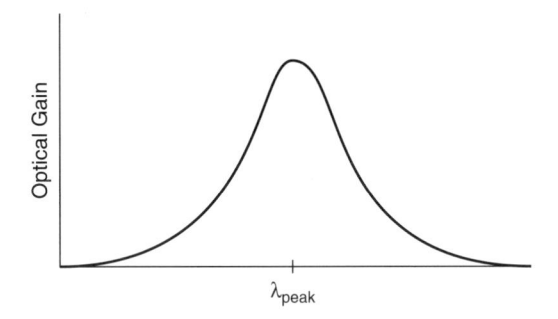

Figure 4.7.3 Gain curve for a practical laser.

to 70 A are common in such a laser. The resulting plasma temperatures can reach 5000 K! Techniques exist to reduce linewidths, as examined in Chapter 6.

The Doppler broadening mechanism depicted in Figure 4.7.1 applies not only to gas lasers where gas molecules may move freely in a tube, but also to solid-state lasers. Although atoms in a solid-state laser crystal will not move as gas molecules do, the lattice of a solid-state laser crystal will vibrate more with increasing temperature. This vibration will affect the system by making the energy band broader, and hence increase the spectral linewidth. The results of broadening are that the gain is not maximum at a specific wavelength and zero at all other wavelengths but rather, is a curve with maximum gain that gradually tapers off to zero at wavelengths around the maximum, as illustrated in Figure 4.7.3.

At all points on the gain curve where the gain is sufficient to overcome losses in the laser (and the laser cavity is resonant), the laser may oscillate and have output.[4]

---

[4] The tone of the discussion so far has been that a narrow gain profile (and hence narrow output linewidth) is always desirable. It is worth mentioning, though, that sometimes a wide gain profile is desirable in order to support ultrashort pulses (those in the pico- and femtosecond range). The broad spectra of these ultrashort pulse lasers are not due to Doppler broadening but to a broad gain curve from the gain medium itself.

104 LASING PROCESSES

There is a further complication in this scenario: the cavity itself, which acts as an interferometer resonant only at certain wavelengths. The configuration of the cavity leads to longitudinal modes: a series of discrete frequencies close together. Collectively, these resemble a broad line much wider, spectrally, than a single line. Both longitudinal and transverse modes are examined in detail in Chapter 6.

4.8 THRESHOLDS FOR LASING

In any medium there will be losses due to scattering, absorption by impurities, and losses caused by the cavity and tube walls itself. In addition, to be practical, we will purposely extract a portion of the laser beam within the laser and use that as our output beam. In a practical laser such as a HeNe, we might extract 1% of the light in the cavity as a beam. That really means a loss of 1%. Obviously, we must have enough gain inside the laser medium to allow us to overcome all losses in the laser as well as to allow us to extract our output beam and still allow laser action to continue. Given all losses, we can calculate a minimum gain that allows laser action. This is the threshold gain of the laser medium. Although some elements of the laser cannot be easily controlled (such as the wavelength of the transition in a gas laser) we *can* control the amount of power we extract through the output coupler, the partially reflecting mirror at the front of the laser. It should be evident that there is a limit on how low the reflectivity of the two cavity mirrors can be for a given laser medium. Remember that regular aluminum mirrors have a reflectivity of only about 85%. Indeed, many lasers, such as the HeNe, lack the gain needed to allow such huge losses. This explains why dielectric mirrors are used for many lasers. These mirrors use multiple thin-film layers to interfere with, and reflect, incoming light. Such mirrors can have reflectivities of up to 99.999%! They are not easy to make, though, and require a high-vacuum thin-film coating apparatus to produce. This also explains why laser mirrors are so expensive as opposed to the common metal film (e.g., aluminum) variety.

Of course, the length of the laser medium also affects threshold gain. Obviously, the longer the tube, the higher the amplification, hence higher losses may be tolerated. Take the argon gas laser as an example. Short argon lasers (30 cm) absolutely require dielectric cavity mirrors, but there are many reports of longer lasers (1 m or more) lasing with aluminum mirrors, especially on the strongest transitions (514 and 488 nm). Many of the weaker transitions simply do not have the gain to lase without dielectric mirrors. The stronger transitions may well tolerate 15% or greater losses in the mirrors.[5]

Some lasers have *huge* gains, so high that light is amplified to a usable level in a single pass down the tube. Such lasers are termed *superradiant* and operate without feedback (i.e., no cavity mirrors). Nitrogen lasers are usually superradiant. It may be noted, though, that including a single rear cavity mirror in this type of laser boosts

---
[5] W. B. Bridges estimated the gain of the 488-nm line in *Applied Physics Letters*, Vol. 4, No. 7 (1964) (the original article detailing the discovery of the argon laser) as >15%, based on the fact that a 1-m tube would lase this line with aluminum mirrors of 90% reflectivity.

power output and creates a beam with better characteristics (such as less divergence of the beam as it exits the laser). Other lasers, such as copper-vapor lasers, require very little feedback to lase, and an uncoated glass window (which reflects 8% of incident light) makes a suitable output coupler for such a laser. Nitrogen-laser-pumped dye lasers frequently have enough gain to operate superradiantly as well. Almost all superradiant lasers are strictly pulsed and will not operate in CW mode.[6]

X-ray lasers are another, if not somewhat extreme, example of superradiant lasers. These are really short cylinders of very dense, very hot plasmas. Since the lifetime of the plasma is very short, typically less than 1 ns, they are mirrorless lasers (since, at most, radiation could traverse only a pass or two in the short time interval; the same situation occurs in nitrogen lasers with short pulses). They amplify spontaneous radiation, and because they cannot utilize mirrors, the gains are typically 500 m^{-1}, enormous by laser standards! This enormous gain does not manifest itself, however, because the plasmas are usually created from a high-intensity laser interaction with a flat metal target (the metal being the lasing material). The density gradients in the gain region (between areas of different refractive indices) are huge, so much of the amplified light is refracted away from the gain region as it propagates along the gain cylinder. At the end, the integrated gain (that which is actually usable) is only about one-tenth of the gain inside the laser medium itself.

Note also that there is a threshold on the pumping rate that must be reached for lasing action to occur. First, until the pump rate (the rate of input power) is sufficient to allow population inversion (higher populations at the upper lasing level than the lower lasing level), laser gain (and hence coherent output) will not begin. But population inversion alone does not guarantee laser output, as the laser will not produce output until the gain overcomes losses in the laser. The situation is outlined in Figure 4.8.1, where gain is seen to increase as pumping power; however, laser output does not begin until gain in the laser is equal to the total losses. This is the threshold pumping power (P_{th}), and any increase in pumping power after that point will lead to an increase in laser output (the dashed line in the diagram). This is easily seen with diode lasers, in which a certain minimum current must be reached before any laser output is seen as in Figure 4.8.2, where the threshold current for this particular laser diode is about 11 mA. Output is emitted from the laser diode below the threshold, but this is primarily incoherent spontaneous emission. Once threshold is surpassed, any increase in current leads to an increase in laser output. There are physical limits on output power, though, and they usually involve the laser medium and how much power it will take per unit volume before saturating or overheating, many times destructively. In the diode laser a high current density will melt the semiconductor crystal and destroy the laser. This sets a practical limit on pumping power for diode lasers (in the case of the diode used to generate Figure 4.8.2, the current limit is about 50 mA; similar limits exist with other types of lasers).

---

[6] Even long argon lasers exhibit superradiance effects: W. R. Bennett et al. describe superradiance effects in a 3-m tube in *Applied Physics Letters*, Vol. 4, No. 10 (1964). The gain of the 476.5-nm line in a pulsed laser is so high that it may well lase without both cavity mirrors in such a long laser!

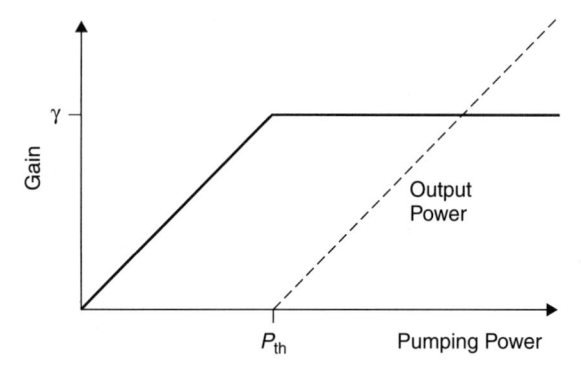

Figure 4.8.1 Gain versus pumping power.

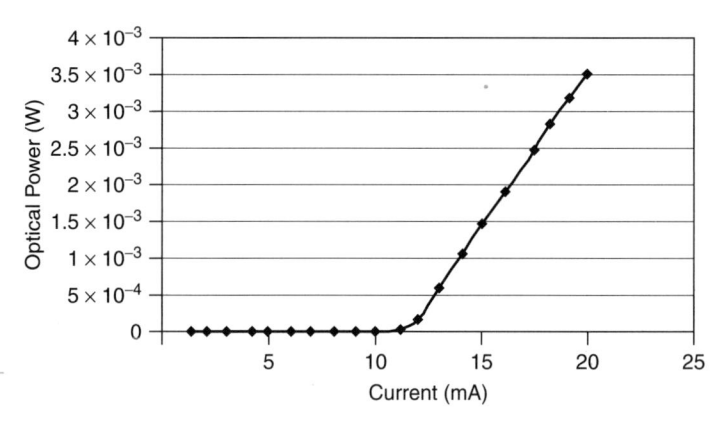

Figure 4.8.2 Output power versus current for a typical diode laser.

4.9 CALCULATING THRESHOLD GAIN

When the laser is operating at steady-state conditions (i.e., constant optical power output), the net gain in the laser must be 1. If the net gain were greater than 1, the output power would increase. Net gains below 1 cause the output power to drop until the laser ceases to operate. So consider, then, a round trip by a stream of photons in the tube (Figure 4.9.1). The power after the round trip must equal the power before the round trip. Sure, the photon stream gains energy (through laser gain), but just as much power as is gained is lost through absorption in the medium, in addition to the portion extracted as the output beam.

The power gained during the round trip is $\exp(g \cdot 2x)$, where $2x$ is the total path length of the laser (x is the length of the gain medium through which the stream of photons passes twice). The power lost during the same trip is $\exp(-\gamma \cdot 2x)$, where γ

On each pass: Gain = exp(gx)
Loss = exp(-γx)

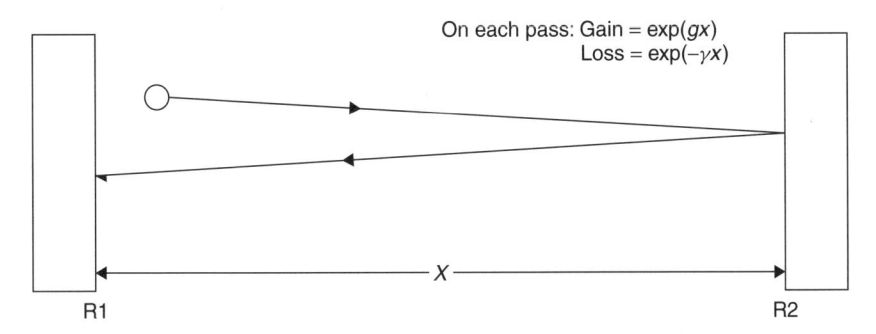

Figure 4.9.1 Gain and loss in a round trip through a laser.

is a new term describing all losses in the cavity due to the lasing mechanisms, with the exception of the mirrors themselves. The losses from the mirrors themselves will be seen as the reflectivities of the two mirrors, labeled R_1 and R_2. A perfect mirror, with 100% reflection, has an R value of 1.

Equating these parameters for a round trip through the tube yields

$$\text{net gain} = \text{laser gain} \times \text{loss} \times \text{loss at mirror 1} \times \text{loss at mirror 2}$$
$$= \exp(2gx)\exp(-\gamma \cdot 2x) R_1 R_2 \quad (4.9.1)$$

Knowing R_1 and R_2 as well as γ, we can solve for gain, which must equal 1 for an operating laser. This gain will be the *threshold gain*, the gain required to allow the laser to operate in this steady-state condition:

$$g_{\text{threshold}} = \gamma + \frac{1}{2x}\ln\left(\frac{1}{R_1 R_2}\right) \quad (4.9.2)$$

where x is the length of the gain medium. In many lasers the gain of the laser medium is proportional to the pump energy (but not without limits, as we shall see). A minimum pump energy will hence be required to generate a gain of at least the threshold value in order for lasing to begin.

Example 4.9.1 Gain and Loss in a HeNe Laser As an application of the threshold gain formula, consider a helium–neon laser in which the loss is known to be 0.05 m^{-1}. The laser has an actual plasma tube length (where gain occurs, not the entire length of the tube between mirrors) of 20 cm. One mirror is 99.9% reflecting, and the output coupler is 95% reflecting. Calculate the threshold gain for the tube.

SOLUTION Using equation (4.9.2), we can substitute as follows:

$$g_{threshold} = \gamma + \frac{1}{2x}\ln\left(\frac{1}{R_1 R_2}\right)$$

$$= 0.05 + \frac{1}{0.4}\ln\left(\frac{1}{0.999 \times 0.95}\right)$$

$$= 0.181 \text{ m}^{-1}$$

This is an excessively high gain requirement for a HeNe laser, and the laser would probably not oscillate. To allow the laser to work, either the plasma tube must be made longer than 20 cm or the reflectivity of the output coupler must be increased to better than 95%.

Threshold gain has many practical applications such as determining the minimum required reflectivities of cavity optics as well as allowed insertion losses in the cavity. One can experimentally measure the gain and loss of a lasing medium directly. One simple method (for a helium–neon laser) is to pass the light from an operating helium–neon laser directly through the bore of a bare plasma tube (windows on each end rather than the usual integral mirrors).[7] As the beam from the first laser passes through the plasma tube, it is amplified. By measuring the power of the exiting beam with the bare tube both energized and not energized (and knowing the length of the gain medium) the gain of the medium can be measured directly, giving a useful value for g. This is called a *MOPA* (master oscillator, power amplifier) *configuration*, the first laser being the oscillator and the second, the amplifier. It is used in many industrial lasers (e.g., Nd:glass) to generate enormous power levels but is useful here for experimentation. Figure 4.9.2 depicts an experimental setup of this type, and Figure 4.9.3 shows an actual experimental setup in which the probe laser is on the bottom of the photograph and the beam is steered by two mirrors to pass through the gain tube. Upon exit from the gain tube, power is measured using an optical power meter.

One can measure the gain of any transition this way simply by using a pump beam at the desired wavelength. Obviously, the oscillator laser must have exactly the same wavelength as the transition whose gain we are attempting to measure, so we would always use the same type of laser (a HeNe laser to measure amplification of a HeNe tube, argon for an argon tube, etc). HeNe lasers have relatively large gains at the red transition (632.8 nm) but much smaller gains at the other allowed lasing transitions, such as the green (543 nm) or orange (612 nm). Not visible in Figure 4.9.3 is the fact that the probe beam is green. This illustrates a major advan-

---

[7]This method is described in an article by S. M. Jarrett and G. C. Barker in the *Journal of Applied Physics*, Vol. 39, No. 10 (1968). This setup allows the effect of many parameters on laser gain to be demonstrated, including tube pressure and magnetic field (important in ion lasers and covered in Chapter 9). The tube can also be used to induce a loss, and although the authors state this, they do not attempt to measure the loss coefficient of the tube in this work.

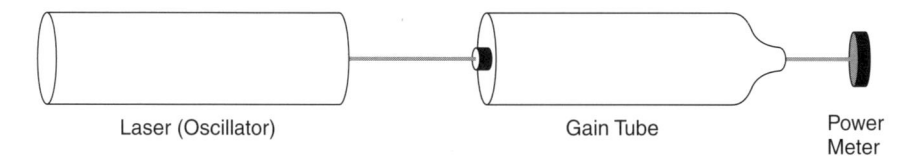

Figure 4.9.2 Experimental setup for measuring the gain of a laser directly.

tage of this technique, as one can measure the gain of any transition by selecting a laser (the oscillator) of the wavelength desired for the transition to be measured. In the case of helium–neon lasers operating at wavelengths other than red (which have lower gains at these wavelengths), the optical elements for these lasers must have extremely low losses. As well, the output coupler has a much lower transmission than a red laser would have. Green helium–neon lasers have much lower output powers than those of similar-size red laser tubes.

The technique just described works with a variety of lasers, including argon-ion and YAG lasers, but a second (and somewhat simpler) method for measuring gain is to insert a variable loss into the cavity of a working laser. In its simplest form, this loss may be a slide of glass at a certain angle. At Brewster's angle, for example, there will be zero loss, as least in one polarization. By varying the angle to the point where lasing is extinguished, one may determine the value of the inserted loss. Summing other losses in the laser, such as that of the output coupler (OC), loss at tube windows, and attenuation in the tube itself, one may then calculate the gain of the lasing medium.

Figure 4.9.3 Actual experimental setup for measuring gain.

110 LASING PROCESSES

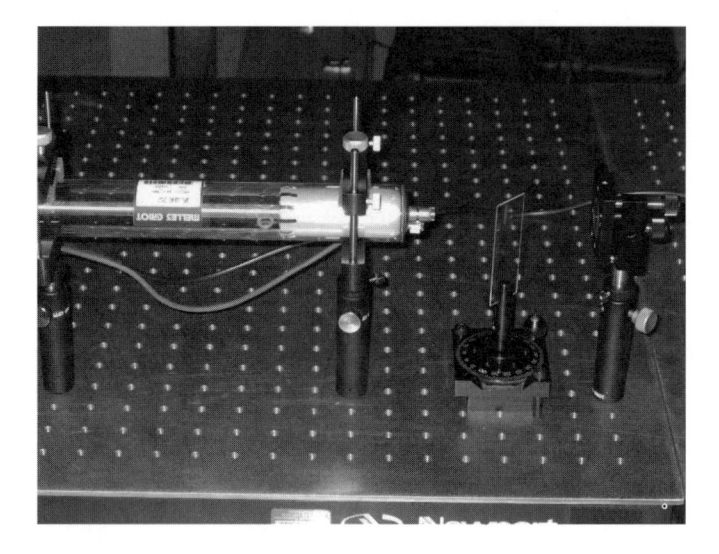

Figure 4.9.4 Second experimental method to determine laser gain.

Figure 4.9.4 shows an experimental setup in the author's lab in which a HeNe tube with external mirrors is set up on an optical breadboard. The photo details the variable loss, which is a glass slide on a rotating stage. The slide is rotated until the laser ceases to oscillate. That angle is now substituted into Fresnel equations to yield an answer in terms of inserted loss. Given the angle θ of the glass slide, the angle of the beam inside the glass itself (with index of refraction n) may be found by using Snell's law as follows:

$$\frac{\sin \theta_i}{n_i} = \frac{\sin \theta_r}{n_r} \tag{4.9.3}$$

where the subscript i represents the incident ray and r the refracted ray. The reflection (loss) in both polarizations may now be found by using the Fresnel equations:

$$R_p = \left(\frac{n \cos \theta_i - \cos \theta_r}{n \cos \theta_i + \cos \theta_r}\right)^2 \tag{4.9.4}$$

$$R_s = \left(\frac{\cos \theta_i - n \cos \theta_r}{\cos \theta_i + n \cos \theta_r}\right)^2 \tag{4.9.5}$$

The polarization R_p is parallel to the Brewster window and so is the polarization of concern here. The other polarization (R_s) yields a minimum loss of about 6%, which is far too high to allow laser oscillation in that plane (at least with a small laser with a small gain such as the one employed here). In the process of inserting this loss, we have highly polarized the output beam.

CALCULATING THRESHOLD GAIN   111

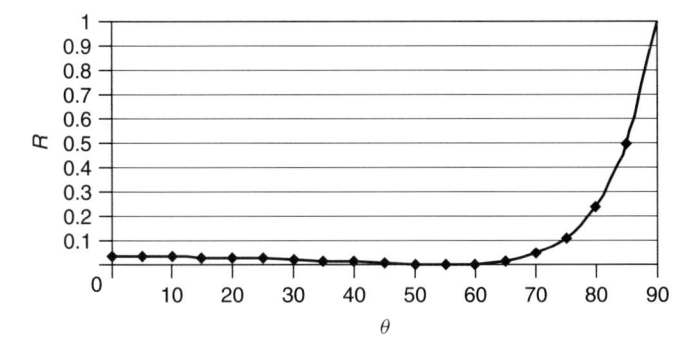

Figure 4.9.5 Reflective loss as a function of angle for an intracavity glass slide.

Using the Fresnel equations, we can determine the loss of the glass slide (the reflectivity) as a function of angle (for one polarization only); this is plotted in Figure 4.9.5. It must be remembered that this loss is for each glass-to-air interface, so these figures must be multiplied by 2 for a single pass through the slide. Inserting this variable loss into the cavity between the plasma tube and any one cavity mirror, we find the output power as a function of θ, portrayed in Figure 4.9.6. Like Figure 4.9.5, the loss increases more rapidly on the large-angle side, so the output power decreases more rapidly on that side. Finally, power output as a function of inserted loss from this slide is plotted in Figure 4.9.7. You are reminded again that the inserted loss is for one surface only: The actual slide has two surfaces, so this value must be multiplied by 2.

Example 4.9.2   Gain of a Real HeNe Laser   Consider the setup in Figure 4.9.4, in which a glass slide with an index of refraction of 1.46 is used as a variable loss. When the glass slide is at Brewster's angle, the maximum laser output will appear, but as the slide is rotated at an angle away from Brewster's angle, lasing is observed to cease at 45.5 degrees. We now find the loss caused by the inserted slide using the Fresnel equation. Using Snell's law ($n_1/n_2 = \sin\theta_1/\sin\theta_2$) the

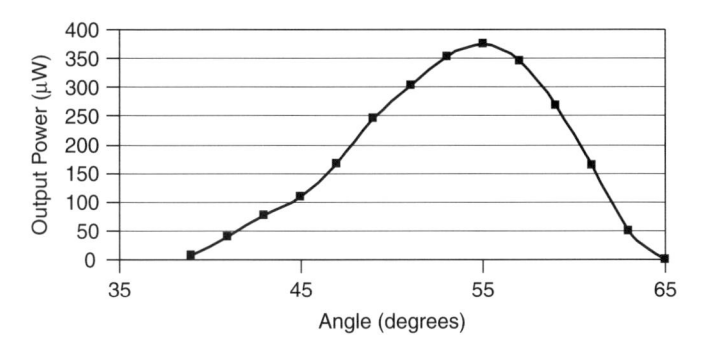

Figure 4.9.6 Power output as a function of angle.

112 LASING PROCESSES

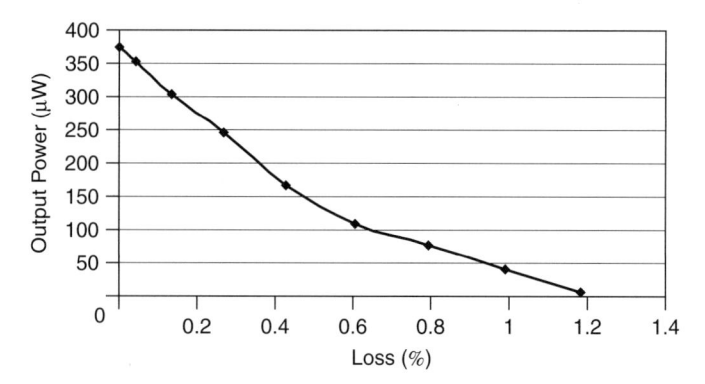

Figure 4.9.7 Power output as a function of inserted loss.

angle of refraction in the glass slide is found to be 29.69 degrees. This is then substituted into the Fresnel equation (with $n_1 = 1.000$ for air) for reflectivity in the parallel plane:

$$r = \frac{n_2 \cos \theta_1 - n_1 \cos \theta_2}{n_2 \cos \theta_1 + n_1 \cos \theta_2} \quad (4.9.6)$$

Squaring the reflection coefficient to get the intensity coefficient, the reflectivity at this angle is found to be 0.00560. This is for a single surface, so the reflection loss for one pass through the slide is 0.0112.

Summing this loss with other losses in the laser, and knowing that gain equals loss at the point where the laser just begins to oscillate, we may determine gain of the laser. Consider a round trip through the laser (as we did in Section 4.6), which consists of two passes through the plasma tube (i.e., two gains), one reflection (loss) off each cavity mirror, four passes through the tube windows (one at each end of the tube), and two passes through the variable loss element:

$$\begin{bmatrix} \text{total} \\ \text{round-trip} \\ \text{loss} \end{bmatrix} = 2 \times \begin{bmatrix} \text{attenuation} \\ \text{in the} \\ \text{tube} \end{bmatrix} + 2 \times \begin{bmatrix} \text{inserted} \\ \text{loss} \end{bmatrix}$$

$$+ 4 \times \begin{bmatrix} \text{loss at the} \\ \text{tube} \\ \text{windows} \end{bmatrix} + \begin{bmatrix} \text{loss at} \\ \text{the HR} \end{bmatrix} + \begin{bmatrix} \text{loss at} \\ \text{the OC} \end{bmatrix}$$

Typical values for loss in a HeNe laser include:

- Loss at the OC of 1.0% (the transmission of the mirror)
- Loss at the high reflector (HR) is negligible (the reflectivity is better that 99.9995% so the loss is less than 0.0005%).

- Attenuation of the laser medium itself is approximately 0.05 m^{-1}, so for a 35-cm tube, it is 0.0175.
- Loss at each window is approximately 0.25% per pass (plane, A/R-coated windows; this figure will change if the tube features Brewster windows, as most argon lasers do).

Total losses sum to 0.0817, and since the gain is equal to this figure (for two passes through the plasma tube), the gain is calculated to be 0.0408 for a single pass through the tube. The gain medium itself, not the length of the tube, but rather that of the actual inner plasma tube where amplification occurs, is 29.5 cm in this particular case, so the gain is determined to be 0.138 m^{-1}. This is typical for a small HeNe laser.

PROBLEMS

4.1 Prove that temperature alone cannot create an inversion. Determine the population of atoms at two energy levels (say, 1 and 3 eV) at low (10 K) and high (5000 K) temperatures. Now determine the negative temperature required for a population inversion between these two energies to exist (this is from a thermodynamics point of view).

4.2 Calculate the temperature required to allow the rate of stimulated emission to exceed the rate of spontaneous emission for a system emitting radiation at 1064 nm (YAG laser output). To do this, set the ratio in equation (4.5.12) equal to 1 and solve.

4.3 To create a population inversion, one must populate the upper lasing level with more atoms than at the lower level. This may be difficult to do, as the lower lasing level is very close to the ground level, where it is populated thermally. The problem can be alleviated by cooling the laser to depopulate the lower energy levels. Imagine an Nd:YAG laser that is operating at a temperature of 1000 K. The laser emits a wavelength of 1064 nm and the lower level of YAG is only 0.2 eV above ground state. If the rod contains a total of 1×10^{19} active lasing ions:

(a) Calculate the number of ions that must be excited at the upper level to ensure that population inversion occurs.

(b) If the rod absorbs 15% of the pump light (in this case, 880 nm), calculate the power input required to allow the rod to lase (in joules; assume that pumping occurs as one fast pulse).

(c) To complete the problem, calculate the thermal population of the lower level and equate this to the population of the upper level. Now calculate the energy required by assuming that 15% of incident photons of 880-nm pump light are converted to excited atoms in the upper level. This is a

114 LASING PROCESSES

simplified problem but does illustrate the problems involved with thermal populations of levels.

4.4 Calculate the threshold gain of two configurations of a helium–neon laser, which each have a loss factor of 0.05 m^{-1}, as follows:

(a) A 50-cm tube with one mirror 99% reflecting and the output coupler 90% reflecting (i.e., $T = 10\%$ for the output coupler)

(b) A 20-cm tube with one mirror 99% reflecting and the output coupler 95% reflecting

(c) A 20-cm tube with one mirror 99% reflecting and the output coupler 97% reflecting

4.5 An argon laser has a 90-cm gain tube and operates at a wavelength of 488 nm. Calculate the Doppler linewidth of the laser. Plasma temperatures in an argon laser run at about 5000 K. Compare this to the linewidth of the HeNe laser.

4.6 A small YAG ($Y_3Al_5O_{12}$) laser rod weighs 10 g and has a dopant (Nd^{3+}) concentration of 1%. The laser is pumped by a flashlamp and is 10% efficient operating at 1064 nm.

(a) Calculate the number of Nd^{3+} ions in the rod itself.

(b) Calculate the number of ions involved in actual laser emission.

(c) Calculate the maximum available energy per pulse from the laser.

(d) Assuming that the rod is from a CW YAG laser and the dopant level is 0.1%, find the maximum energy available per pulse.

4.7 A superradiant nitrogen laser (i.e., no cavity optics) operates at 337.1 nm. It is a pulsed laser with 10-ns pulses, a tube 30 cm in length, and fires 60 times per second. Ignoring loss in the laser, calculate the gain of the laser (in m^{-1}) if it produces a 1-mW average output. To do this problem, calculate the energy of a single photon, then the energy required per pulse to give an average power of 1 mW (i.e., 1 mJ/s). Finally, consider the number of photons required per pulse. Assume that the gain required is that sufficient to amplify a single photon at the far end of the tube to the required number of photons after a length of 30 cm.

4.8 In a method similar to Example 4.9.1, an inserted loss created by a glass slide of $n = 1.37$ is used to determine the gain of a HeNe laser. The laser ceases to oscillate when the glass slide is at 60.5 degrees, but the uncertainty in measuring this angle is ± 1.0 degree. Calculate the gain of the laser (assuming the same losses as in Example 4.9.1) as well as uncertainties. Express the answer as a gain in m^{-1} with a "$\pm$" uncertainty figure.

4.9 The gain of a HeNe laser tube is measured using the method outlined in Figure 4.9.3. When using the 612-nm line in the orange, it is found to have a measured net gain of 0.035 m^{-1} (this figure includes losses from the Brewster windows and absorption by the medium itself). Considering that the HR for this laser has a reflectivity of 99.9999% (essentially, 1), calculate the minimum reflectivity of the OC to allow this laser to operate. If the same laser shows a net gain of 0.135 m^{-1} when the red 632.8-nm line is used, calculate the minimum reflectivity of the OC for this laser when operating on that line.

CHAPTER 5

Lasing Transitions and Gain

Despite the fact that the ruby was the first laser, it is amazing that it worked at all. Consider that there are only three energy levels in the system: a pump level, an upper-lasing level, and a ground state which serves as the lower-lasing level. Population inversion, of course, occurs between the upper and lower lasing levels of the system. The ramifications of this are that the upper lasing level must be populated to a level exceeding that of the ground state, a difficult feat requiring an enormous amount of pump energy. Later it was discovered that other energy systems exist which utilize a discrete lower lasing level. Such systems, in general, allow generation of a population inversion with much less pump energy and are usually more efficient than three-level systems. An example is the YAG laser, which is much more efficient than ruby and can easily operate in continuous-wave (CW) mode. Today, four-level lasers are by far the most prolific type.

5.1 SELECTIVE PUMPING

It is evident that the pumping method must ensure that the upper lasing level is populated while the lower level is relatively empty. For this reason, pumping mechanisms must be designed with forethought. Consider the helium–neon laser described in Chapter 4. Helium is used as an intermediary that absorbs energy from the electrical discharge (by electron collision) and transfers it, quite selectively, to the upper lasing level of neon, allowing that level to populate while the lower level receives no energy from helium. The lower level will certainly be populated thermally (and to a larger extent than the upper level), but at over 18 eV above ground level, the thermal population of the level will be negligible.

Other examples of selective pumping exist in solid-state lasers such as the YAG. Examining the absorption spectrum of YAG in Figure 5.1.1, we see a large absorption band in the red and near-infrared regions of the spectrum around 750 and 800 nm. High-powered lamps using inert gases are employed to pump this rod optically. Consider the output spectra of various lamps that may be used for pumping, as shown in Figure 5.1.2. Xenon lamps, the most common used for optically pumped

Fundamentals of Light Sources and Lasers, by Mark Csele
ISBN 0-471-47660-9   Copyright © 2004 John Wiley & Sons, Inc.

118 LASING TRANSITIONS AND GAIN

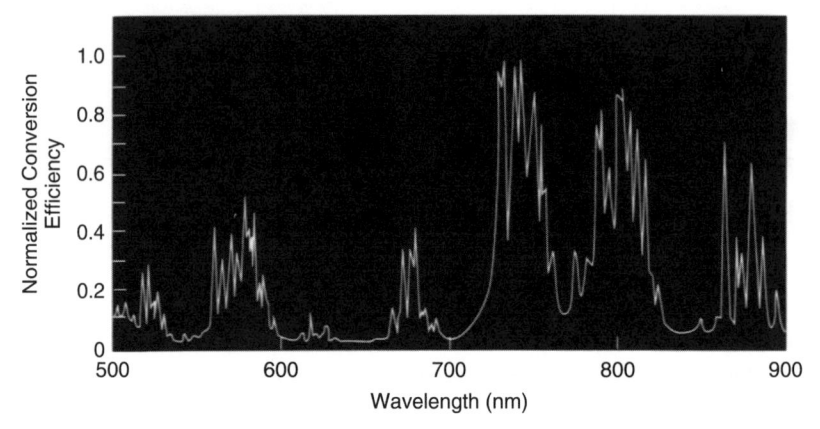

Figure 5.1.1 Absorption spectrum of YAG. (Courtesy of Perkin-Elmer OptoElectronics.)

lasers, show an output rich in the blue and green end of the spectrum, where YAG does not absorb light particularly well. Most of the energy output by the xenon lamp in this region is wasted since the YAG rod cannot absorb it. Krypton lamps, on the other hand, have a rich output precisely in the region where YAG absorbs, around

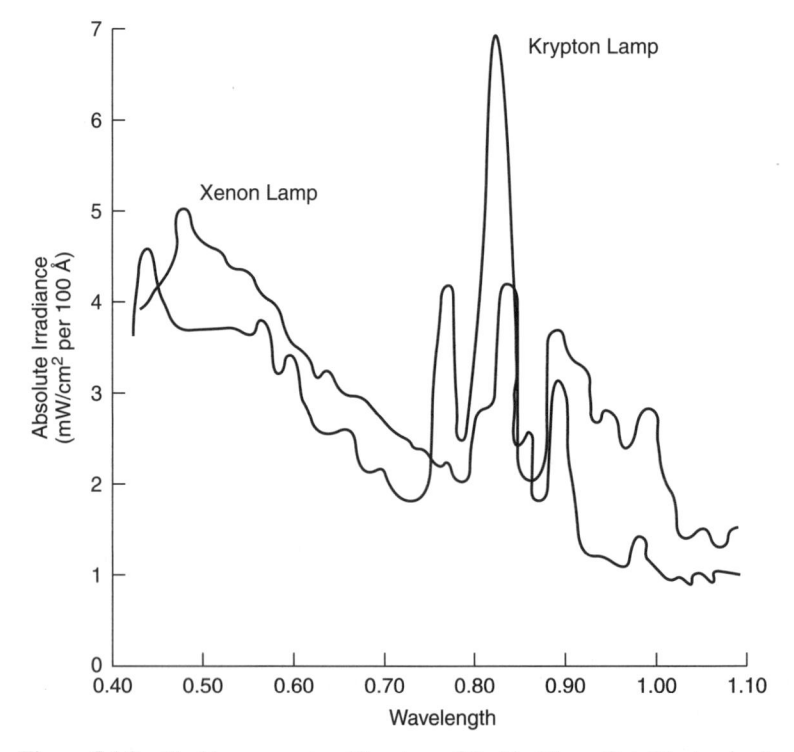

Figure 5.1.2 Flashlamp spectra. (Courtesy of Perkin-Elmer OptoElectronics.)

750 to 850 nm. Krypton lamps are a more efficient pump source than xenon lamps for YAG rods and are hence used in most commercial lasers of this type. Xenon lamps may still be used as a pump source, but are not as efficient.

Like the YAG, the ruby laser has defined absorption bands, but these are in the yellow–green region of the spectrum, where xenon lamps have much higher output than krypton lamps, hence xenon lamps are a better match as a pump lamp. Later in the chapter we revisit YAG and ruby lasers.

5.2 THREE- AND FOUR-LEVEL LASERS

Lasers are classed by the number of energy levels involved in the actual lasing process as three- or four-level lasers. In a three-level system (the simplest), energy injected into the gain medium excites atoms to a pump level above the upper lasing level. From there atoms decay to the upper lasing level. This decay to the upper lasing level usually occurs by emitting heat, not photons. It is rapid and quickly populates the upper energy level. This upper level often has a long lifetime, so a healthy population of atoms builds in that level. Lasing transitions now occur between the upper level and the ground state, emitting laser light in the process. This system is characterized by the lack of a discrete lower lasing level; the ground state serves that purpose. The left side of Figure 5.2.1 shows the energies involved in a three-level laser, including the pump, upper lasing level, and lower-lasing levels.

Four-level systems feature a discrete lower lasing level between the upper and ground states. Atoms making a laser transition to the lower state decay further to the ground state, in some cases by emitting a photon (in the case of an argon-ion laser, a 74-nm photon is emitted during this decay to the ground state). Four-level lasers are by far the most common. In a helium–neon laser, for example, the pump level is in helium (it need not be in the same species as the lasing atom), while the upper and lower lasing levels are in neon. A carbon dioxide laser has a similar situation, where the pump level is in nitrogen gas, which like helium in a

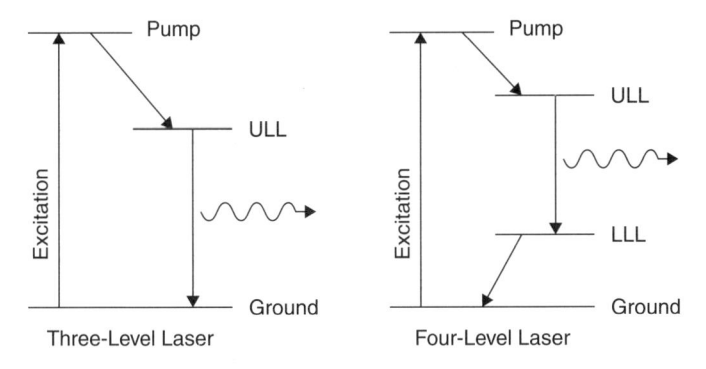

Figure 5.2.1 Ideal three- and four-level lasers.

helium–neon laser, is in larger proportions than the lasing species. It is also possible in some lasers, such as metal-vapor lasers, to pump directly to the upper lasing level (i.e., there is no pump level). In this case, atoms are excited to the upper lasing level, where they make a transition to the lower lasing level, finally decaying to the ground state. These are particularly efficient systems since they lack an energy loss when atoms in the pump level decay to the upper lasing level (which, in most cases, shows up as heat in the laser). Although the metal-vapor system has only three levels, it has more in common with a four-level system than does a classic three-level system.

In a four-level laser, laser gain is realized as soon as pump energy is applied to the system. Pump energy is injected into the pump level, where it decays, in most cases almost instantaneously, to the upper lasing level. Assuming that the upper level has a longer lifetime than the lower level (and in most four-level lasers it does), a population inversion occurs almost immediately after pump energy is injected. Although there may not be a usable output beam (since any small gain produced will be lost to absorption and other losses within the system until threshold is reached), there is a population inversion and hence a gain. Injecting a little more energy into the system will raise the gain to a level where it exceeds lasing threshold and a usable output beam appears.

So the nature of a four-level laser is such that a population inversion is easy to achieve—but what happens in a three-level laser where the lower lasing level is the ground level? We know from Boltzmann statistics that at a given operating temperature there will be an enormous population of lasing atoms at ground state, and that to achieve a laser gain we must raise over half of these to the upper lasing level, which in many cases is far above ground. Consider the dynamics of the ruby laser system in Figure 5.2.2, in which the upper lasing level is about 1.8 eV above ground. The thermal population of this level at 300 K is negligible. Also evident in the figure are two pump bands in the blue and green regions of the spectrum.

Strong pumping is required for such a laser to achieve a population inversion, and these lasers exhibit a high threshold of pump energy. A considerable amount of pump energy is required just to achieve inversion, let alone lasing threshold! There is also a time delay between the onset of pumping power and lasing output. As pump energy is injected into the ruby, one must wait until a significant population of atoms (certainly, over half of the chromium ions in the rod) reaches the upper lasing level before lasing can begin. So why does ruby work quite well as a laser material (after all, it was the first)? The reason lies in the two broad pump bands, which readily absorb energy from a flashlamp (xenon flashlamps, you will recall from an earlier discussion, emit a good portion of their radiation in the violet-to-green region, so are commonly used with ruby lasers). As well as being broad, the pump bands have incredibly short lifetimes (on the order of 1 µs), which causes energy absorbed into these pump bands to relax almost immediately to the upper lasing level. Finally, the lifetime of the upper level, 3 ms, is quite long, allowing the excited ions to remain there long enough to have a good chance of emitting light by stimulated (as opposed to spontaneous) emission. Although most ruby lasers are pulsed, it is possible to operate the

THREE- AND FOUR-LEVEL LASERS **121**

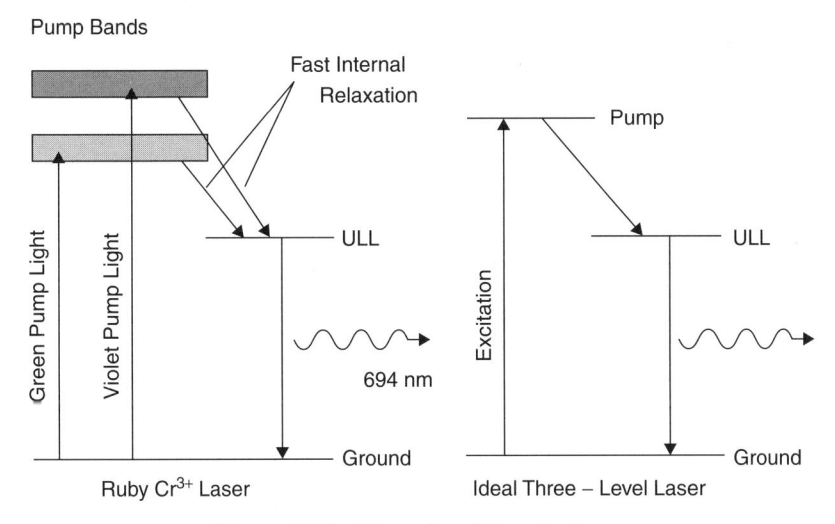

Figure 5.2.2 Dynamics of the ruby laser.

material as a CW laser. Special conditions are required, however, such as extreme cooling to depopulate a portion of the ground level; this is, not surprisingly, a collection of closely spaced levels. Doing so allows the system to operate more like a four-level system, as described next.

Now, in contrast to the three-level ruby laser, consider a four-level laser such as the YAG (also a solid-state laser), with energy levels as depicted in Figure 5.2.3. The YAG system is characterized by a cluster of pump levels from which excited atoms decay rapidly to the upper laser level. There are multiple pump levels, allowing the

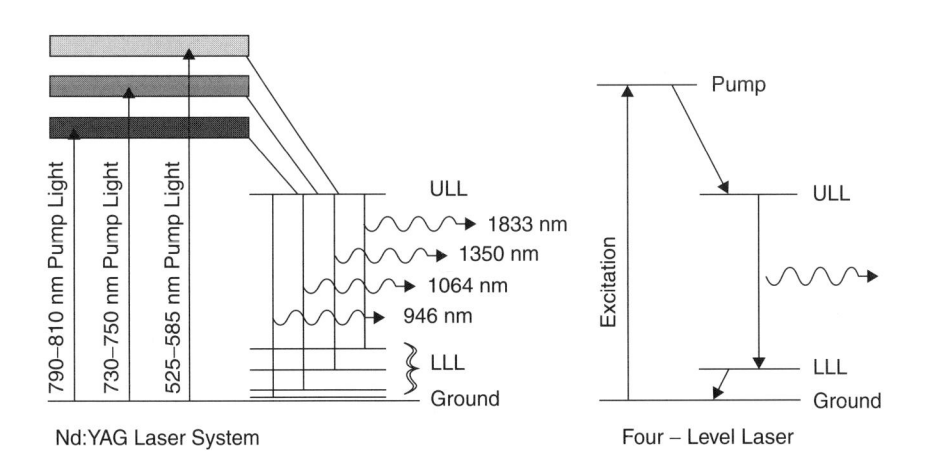

Figure 5.2.3 Energy levels in the YAG laser.

122   LASING TRANSITIONS AND GAIN

system to absorb energy at a variety of wavelengths, including the important band at 790 to 810 nm, useful for pumping via a semiconductor laser. All pump levels have short lifetimes, around 100 ns, and decay rapidly to the upper lasing level. The upper level has a very long lifetime of 1.2 ms compared to the lower level, which decays to ground in 30 ns. Many lasing transitions are possible in this system, including the most powerful, at 1064 nm. The lower lasing level of the transition at 946 nm is extremely close to the ion ground state. Because of the large thermal populations of the lower lasing level, lasing on this transition occurs in a mode resembling a three-level laser, and strong pumping is required to allow this transition to oscillate. For other transitions such as 1064 nm, the situation is more favorable and a population inversion will occur quickly.

Most lasers are four-level lasers: for example, the familiar HeNe laser with energy levels pictured in Figure 5.2.4. The pumping mechanism, involving resonant energy levels between helium and neon, has been described previously, but an examination of the lower levels (of which there are many) reveals an intervening *metastable state* (so-called because it has a relatively long lifetime, making it almost stable) to which atoms in one of the lower lasing levels can rapidly decay. This pathway allows depopulation of the lower lasing levels, making population inversion easy to achieve. This rapid decay is radiative; it is accompanied by spontaneous emission of light at about 600 nm. This light does not contribute to laser action but is seen as an orange line in the spectrum of the tube emission as viewed through a spectroscope. It is part of the pink glow coming from the length of the tube. Not

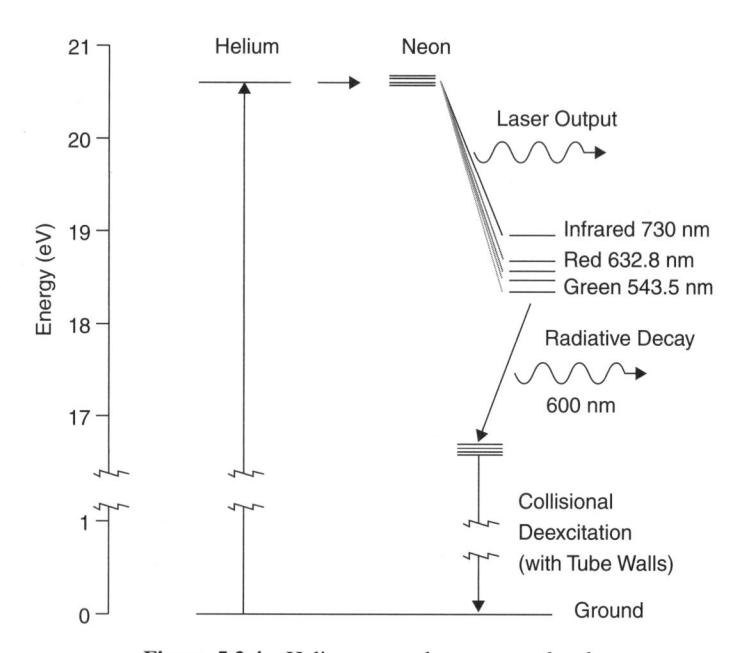

Figure 5.2.4 Helium–neon laser energy levels.

surprisingly, atoms at the lower level must emit the energy somewhere when making the transition to a yet lower energy level, and the 600-nm spontaneous emission provides that outlet. From that metastable state they must collide with the walls of the tube to return to ground state.

This scenario illustrates that the HeNe laser involves four energy levels in the active lasing atom itself (i.e., the neon atom) plus a pump level in helium. The neon levels are the ground state, the upper lasing level, the lower lasing level, and the metastable state, from which it returns to ground state. Although it is tempting to call it a *five-level laser*, it is classed as a four-level system since only four states are really involved with the laser process. Other levels are simply there as stepping stones. The process begins again when ground-state neon atoms are again pumped by collision with energetic helium atoms. Some lasers feature hybrid-level systems. One example is the copper-vapor laser, which has a three-level system which more closely resembles a four-level system since it lacks a pump level (Figure 5.2.5). In this system, energy is injected directly into the upper lasing level (ULL) (by direct collision of electrons with copper atoms), but unlike a true three-level system, it does have a discrete lower lasing level (LLL).

Although a system such as this could theoretically offer high efficiencies since no energy is wasted in the decay of energy from the pump level to the upper lasing level (i.e., the system has a high quantum efficiency, as we examine later in the chapter), this particular laser does not operate in CW mode, due to an unfavorable situation in which the lifetime of the ULL is shorter than that of the LLL (as discussed in the next section).

Perhaps the most dramatic effect of the type of energy-level system is the amount of pump energy required to generate laser action. In any laser, output can begin as soon as population inversion occurs, followed by stimulated emission. Gain increases until it exceeds total losses in the system and laser output occurs. In a three-level laser it takes a large amount of pump energy just to generate an inversion, since the population of the ULL must exceed that of the ground state. In a four-level

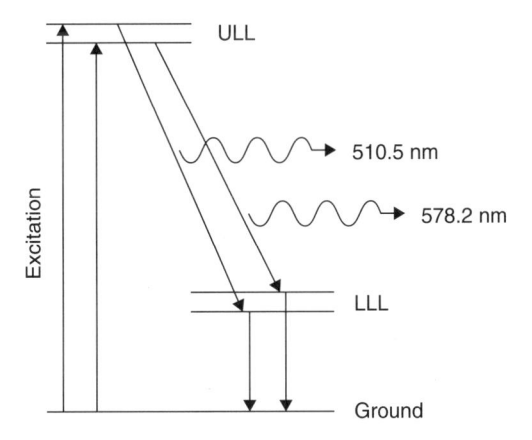

Figure 5.2.5 Copper-vapor laser energy levels.

laser, inversion can be achieved with much lower pump energies since the LLL is a discrete level that is not populated to nearly the level of the ground state. Assuming that the LLL for a four-level laser is a few electron volts above ground state, it will not be thermally populated to any large extent—unlike the three-level laser, in which the ground state has a large population, making inversion difficult to achieve. We examine this is detail in Section 5.7.

5.3 CW LASING ACTION

Not all lasers can operate in CW mode; many lasers operate strictly in pulsed mode. Most of these lasers never became commercially feasible because industry demand is considerable higher for CW lasers. However, nitrogen and excimer pulsed lasers are popular. For example, molecular nitrogen, with a UV output at 337.1 nm, is a self-terminating laser transition that cannot operate in CW mode.

In any laser, the probability that a particular system will lase in CW mode (i.e., produce continuous laser output) is determined in large part by the relative lifetime of the levels involved in the laser transition. If the lower level has a relatively long lifetime, atoms in that lower energy state stay there longer, giving them a good chance of absorbing photons as well as of violating the population inversion criteria. In this situation a pulsed laser may still be possible where the upper level is filled quickly and preferentially over the lower level, but eventually the population of the lower level will exceed that of the upper level and lasing will cease. On the other hand, if the lower level has a short lifetime, atoms in that state decay quickly to another state (ground state in a model four-level laser, but quite possibly an intervening level), where they will not be available to absorb the newly emitted photons in the laser. CW gas lasers invariably possess the latter characteristic.

The lifetime, or to be more specific, the spontaneous radiative lifetime of a lasing species is the time that an atom will stay at a particular energy level before spontaneously losing energy. An atom at an upper lasing level, for example, can lose energy by emitting a photon, either spontaneously or through stimulated emission. You will recall from Chapter 4 [equation (4.5.3)] that the rate of spontaneous emission is defined by the product of the Einstein coefficient and the number of atoms at the upper energy level N_2:

$$r_{\text{spontaneous}} = A_{21} N_2$$

where A_{21} is the Einstein coefficient for spontaneous emission and N_2 is the number of atoms at the upper energy state. It might be noted that the A coefficient for absorption is related to the spontaneous lifetime as

$$A_{21} = \frac{1}{\tau}$$

where τ is the spontaneous radiative lifetime of the upper state. So if the upper level has a long spontaneous lifetime, the A coefficient will be small, as will be the rate of spontaneous emission. With a low spontaneous emission rate, stimulated emission is given a better chance of occurring. There are also implications here regarding the storage of energy in a particular energy level, which allows the production of giant pulses using a technique called Q-switching (which we examine in Chapter 7).

In a four-level laser the ability to lase in CW mode is not only defined by the lifetime of the upper lasing level but also by the lifetime of the lower lasing level. If the lower level has a relatively long lifetime, atoms in that (lower) energy state stay there longer, giving them a good chance of absorbing photons. This, of course, would serve to hinder laser action since (1) the laser medium will strongly absorb photons produced by the laser, and (2) this allows the atomic population of the lower level to build violating the required inversion criteria. On the other hand, if the lower level has a short lifetime, atoms in that state decay quickly to another state (the ground state in a model four-level laser system but possibly to an intermediate level), where they will not absorb the newly emitted photons in the laser. CW gas lasers (those that produce continuous lasing output) invariably possess the latter characteristic since a long lower lifetime (i.e., longer than the lifetime of the upper state) would probably not allow a population inversion to be maintained. We expand on the depopulation mechanism in Section 5.5.

A nonfavorable condition (for CW lasing action) exists in the nitrogen laser where the upper lasing level is *very* short compared to the lower level. The upper level has a lifetime of about 10 ns, whereas the lower level has a lifetime of about 10 ms.[1] Assuming that a mechanism exists to pump energy quickly into the upper level (and it does, in the form of a high-current transverse discharge), the laser can operate for a maximum of 10 ns before the lower level is populated, inversion is violated, and lasing ceases. The dynamics of the nitrogen laser are depicted in Figure 5.3.1, with important energy levels at various times in the lasing process highlighted. At $t = 0$ s, pumping begins and the upper lasing level fills. Laser action ensues until at about 10 ns, the population of the lower lasing level exceeds that of the upper level and lasing action ceases since population inversion is no longer maintained. This illustrates the self-terminating nature of the species. About 10 ms later the lower level depopulates via a transition to a metastable energy state that has a longer lifetime yet. The situation also illustrates that it is not possible to operate such a laser in CW mode, since once the lower level has filled, these molecules of nitrogen are no longer available to be pumped to the upper lasing level.

The only reason the nitrogen laser operates at all is that a fast pump mechanism exists in which a current of thousands of amperes is made to pass through a small volume of gas. With fast rise times on the current, inversion can be achieved and lasing ensues (it is short-lived, however, due to the short ULL lifetime). Achieving a fast discharge such as this requires careful laser design, as we shall see in Chapter 10. Another example of a pulsed laser system with unfavorable lifetimes

---

[1] The figures quoted are for a low-pressure nitrogen laser operating at pressures around 30 torr. The lifetime situation is worse in TEA nitrogen lasers, which operate at many atmospheres of pressure. The upper lasing level lifetime of these lasers can be as short as 1 to 2 ns. In Chapter 10 the lifetime situation is described in detail.

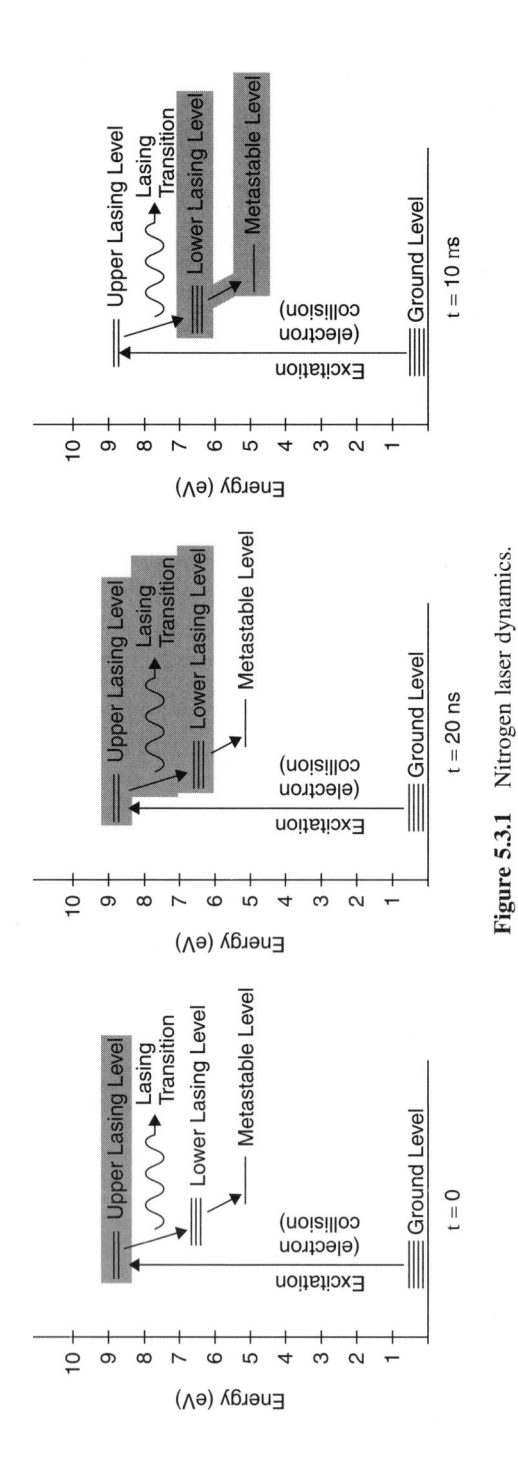

Figure 5.3.1 Nitrogen laser dynamics.

is the copper-vapor laser. This laser has an upper-level lifetime of about 10 ns and a lower-level lifetime of 270 μs. An effect called *resonance trapping* can be used to increase this upper-level lifetime. In this effect the ground-state population of copper atoms influences population inversion by trapping radiation emitted from a transition between the upper level and the ground state by spontaneous emission. This radiation is emitted spontaneously by atoms at the ULL as opposed to the lasing transition to the LLL, which is brought about by stimulated emission. The spontaneously emitted radiation is reabsorbed by ground-state atoms (or *trapped*) and serves to pump copper atoms at the ground state back into the upper level. This serves to extend the effective upper-level lifetime from 10 ns to 370 ns.[2] This improves the situation and greatly reduces the discharge pumping rate required, which simplifies the design of the laser since the discharge may be much slower. (Nitrogen lasers, which have an ULL lifetime of about 20 ns, utilize extremely fast discharge electronics; see Chapter 10 for details.) Still, the upper-level lifetime of 370 ns is much shorter than the lower-level lifetime, so this laser operates strictly in pulsed mode.

5.4 THERMAL POPULATION EFFECTS

As mentioned previously, the effect of thermal energy on the population of energy levels is negligible in most cases. For a helium–neon laser, the lower lasing level is almost 19 eV above ground, so thermal populations are essentially zero at any practical operating temperature. The situation does exist, though, in some laser systems where the lower lasing level is very close to the ground level and thermal effects cannot be ignored. Where two or more transitions are possible, with one having a lower level close to ground state, a transition with a higher lower level may be favored.

Consider a common YAG laser in which neodymium ions are the active lasing species. Examining the energy levels and transitions of this laser in Figure 5.2.3, we see that several lasing transitions are possible, yet 1064 nm is the most powerful and the others rarely appear. As expected, each lower level is actually a cluster of levels tightly clumped together. The lower level for the 946-nm transition, for example, is part of a cluster of levels situated below 0.1 eV, referred to as *ground state*. Indeed, even ground state for an ion may be composed of a number of tightly clumped levels. Unlike the 1064-nm transition, where the lower level is 1.2 eV above ground state and where the population of this level from temperature alone is negligible, the population of the lower level for the 946-nm transition is very much affected by thermal population effects. Note that in a hot YAG rod (i.e., around 1000 K), the thermal effects on the 1064-nm transition *can* become

---

[2] The effects of resonance trapping are described by L. A. Weaver et al. in the *IEEE Journal of Quantum Electronics*, Vol. QE-10, No. 2, p. 140 (Feb. 1974). The copper-vapor laser system has two visible transitions, each between distinct upper and lower levels. The 578.2-nm transition features an upper lasing level with a lifetime of 10.24 ns, extended to 370 ns by resonance trapping. The 510.6-nm transition features an upper lasing level with a lifetime of 9.60 ns, extended to 615 ns by resonance trapping. In both cases the lower-level lifetimes are 270 μs.

significant (although in practice a rod would never be allowed to reach so high a temperature).

Example 5.4.1 Thermal Populations in a YAG Rod Consider a YAG rod doped with a total of 1.01×10^{20} neodymium ions. At a temperature of 300 K, the population of the lower level for the 946-nm transition can be calculated using Boltzmann statistics to be

$$N = N_0 \exp\left(-\frac{E}{kT}\right)$$

$$= 1.01 \times 10^{20} \exp\left[\frac{-1.6 \times 10^{-20}}{1.38 \times 10^{-23}(300)}\right]$$

$$= 2.1 \times 10^{18}$$

This means that the upper lasing level for that transition must be filled to at least that quantity in order to have 946-nm emission. Contrast this to the population of the lower level for the 1064-nm transition at the same temperature as that of one (yes, only one) of the 1.01×10^{20} neodymium ions. It is fair to say that thermal effects on the 1064-nm transition's lower level are negligible, but clearly that is not the case for the 946-nm transition, which has a lower level much closer to ground. All things being equal, we need a much higher pump power to excite the 946-nm transition. Like a three-level laser, this transition requires a large population to be excited to the upper level to ensure an inversion. Despite strong pumping, it is possible in some cases that the maximum number of ions pumped to the upper level for this transition will *still* be insufficient to ensure an inversion. The option, then, is to cool the rod (e.g., with liquid nitrogen at 77 K) to lower the thermal population of this level and allow a population inversion to occur.

Other systems exist with similar energy levels, such as erbium (Er^{3+} with a 1.55-μm emission) as used in fiber amplifiers. At a temperature of 300 K the system is essentially a three-level system since the lower lasing level is quite close to the ground state. At a temperature of 77 K (brought on by liquid-nitrogen cooling), the system is a four-level system, and population inversion is much easier to achieve.

5.5 DEPOPULATION OF LOWER ENERGY LEVELS IN FOUR-LEVEL LASERS

In a previous example we saw how the lower lasing levels of the nitrogen laser are depopulated as molecules drop in energy to a metastable state with a lower energy. In any four-level laser, some method of depopulation of the lower lasing level exists. These methods may result in the production of a photon of light (although not one that contributes to lasing action) or as the result of a nonradiative process of the type seen in semiconductor diodes in Chapter 2. The argon laser provides us with a straightforward example of depopulation via radiative energy loss, in which argon ions at the lower lasing levels (around 33 eV above the ion ground-state energy) drop to a lower level by emission of a photon as shown in Figure 5.5.1. This drop is quite large and results in the release of a photon with 18 eV of energy, corresponding to an extreme-UV wavelength of 74 nm. This photon does not contribute to laser action, nor is it normally observed as a spontaneous emission from the tube since few materials (including quartz windows) are transparent to this extremely short wavelength.

Atoms of neon in a helium–neon laser undergo a two-stage process, outlined in Figure 5.5.2, to relax to the ground state, from where they can again be pumped to the upper lasing level to take part in coherent light emission. From the lower lasing level neon atoms first undergo a radiative transition to an intermediate level by emission of a photon around 600 nm. This emission contributes to the pink-orange glow (spontaneous emission) seen from an operating HeNe tube. From that intermediate level, radiative emission is not possible, due to quantum-mechanical selection rules for allowed transitions (Chapter 3), so the only avenue available is collisional transfer of energy between the excited neon atoms and the walls of the laser tube. This also explains a peculiarity with HeNe lasers in that a larger-diameter plasma tube results in the reduction of laser power. In many lasers, such as the carbon dioxide laser, use of a larger-diameter plasma tube results in more gain volume and hence in higher energies being available for a given length of tube, but this is not the case with HeNe lasers. Most HeNe tubes have an inner plasma tube bore of only 1 mm. Use of a larger-diameter tube results in a buildup or *bottleneck* of neon atoms residing at the intermediate (metastable) energy level. A metastable state has a long lifetime relative to the lifetimes of other levels in the system. In some cases it is an energy level from which a radiative transition cannot occur (due to the selection rules of quantum mechanics), so photon emission as a simple way to

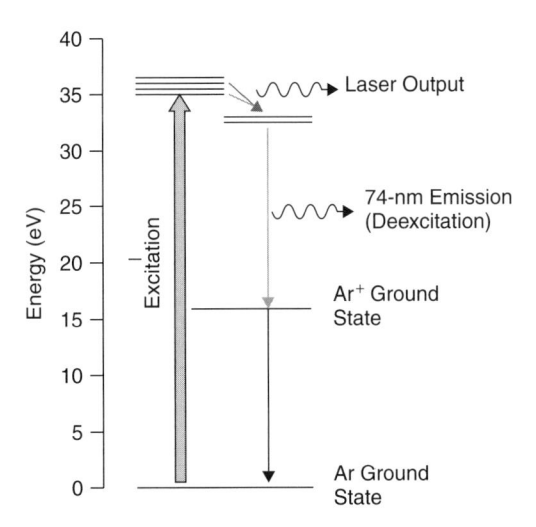

Figure 5.5.1 Depopulation of the argon-ion lower lasing level.

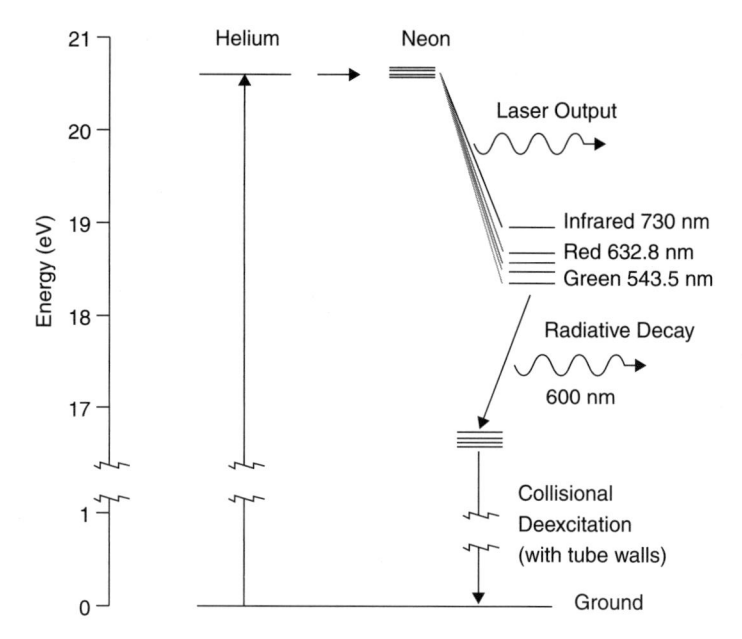

Figure 5.5.2 Depopulation of neon atoms in the HeNe laser.

return to ground state is not possible. Many atomic systems, including neon and nitrogen, have such states. In the case of the HeNe laser, neon atoms in the metastable state are no longer available to contribute to laser action, so the gain is reduced and consequently, the output power is also reduced. Small tube diameters are required to allow excited neon atoms to lose energy quickly to the tube walls and again enter the laser process.

The final example of a unique depopulation process is in the semiconductor laser (a four-level laser), in which a nonradiative process occurs. In this laser the ground and lower lasing levels exist within the valence band, and the pump and upper lasing levels exist within the conduction band, as shown in Figure 5.5.3. Excitation promotes an electron in the semiconductor crystal to a pump level. From that level it falls to the lowest level in the conduction band, which serves as the upper lasing level. Emission occurs between that level and the highest level in the valence band, which serves as the lowest lasing level. This level must depopulate to maintain a population inversion, and does so by falling to a lower level in the valence band. This drop in energy is accompanied by the production of a phonon, a vibration in the crystal lattice of the semiconductor.

5.6 RATE EQUATION ANALYSIS FOR ATOMIC TRANSITIONS

Revisiting Section 4.5, we now take a more exhaustive approach to the analysis of laser rate equations based on the nature of the energy levels themselves. Before

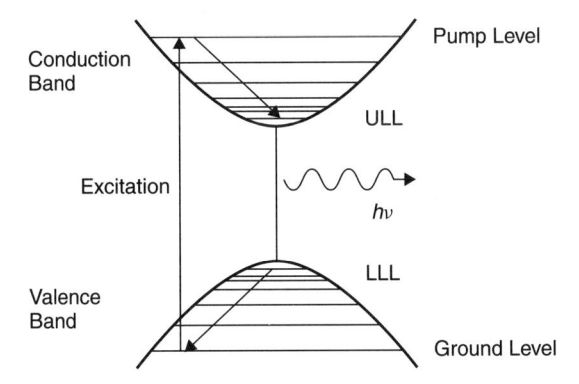

Figure 5.5.3 Depopulation of semiconductor atoms.

we examine laser systems, consider a simple atomic system involving only two energy levels, N_1 and N_2, where N_1 is the lower level for the transition and N_2 the upper level. An atom at the lower level can absorb a photon of a specific energy and be promoted to the upper level. Similarly, an atom at the upper level can jump to the lower level and in the process emit a photon of light with energy corresponding to the energy difference between the two levels. In addition to displaying the spontaneous emission of a photon, the same atom at the upper level can be stimulated to emit a photon of light in the process of stimulated emission.

We begin by defining the *cross section* of an atom, which specifies the effective size of an atom. The larger the atomic cross section, the larger the target to interact with other atoms or particles or an optical wave. In a physical sense, one would expect a real atom to have a cross section of only a few atomic radii in diameter. Referring back to Chapter 2, we find that Bohr postulated that the radius of an orbit was given by

$$r = n^2 a_0$$

where n is the quantum number and a_0 is the Bohr radius (5.29×10^{-11} m). Following this logic, for an atom with an $n = 2$ outer orbit, we expect the cross section to be only two 2 Å (2×10^{-10} m) radius. The odds of such a small object interacting with anything would be slim (unless the pressures involved were extremely high, so as to increase the number of atoms available for interaction). In reality, though, atoms are found to have cross sections of 100 nm or more! The large cross section is caused by the electric dipole of the atom, which acts as a tiny antenna capturing signals.

Cross section is important because it gives rise to a figure called the *transitional probability*. For a given transition between two energy levels, this figure represents the odds that a particular atom will make a particular transition and is denoted W. It has units of probability per atom per second (measured in units of s^{-1}) and is a real physical value, which may be determined experimentally by examining the absorption of an atomic transition across the entire bandwidth of the transition to determine time constants (which also determine the Einstein A coefficient, important in rate equations). Experimental values may be referenced using a source of atomic data such as the *CRC Handbook of Chemistry and Physics* (Boca Raton, FL: CRC Press), a standard source for data such as these. The *CRC Handbook* lists transitional probabilities for various atomic and ionic species in units of $10^8 \, s^{-1}$ and gives statistical weight and upper energy levels allowing the calculation of atomic oscillator strength and transition strength, two useful figures when considering the usefulness of a particular transition for lasing action. Transition strength, denoted S, is simply the integral of cross section across all frequencies. It is related to the radiative lifetime and wavelength of the transition itself and may be calculated from those parameters.

As an example, consider the transitional probabilities of the red (632.82 nm) and green (543.37 nm) transitions of neon responsible for two popular lines of the helium–neon laser. The red line has a transitional probability listed as $0.0339 \times 10^8 \, s^{-1}$, and the green line, as $0.00283 \times 10^8 \, s^{-1}$. It is immediately apparent that the red transition (which shares exactly the same upper energy level as the green transition) is much more probable, so that if both transitions are allowed to oscillate (i.e., through the use of broadband mirrors in a laser cavity) the red transition will probably overtake the green. Conversion to line strength also reveals why the red transition is much more popular in the HeNe system.

Now that we have defined probability (and cross section), we can examine an atomic system by defining the rate of energy flow into or out of a particular level as the change in population of the level (ΔN) over a given time period (Δt). In proper calculus notation (which we specifically avoided in Chapter 4) we express a rate as dN/dt, which has units of atoms per second. Simply put, the rate equation for any given level is

$$r = dN/dt = WN \qquad (5.6.1)$$

where W is the probability of a transition and N is the population of the level.

For an atom at the lower energy level (N_1) the probability of absorbing a photon and being promoted to the upper level is W_{12}, where the subscripts indicate the transition and the direction in which it is occurring (in this case, from level 1 to level 2). In the case of the upper level, though, there are two probabilities: the probability of spontaneous emission occurring (related to the lifetime of the level) as well as the probability of stimulated emission occurring. We can define, then, these two probabilities as W_{21}, the stimulated emission probability, and $1/\tau_{21}$, the spontaneous emission probability (where τ_{21} is the spontaneous decay time for the upper level, which includes decay by radiative and nonradiative means. You will recall from Chapter 4 that this is the inverse of the Einstein A coefficient). Spontaneous emission does not depend on incident light but solely on population at an upper energy level available to make the downward transition. The situation in a two-level system is outlined in Figure 5.6.1.

It must be noted that W (the pumping transition probability measured in units of s^{-1}) is a function of the intensity of incident light. For an absorption process it is obvious that more light intensity results in a higher probability of absorption of a photon. The same holds true for stimulated emissions, where a higher intensity of light results in a higher probability of a stimulated emission occurring. The probability of a single event (absorption or stimulated emission) occurring is related to the cross section (defined earlier) by

$$W = \frac{\sigma I}{h\nu} \quad (5.6.2)$$

where σ is the cross section of the transition (in cm^2), I the intensity of the incident photon stream (in W/cm^2), and $h\nu$ the energy of a single photon in the stream (in joules).

The cross section of real gas atoms at low pressures (conditions normally encountered in gas lasers) usually fall in the range of 1×10^{-11} to 1×10^{-13} cm^2. For example, the neon transition that produces the HeNe gas laser output at 632.8 nm has a cross section of 1×10^{-13} cm^2. We may also relate this probability to the Einstein B coefficient, which we used in Chapter 4, as

$$W_{12} = B_{12}\rho \quad (5.6.3)$$

where ρ is the energy density. Multiplying this probability parameter by the number of atoms at a specific level results in the rate at which atoms make the transition (in units of atoms per unit time). This relationship may be derived by combining equation (5.6.3) with equation (4.5.1).

We now begin an analysis of a two-level system by expressing the rates at which atoms populate each level. In a two-level system we can express the rate of each

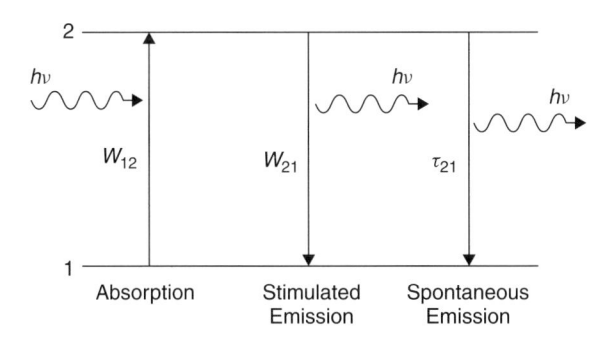

Figure 5.6.1 Processes in a two-level atomic system.

134 LASING TRANSITIONS AND GAIN

level[3] by summing mechanisms that serve to add to the level and subtracting mechanisms that serve to depopulate the level as follows. For the upper level:

$$\frac{dN_2}{dt} = \text{(absorption from } 1 \to 2\text{)}$$
$$- \text{(stimulated emission from } 2 \to 1\text{)}$$
$$- \text{(spontaneous decay from } 2 \to 1\text{)}$$
$$= W_{12}N_1(t) - W_{21}N_2(t) - \frac{N_2(t)}{\tau_{21}} \quad (5.6.4)$$

Similarly, for the lower level:

$$\frac{dN_1}{dt} = \text{(stimulated emission from } 2 \to 1\text{)}$$
$$+ \text{(spontaneous decay from } 2 \to 1\text{)}$$
$$- \text{(absorption from } 1 \to 2\text{)}$$
$$= W_{21}N_2(t) + \frac{N_2(t)}{\tau_{21}} - W_{12}N_1(t) \quad (5.6.5)$$

In these equations $N_1(t)$ and $N_2(t)$ represent the population of the two levels at any time t. Assuming that we have a total number of N atoms in the system, $N = N_1(t) + N_2(t)$ remains constant. This also implies that the rates above are identical but opposite in sign (you can check this in the equations)! Since a laser requires a population inversion to work, it is most useful to express the population difference at any time t as $\Delta N = N_1(t) - N_2(t)$ (redefined later in this chapter). When this number is less than zero, we have an inversion. We can now solve for the rate of change of the population difference as

$$\frac{d\Delta N(t)}{dt} = \frac{dN_1(t)}{dt} - \frac{dN_2(t)}{dt}$$
$$= -2W_{12}N_1(t) + 2W_{21}N_2(t) + \frac{2N_2(t)}{\tau_{21}} \quad (5.6.6)$$

But the probability of an absorption is the same as the probability of a stimulated emission (i.e., $W_{12} = W_{21}$ at the same intensity of incident photons),[4] so that we

---
[3] Be forewarned that simplifications have been made here to reflect the majority of real atomic systems. For example, the upward decay term W_{12}, which represents the possible thermal pumping of an upper lasing level, has been omitted from the equations. A more rigorous treatment of such rate equations can be found in *Lasers* by A. E. Siegman (Sausalito, CA: University Science Books, 1986, pp. 204–207).

[4] This assumes that there is no degeneracy in the levels, degeneracy being a condition where two levels have the same energy. If degeneracy exists, these probabilities must be ratioed according to $g_1 W_{12} = g_2 W_{21}$, where g_1 and g_2 are the degeneracies of each corresponding level.

may now express the equation as

$$\frac{d\Delta N(t)}{dt} = -2W_{21}[N_1(t) - N_2(t)] + \frac{2N_2(t)}{\tau_{21}}$$

$$= -2W_{21}\Delta N(t) + \frac{2N_2(t)}{\tau_{21}} \quad (5.6.7)$$

Furthermore, the total population of the system (N_0) remains constant with $N_0 = N_1(t) + N_2(t)$, so we may substitute $\Delta N(t) = N_0 - 2N_2(t)$, where the "×2" factor arises from the fact that a decrease of one atom from N_2 results in an increase of one atom in level N_1, so that the total change in the difference ΔN two atoms.[5]

$$\frac{d\Delta N(t)}{dt} = -2W_{21}\Delta N(t) + \frac{N_0 - \Delta N(t)}{\tau_{21}} \quad (5.6.8)$$

The rate at which the population difference occurs, then, is proportional to W_{21}, which in turn is proportional to the incident light intensity (a higher incident stream of photons leads to a higher rate of stimulated emission). At steady state, though, the rate of change of population difference is zero, so equation (5.6.8) may be solved as

$$\frac{d\Delta N(t)}{dt} = 0 = -2W_{21}\Delta N(t) + \frac{N_0 - \Delta N(t)}{\tau_{21}}$$

$$-2W_{21}\tau_{21}\Delta N(t) = \Delta N(t) - N_0$$

$$(1 + 2W_{21}\tau_{21})\Delta N(t) = N_0$$

$$\Delta N(t) = \frac{N_0}{1 + 2W_{21}\tau_{21}} \quad (5.6.9)$$

At low pump powers W_{21} tends toward zero, so the population difference is simply that of the number of atoms in the system (i.e., there are no excited atoms and the entire population is at level 1^5). At high pump powers the W_{21} term becomes large, so $\Delta N(t)$ tends toward zero. In other words, strong pumping will bring one-half of the atomic population to the upper level and one-half to the lower level, so that there is no population difference. The atomic population of each level at various pump power levels is plotted in Figure 5.6.2, with N_1 starting at N_0 (the total

---

[5] In another simplification of the problem, the total population is assumed to be N. The correct approach would be to use the fact that the population difference is $2\Delta N(t) = 2N_2(t) - N_0$, where N_0 is the thermal population difference equal to the population of level 1 minus the population of level 2 at thermal equilibrium. For most lasers, especially those in the visible region, the energy difference between the two levels is large, so Boltzmann statistics will show that the thermal population of level 2 is essentially zero. We hence arrive at this simplification using the total population N and ignoring thermal populations of the upper level.

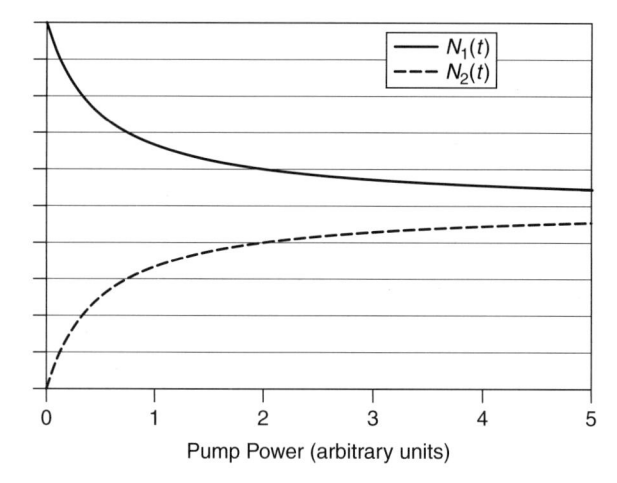

Figure 5.6.2 Energy-level populations for a two-level system.

number of atoms in the system) at zero pump power. In other words, there is no population at level N_2; in reality, N_2 would be thermally populated according to Boltzmann statistics, but for most systems this population would be negligible. Under no conditions can an inversion be achieved in this two-level system, so laser action is not possible with a two-level atomic system[6]; to obtain laser action, a system with more than two related energy levels is required. Another useful parameter to define is the *saturation intensity*, the intensity of the incident photon stream at which the rate of stimulated emission is equal to the rate of spontaneous emission.

5.7 RATE EQUATION ANALYSIS FOR THREE- AND FOUR-LEVEL LASERS

Most practical lasers feature three- and four-level atomic systems, so we shall apply the same technique used in the analysis of the two-level system to these systems. Bear in mind that many simplifications will be made to suit the model of an ideal laser, and that real lasers may have considerably more complex equations. Furthermore, we will use the concept of an optically pumped laser for simplicity (i.e., a stream of incident pump photons is used to excite lasing atoms). For other types of lasers, pump energy would be supplied in various forms, such as a DC current for a semiconductor laser diode or a flow of excited electrons for a gas laser.

---

[6]A generalization since the excimer laser appears to be a two-level laser, with the upper lasing level created, transiently, in the laser itself when two atoms are forced to bind to form an excited molecule (the excimer). The lower lasing level exists when the atoms are not bound together (the normal situation) since the lasing molecule cannot exist in a nonexcited state (which would be the expected LLL). These levels, then, are distinct species (one a bound molecule and the other two discrete atomic species) and not simple energy levels in a single species as the general examples of this section imply. See Chapter 10 for details.

RATE EQUATION ANALYSIS FOR THREE- AND FOUR-LEVEL LASERS

The analysis of a three-level laser, the most basic system, is methodical and progresses with an examination of energy flow into and out of each level in the system. We begin with a look at the pump level, which is populated via upward transitions from the ground state and loses population via downward decay to the ULL (denoted level 3 in Figure 5.7.1). It is most useful to express each level as a rate so that these may be equated for the equilibrium condition that exists when the laser is operating.

The rate equation for the pump level can be expressed as the change of population of atoms in the pump level as follows:

$$\frac{dN_3}{dt} = W_{13}(N_1 - N_3) - \frac{N_3}{\tau_3} \tag{5.7.1}$$

where W_{13} is the probability of an atom making the transition from level 1 to level 3 and τ_3 is the decay lifetime of the pump level. The flow of energy into the pump level is described by the first half of the equation, in which $(N_1 - N_3)$ represents the number of atoms available at the ground state to be pumped to level 3 by W_{13} (this is the pump which, as we stated earlier, is proportional to the intensity of the incident photon stream pumping the system). Energy from the pump level flows out downward by decay to the ULL, as described by the second half of the equation (N_3/τ_3) in units of number of atoms per second. Note that the decay lifetime τ_3 is a total *lifetime* representing both the decay from level 3 to level 2 and the decay from level 3 to level 1. If each decay path $3 \to 2$ and $3 \to 1$ has time constants τ_{32} and τ_{31}, respectively, $1/\tau_3 = 1/\tau_{32} + 1/\tau_{31}$.

Following a similar line of reasoning, the rate equation for the upper lasing level can be expressed as

$$\frac{dN_2}{dt} = \frac{N_3}{\tau_{32}} - \frac{N_2}{\tau_{21}} \tag{5.7.2}$$

where τ_{32} is the decay lifetime from the pump level to the ULL and τ_{21} is the lifetime of the ULL. Similar to the example with the pump level, energy flows into the ULL from above via decay from the pump level, the number of atoms per second being

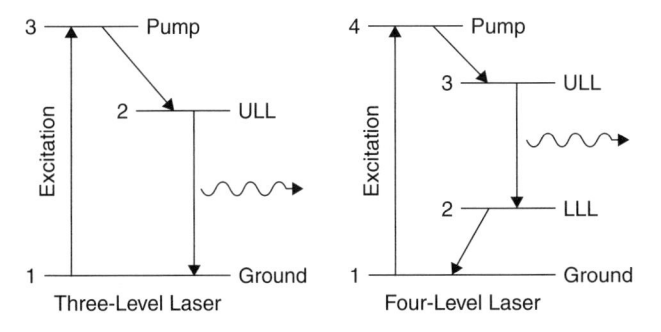

Figure 5.7.1 Energy levels in three- and four-level lasers.

the population of the pump level (N_3) divided by the decay lifetime from the pump level. Note that we use the specific decay rate $1/\tau_{32}$ to describe the flow of energy from the pump level only to the ULL. The other path that may be possible is from the pump level directly to ground (τ_{31}), but that is undesirable and in any case, does not affect the rate equation for the ULL (except that the process may serve to depopulate the pump level, which would otherwise decay to the ULL). Energy flows out of the ULL via the transition to ground (which serves as the lower-lasing level in a three-level system such as this), which is modeled by the second part of the equation.

The usefulness of these developed rate equations becomes apparent when we realize that under a steady-state condition these rates must both be equal to zero; that is, the population of each level remains constant, so the flow of energy into the level is the same as the flow of energy out of the level (this follows the logic used to analyze the two-level system in Section 5.6). By equating the rate equation for the pump level (5.7.1), to zero, we obtain

$$W_{13}(N_1 - N_3) - \frac{N_3}{\tau_3} = 0$$

$$W_{13}(N_1 - N_3) = \frac{N_3}{\tau_3}$$

$$W_{13}N_1 - W_{13}N_3 = \frac{N_3}{\tau_3}$$

$$W_{13}N_1 = N_3\left(\frac{1}{\tau_3} + W_{13}\right)$$

This may be simplified mathematically, though, since τ_3 is a small quantity (indeed, for a practical three-level laser the decay from the pump level to the upper level must be much faster than the decay from the upper to the lower lasing levels) and hence, numerically speaking, $1/\tau_3$ is much greater that W_{13} so that the latter term can essentially be ignored, giving us an expression for N_3:

$$W_{13}\tau_3 N_1 = N_3 \qquad (5.7.3)$$

Not surprisingly, the population of the pump level depends on the rate of pumping itself and on the total lifetime of that level (a longer lifetime allowing higher populations to build up). We may now equate the rate of the ULL (5.7.2) to zero at steady state as well, stating that the population of the ULL remains constant (inversion, and hence gain, does not increase or decrease under steady-state conditions).

$$\frac{dN_2}{dt} = 0 = \frac{N_3}{\tau_{32}} - \frac{N_2}{\tau_{21}}$$

which yields the solution

$$N_2 = N_3 \frac{\tau_{21}}{\tau_{32}} \tag{5.7.4}$$

Finally, combining equations (5.7.3) and (5.7.4) and simplifying equation (5.7.4) by assuming that $\tau_{32} = \tau_3$ (i.e., assuming that there is no leakage from the pump level to ground and that all atoms in the pump level decay solely to the ULL, we yield an expression for inversion now defined as $N_2(t) - N_1(t)$, with $\Delta N > 0$ to signify an inversion:

$$\begin{aligned} \Delta N &= N_2(t) - N_1(t) \\ &= N_3 \frac{\tau_{21}}{\tau_3} - N_1(t) \\ &= W_{13} \tau_3 N_1 \frac{\tau_{21}}{\tau_3} - N_1(t) \\ &= N_1(W_{13} \tau_{21} - 1) \end{aligned} \tag{5.7.5}$$

In terms of the pump rate required in a three-level laser to obtain an inversion, we note that inversion does not occur with the onset of pumping. A minimum pumping rate equal to $W_{13}\tau_{21}$ (where W_{13} is proportional to pumping rate) is needed just to get half of the population at ground state to the upper lasing level (i.e., when $\Delta N = 0$). The population of the upper and lower lasing levels is plotted in Figure 5.7.2, with inversion indicated in the shaded area. The population of the ground state (N_1) starts at N_0 at zero pump power, but as in the two-level system, each level above ground will be thermally populated according to Boltzmann statistics—these populations would be negligible for most systems.

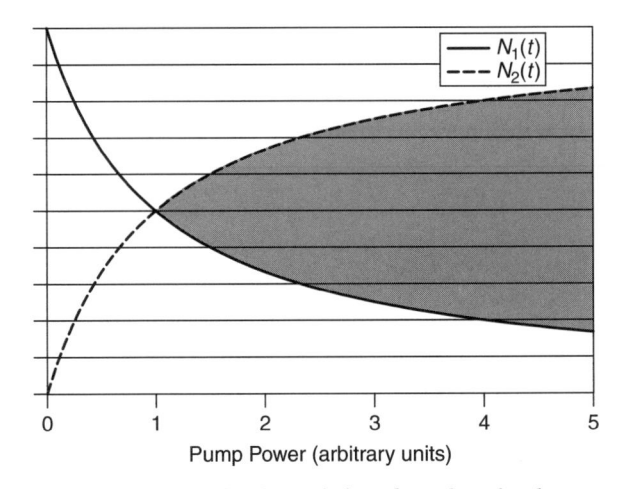

Figure 5.7.2 Energy-level populations for a three-level system.

Although inversion begins at a certain pump power, even more pumping will be required to generate a sufficient inversion to create a high enough gain to overcome losses in the system (which are always significant). As we had seen earlier in the general discussion of three-level lasers (and as a specific example, the ruby laser), as a rule the lifetime of the ULL (τ_{21}) must be much longer than the decay rate from the pump level (τ_{32}) so that the population of the pump level will remain very low while the population of the ULL will remain very large (and hence inversion will be maintained). In the case of ruby, the 3 ms lifetime of the ULL is far longer than that of the pump level, allowing an inversion to occur (especially important in a three-level laser since the LLL is ground state, which usually has a large population).

Analysis of a four-level laser now progresses in much the same manner as that of the three-level laser, with an examination of energy flow into and out of each level in the system. We begin with a look at the pump level (denoted as level 4 in Figure 5.7.1). The rate equation for the pump level can be expressed as the change of population of atoms in the pump level as follows:

$$\frac{dN_4}{dt} = W_{14}(N_1 - N_4) - \frac{N_4}{\tau_4} \quad (5.7.6)$$

where W_{14} is the probability of an atom making the transition from level 1 to level 4 and τ_4 is the total lifetime of the pump level. The flow of energy into the pump level is described by the first half of the equation, in which $(N_1 - N_4)$ represents the number of atoms available at the ground state to be pumped to level 4. Energy from the pump level flows out downward by decay to the ULL. This is described by the second half of the equation in units of number of atoms per second. Like the three-level example, note that τ_4 represents the total lifetime of the pump level (4), which can decay to levels 3, 2, or 1 (i.e., $1/\tau_4 = 1/\tau_{43} + 1/\tau_{42} + 1/\tau_{41}$). In a real laser one would hope that the majority of pump atoms decay to level 3, the ULL, in order to generate an inversion; however, other paths are possible (although clearly undesirable).

The equation for the upper lasing level can be expressed as

$$\frac{dN_3}{dt} = \frac{N_4}{\tau_{43}} - \frac{N_3}{\tau_3} \quad (5.7.7)$$

where τ_{43} is the decay lifetime from the pump level to the ULL and τ_3 is the lifetime of the ULL. Similar to the example with the pump level, energy flows into the ULL from above via decay from the pump level, the number of atoms per second being the population of the pump level divided by the decay lifetime from the pump level. Energy flows out of the ULL via decay, hopefully to level 2, by producing a photon of laser light, but also possibly to level 1. Finally, we can express the rate equation of the lower lasing level as

$$\frac{dN_2}{dt} = \frac{N_4}{\tau_{42}} + \frac{N_3}{\tau_{32}} - \frac{N_2}{\tau_{21}} \tag{5.7.8}$$

where τ_{42} is the decay lifetime from the pump level directly to the LLL, τ_{32} the lifetime of the laser transition (ULL to LLL), and τ_{21} the lifetime of the LLL. The first term represents a population of the LLL directly from the pump level. As mentioned previously, this is not a desirable transition but represents a loss of population from the pump level which does not contribute to laser action; worse yet, it reduces the population inversion necessary for lasing action (which is why it is shown here, as a reminder that such paths are possible). Luckily, in a practical laser system (those that work for real lasers), this term is negligible, so we shall ignore it here. The second term represents the expected route in which atoms decay from the ULL to the LLL, and the final term represents the decay from the LLL to the ground level.

Once again, the rate equations are most useful when equated to zero under steady-state conditions. Like the three-level laser, we begin with the pump level:

$$W_{14}(N_1 - N_4) - \frac{N_4}{\tau_4} = 0 \tag{5.7.9}$$

which may be solved algebraically in terms of N_4 as

$$\frac{W_{14}\tau_4}{1 + W_{14}\tau_4} N_1 = N_4 \tag{5.7.10}$$

But the population of N_4 in a real laser will be much smaller than that of N_1, so we can simply alter equation (5.7.9) by eliminating one term from the left side of the equation:

$$W_{14}N_1 - \frac{N_4}{\tau_4} = 0 \tag{5.7.11}$$

yielding the solution

$$W_{14}\tau_4 N_1 = N_4 \tag{5.7.12}$$

which states basically that as long as the excitation is not large enough to completely drive a large proportion of ground-state atoms to higher-energy states (the assumption we made to simplify the solution), the population of the pump level is proportional to the pump rate. Similarly, the rate equation for the ULL can be equated to zero to solve for a steady-state condition:

$$\frac{N_4}{\tau_{43}} - \frac{N_3}{\tau_3} = 0$$

which yields the solution

$$N_3 = N_4 \frac{\tau_3}{\tau_{43}} \quad (5.7.13)$$

As we expressed earlier, in a practical four-level laser system, τ_3 (the lifetime of the ULL) is much longer than τ_{43} (the decay rate from the pump level), so that the population of the pump level will remain very low while the population of the ULL will remain very large. Finally, we can equate the rate of the LLL to zero at steady state for the same reasons (remembering that the term $N_4/\tau_{42} = 0$), as follows:

$$\frac{N_3}{\tau_{32}} - \frac{N_2}{\tau_{21}} = 0$$

which yields the solution

$$N_2 = N_3 \frac{\tau_{21}}{\tau_{32}} \quad (5.7.14)$$

Finally, combining equations (5.7.10), (5.7.13), and (5.7.14), we obtain an expression for inversion (again, with $\Delta N > 0$ for an inversion):

$$\Delta N = N_3(t) - N_2(t)$$

$$= N_4 \frac{\tau_3}{\tau_{43}} - N_3 \frac{\tau_{21}}{\tau_{32}}$$

$$= W_{14}\tau_{43}N_1 \frac{\tau_3}{\tau_{43}} - W_{14}\tau_{43}N_1 \frac{\tau_{32}}{\tau_{43}} \frac{\tau_{21}}{\tau_{32}}$$

$$= W_{14}N_1\tau_3 - W_{14}N_1\tau_{21}$$

But in a practical laser the decay time from the lower lasing level ($2 \rightarrow 1$) is very fast, so this term is insignificant against the first term (where τ_3 is much longer), yielding the result

$$\Delta N = W_{14}\tau_3 N_1 \quad (5.7.15)$$

In other words, a population inversion results when any pump energy is supplied, as plotted in Figure 5.7.3, where the inversion is indicated as the shaded area.

Unlike the three-level system, where inversion occurs only after a threshold pump rate is reached, inversion occurs immediately in a four-level system. Most

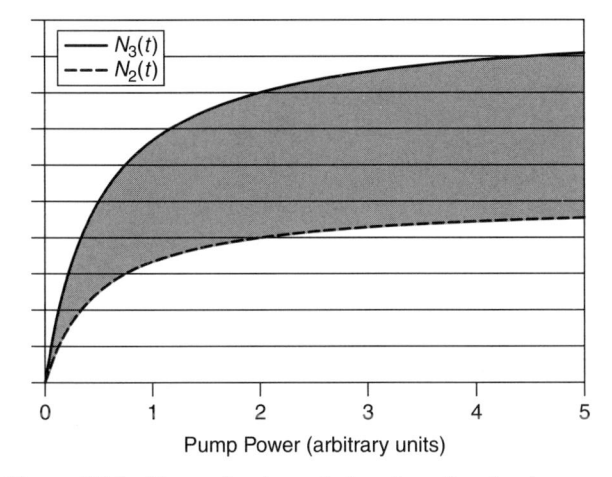

Figure 5.7.3 Energy-level populations for a four-level system.

common laser systems, such as the HeNe, argon, CO_2, and YAG lasers, are four-level systems. Although it may appear that lasing will occur with even a small pump energy, remember that a sufficient inversion and hence gain must be built up to overcome losses in the laser before laser output will occur.

5.8 GAIN REVISITED

Although we have defined gain in Section 4.6 in a practical manner, we may now examine how gain relates to energy-level populations in the atomic system. We begin by expressing the gain coefficient of the system as proportional to the population inversion according to

$$g = (N_2 - N_1)\sigma_0 \tag{5.8.1}$$

where $N_2 - N_1$ is the population difference (ΔN) and σ_0 is the cross section of the stimulated emission process. This states, not surprisingly, that gain is proportional to the population inversion as well as the cross section. Larger cross sections present more opportunity for stimulated emissions to occur and hence render larger gains. Cross section, as defined earlier, can be expressed mathematically as the product of the transition strength (or oscillator strength), S, and the lineshape function of gain, $g(v)$, which gives the normalized gain of the stimulated emission process at any given frequency. Graphically, transition strength is seen as the area under the curve, which defines the cross section as a function of frequency, as depicted in Figure 5.8.1.

144 LASING TRANSITIONS AND GAIN

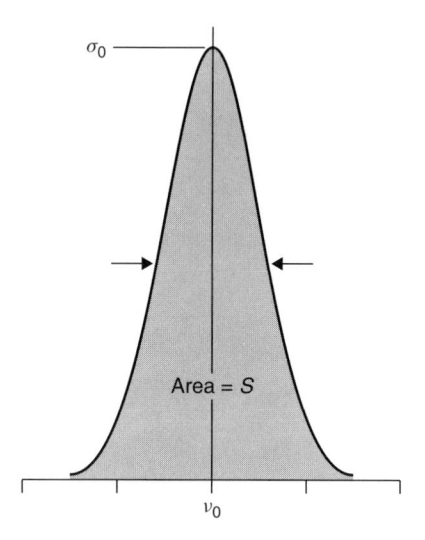

Figure 5.8.1 Transition cross section and strength.

Transition strength is related to the spontaneous lifetime of the atomic species, so we may express cross section mathematically as

$$\sigma(\nu) = Sg(\nu)$$

$$= \frac{\lambda^2}{8\pi t_{sp}} g(\nu) \qquad (5.8.2)$$

where $g(\nu)$ is the lineshape of the gain function and t_{sp} is the spontaneous lifetime of the species. So the cross section of the species is related to the spontaneous lifetime, an observable quantity in the laboratory. It may also be found in tables of physical constants for many species. This leaves the final term, called the *lineshape function for gain*, to be defined.

Gain is actually a function of frequency (or wavelength, if you prefer). Since the laser transition is not infinitely sharp, it has a finite linewidth, so gain is spread out over a range (albeit a narrow range) of frequencies. The lineshape of the gain function, for many lasers, is approximately Gaussian in shape, reaching a peak value of $g(\nu_0)$ at a center frequency ν_0 and tapering off at frequencies away from the center, as depicted in Figure 5.8.2.

The gain can also be expressed as a function of wavelength (useful since linewidths are often measured experimentally in terms of wavelength) as

$$g(\lambda) = g(\lambda_0) \exp\left[-\frac{(\lambda - \lambda_0)^2}{2\delta^2}\right] \qquad (5.8.3)$$

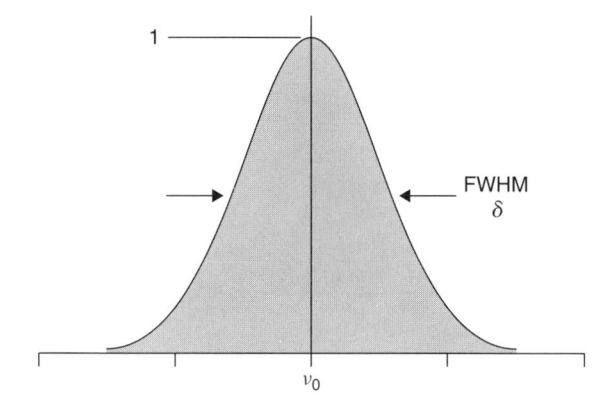

Figure 5.8.2 Typical gain lineshape.

where δ is the FWHM width of the gain spectrum (in nm), which models, mathematically, the Gaussian shape of a typical gain curve. For many purposes it is sufficient to use a gain value at the maxima on the curve, at the center wavelength or frequency (denoted v_0). A reasonable approximation for the gain curve is

$$g(v_0) = \frac{2}{\pi \Delta v} \quad (5.8.4)$$

where Δv is the linewidth (FWHM) of the transition, which applies for most materials that feature what is called *homogeneous broadening*. For gases, which broaden with an inhomogeneous mechanism such as Doppler shifting, a better approximation is

$$g(v_0) = \frac{1}{\Delta v} \quad (5.8.5)$$

where Δv is the linewidth (FWHM) of the transition.

By substituting these approximations for the gain lineshape into equation (5.8.2), we can calculate the maximum value of the cross section to yield $\sigma(0)$ or σ_0. Finally, using σ_0 in equation (5.8.1), we may calculate the gain of the laser (or at least what is called the *small-signal gain* of the laser) for a given inversion. It is further possible to express the threshold gain of a laser in terms of the inversion required to obtain that gain by combining equation (5.8.1) with the equation for threshold gain. In Section 4.9 we examined the threshold gain for an operating laser, and by equating the net gain equation (4.9.1) to unity we yielded a simple equation to calculate threshold gain:

$$g_{th} = \gamma + \frac{1}{2l} \ln \frac{1}{R_1 R_2} \quad (5.8.6)$$

146 LASING TRANSITIONS AND GAIN

We may simply combine this (multiplied by 2 to render an expression for round-trip loss in the laser) with equation (5.8.1) (by equating g_{th} to g) to yield an answer for threshold in terms of the population inversion required to allow a laser to oscillate as follows:

$$2\gamma + \frac{1}{l}\ln\frac{1}{R_1 R_2} = \Delta N_{th}\sigma_0$$

$$\frac{2\gamma l + \ln(1/R_1 R_2)}{\sigma_0 l} = \Delta N_{th} \quad (5.8.7)$$

The numerator term in the equation represents round-trip losses, including absorption by the laser medium itself ($2\gamma l$) as well as losses at both cavity mirrors [$\ln(1/R_1 R_2)$].

It is important to remember that as pump power to the laser is increased from zero, the inversion (ΔN) becomes progressively larger in proportion to pump power, as does laser gain. Since the gain is insufficient to overcome losses in the cavity, there is a net loss in the cavity and no laser action is observed. Eventually, pump power is sufficient to allow the inversion to reach the threshold value, at which point gain is equal to losses and laser action begins. At this point both gain and the amount of inversion are clamped at those threshold values, and a further increase in pump power to the laser does not increase gain but rather is manifested as an increase in laser output.

5.9 SATURATION

The rate equation analysis given in Section 5.7 is a steady-state analysis assuming no (or very little) radiation in the cavity. Consider the rate equation for the ULL of a three-level laser given by equation (5.7.2):

$$\frac{dN_2}{dt} = \frac{N_3}{\tau_{32}} - \frac{N_2}{\tau_{21}}$$

This equation factors in decay from the pump level (N_3/τ_{32}) as a source of positive change (a gain in population at this level) as well as decay via the lasing transition (N_2/τ_{21}), but nowhere does it account for stimulated processes.

If radiation is present in the cavity, stimulated emission will occur at the rate of $N_2 W_{21}$, and this will represent a loss. There will also be a process of absorption (really, a stimulated process opposite to stimulated emission, since it required a photon flux to be present) at a rate of $N_1 W_{12}$ in the opposite direction as photons are absorbed and the energy is used to pump atoms in the LLL back to the ULL. Rate equations must now be rewritten to include these new terms, and it will be noted that as the photon flux increases, the amount of inversion (i.e., the difference in population between the two lasing levels) will decrease, as will gain of the laser!

In a laser, then, gain is not a constant value but rather varies with incident (or, in a practical laser, circulating) power. Imagining the laser gain medium as an amplifier, the gain is quite large. This is termed the *small-signal value*, in which the ULL is well populated and is replenished continually by pumping, so that the population does not change appreciably (in other words, the rate of pumping is large enough to keep the level ULL populated and to keep inversion large. As the input signal to the amplifier reaches a large value, the photon flux is large enough to depopulate the ULL, which then lowers the overall gain in the device since fewer excited atoms will be available at that level to contribute to the stimulated emission process. This results in a saturated gain figure in which the gain is reduced by large photon fluxes in the cavity.

We recall from Section 4.6 that gain is an increase in power for a given distance traveled through the laser amplifier medium. In a normal laser amplifier we would expect power to increase exponentially with length according to

$$P_{output} = P_{input} \exp(g_0 l) \tag{5.9.1}$$

where g_0 is the small-signal gain of the laser amplifier and l is the length of the amplifier. This represents the maximum gain that a laser amplifier can deliver, but as the amplifier saturates, eventually the power increase is a linear function of power according to

$$P_{output} = P_{input} + g_0 l \tag{5.9.2}$$

Clearly, a saturated amplifier delivers less "bang for the buck" than is delivered by an unsaturated amplifier! Also, a high-gain laser (e.g., a YAG or CO_2) suffers a greater decrease in potential output power due to saturation than a low-gain laser such as a HeNe does (but luckily, it takes more power to saturate many of these amplifiers, as we shall see in this section).

The gain of a saturated system is obviously dependent on the photon flux inside the cavity since it is these photons that generate the stimulated emissions which serve to deplete the ULL and hence decrease gain. The saturated gain is then

$$g_{sat} = \frac{g_0}{1 + \rho/\rho_{sat}} \tag{5.9.3}$$

where g_0 is the unsaturated gain of the amplifier, ρ the photon flux in the system, and ρ_{sat} is called the saturation flux. Either photon flux may be calculated from intensities (in W/m^2) measured experimentally or calculated (as we shall do in this section).

The input intensity required to reduce the gain to one-half the initial (unsaturated) value is termed the *saturation intensity*. In terms of intensity, we know the energy of each photon and so may calculate the intensity of the input to the amplifier to decrease gain to one-half of the small-signal value as

$$I_{\text{sat}} = \frac{h\nu}{\sigma\tau} \qquad (5.9.4)$$

where $h\nu$ is the energy of an individual photon, σ the cross section of the transition, and τ the effective or saturation lifetime of a photon. The saturation time constant τ is really the effective lifetime or recovery time of the species and represents the time for the species to become excited and to decay again. It can be approximated by t_{sp} (the spontaneous decay lifetime) for a four-level laser or $2t_{\text{sp}}$ (twice the spontaneous decay lifetime) for a three-level laser. You will note that this relationship is simply that of equation (5.6.2) rearranged so that the transitional probability is replaced with the effective lifetime. Intensity has units of J/cm^2 per second or simply, W/cm^2.

Although the calculation of I_{sat} looks simple enough, equation (5.9.4) assumes a gain medium in which the linewidth is broadened solely by the distribution of energy in what is termed a homogenously-broadened medium—it does not take into account the broadening of linewidths due to Doppler or other mechanisms. Consider that in such a medium the linewidth is predicted by classical mechanics to be:

$$\Delta\nu = \frac{1}{2\pi\tau} \qquad (5.9.5)$$

where $\Delta\nu$ is the expected linewidth (in Hz) and τ is the time constant for the upper level. For a HeNe laser in which the lifetime of the upper-lasing level is 30 ns, this corresponds to a linewidth of only 5.2 MHz; however, in the example of Section 4.7 we found the Doppler-broadened HeNe linewidth to be 1.5 GHz!

While it is difficult to calculate the expected saturation intensity for a given amplifier, it is easy enough to measure it experimentally as depicted in Figure 5.9.1 in which the gain of a HeNe amplifier tube is measured on the 612 nm (orange) transition.

The arrangement is one described in Section 4.9, in which the beam from a laser (the probe beam) is passed through a second (amplifier) tube. Because of the oscillatory nature of the probe laser (discussed in Chapter 9) and limits on the resolution of the power meter employed, there is a large uncertainty in measurements taken at low power levels, as indicated by the error bars on the graph. Regardless, it can clearly be seen that gain is much higher at low intensities and drops as incident power increases. At about 1.6 mW of incident power the gain of the tube is measured to be only a half of what it was when the incident power was 0.4 mW. This is a smaller saturation intensity than expected; however, an estimate would not account for utilization of the lasing volume, and as we shall see in Chapter 6, a laser operating in the normal Gaussian mode (TEM$_{00}$) does not use the lasing volume particularly efficiently.

In contrast to this example, consider a solid-state YAG laser in which the cross section is smaller but the spontaneous lifetime of the medium is much longer.

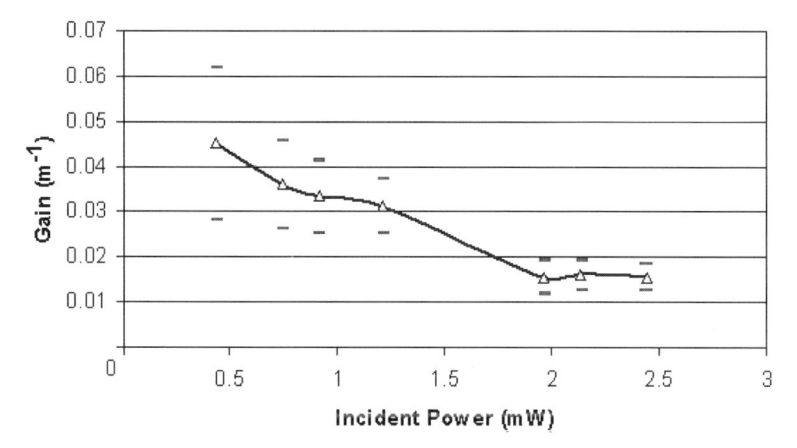

Figure 5.9.1 Saturated gain in a laser.

The intensity required to saturate this medium is about 100 times larger than that required for the HeNe gas laser.

5.10 REQUIRED PUMP POWER AND EFFICIENCY

At this stage of the analysis we can model a simple (ideal) laser system mathematically and calculate key parameters for operation, including the minimum pump power required to generate an inversion (and hence a gain) sufficient to overcome losses in the laser system. We begin by expressing the minimum pump energy that must be delivered to the system:

$$P_{minimum} = \frac{dN_{ULL}}{dt} V h \nu_{mp} \qquad (5.10.1)$$

where V is the active volume of the amplifier and $h\nu_{mp}$ is the photon energy required at the ULL. The photon energy in this case is the energy of the ULL above ground state; in other words, the energy that a photon at the ULL must possess to emit a lasing photon and decay further to ground state. In a three-level laser this is equal to the photon energy of the lasing transition itself, but in a four-level laser this energy is equal to the energy of the lasing transition plus the energy of the LLL above ground.

The rate dN_{ULL}/dt in the equation is the rate at which energy decays from the pump level to the upper lasing level and so represents the energy flowing into the ULL from the pump level above. A substitution for this rate may be determined by solving the equation for decay of the level, which expresses the population of the level at any time t as

$$N_{\text{ULL}}(t) = N_{\text{ULL}}(0)\exp\left(-\frac{t}{\tau}\right) \quad (5.10.2)$$

where $N_{\text{ULL}}(0)$ is the population of the ULL at time $t = 0$ and τ is the spontaneous lifetime of the level. This is the exponential decay rate of the upper lasing level as a function of time. With no external input of energy, the level decays in population to 37% of its initial value within the time period of one spontaneous lifetime (this is, in fact, the definition of spontaneous lifetime). Taking the differential of the equation with respect to time yields an answer of

$$\frac{dN_{\text{ULL}}}{dt} = -\frac{1}{\tau} N_{\text{ULL}} \quad (5.10.3)$$

which describes the first term of our equation. Further substitution may now be made by realizing that in a four-level laser, $N_{\text{ULL}} \simeq \Delta N$, where ΔN is the population inversion of the laser (which also represents the number of coherent photons in the laser itself since we assume that in an ideal laser each excited atom produces one photon of coherent light). This substitution assumes a negligible population of atoms at the LLL since the decay rate from the LLL to ground would be much faster than that of the lasing transition in an ideal laser. For a three-level laser the population of the LLL is far from negligible (since the LLL is the ground state itself), so the number of coherent photons is equal to the population difference between the two lasing transition levels ($\Delta N \simeq N_{\text{ULL}} - N_1$ [7]), so $N_{\text{ULL}} \simeq \Delta N + N_1$ (where $N_{\text{ULL}} = N_2$ for a three-level laser).

Consider now the threshold population inversion as stated in equation (5.8.7). Knowing the reflectivities of both cavity mirrors as well as loss in the lasing medium, we may calculate the threshold population inversion and further equate this to N_2 using the approximation stated. Furthermore, by determining the spontaneous lifetime of the lasing species, we may solve the required rate of transitions from the ULL to the LLL (i.e., dN_{ULL}/dt). Finally, computation of the volume and photon energy of the system allows solution of the minimum required pump power.

Example 5.10.1 **Minimum Required Pump Power** Consider the pump power required for an optically pumped Nd:YAG laser with a rod 4 mm in diameter and 5 cm in length. The material has a transition cross section of 6.5×10^{-19} cm^2 and a spontaneous lifetime of 1.2 ms. The laser cavity has one mirror 100% reflecting and the OC 90% reflecting. Loss in YAG is approximately 0.01 m^{-1}. The rate of transitions required is $dN_{\text{ULL}}/dt = -1/\tau N_{\text{ULL}}$, for which we substitute

---

[7] This assumes no degeneracy. If degeneracy exists, then $\Delta N \simeq N_2 - g_2/g_1 N_1$, where g_1 and g_2 are the degeneracies of each corresponding level.

REQUIRED PUMP POWER AND EFFICIENCY   151

$N_{ULL} = \Delta N_{th}$ from equation (5.8.7) as follows:

$$\frac{dN_{ULL}}{dt} = -\frac{1}{\tau}\Delta N_{th}$$

$$= \frac{2\gamma l + \ln(1/R_1R_2)}{\sigma_0 l \tau}$$

$$= \frac{0.001 + 0.105}{(6.5 \times 10^{-19})(0.05)(1.2 \times 10^{-3})}$$

$$= 2.73 \times 10^{21} \text{ cm}^{-3}/\text{s}$$

So to oscillate a laser with the losses specified requires 2.73×10^{21} transitions per second per cubic centimeter of lasing medium. We may also easily calculate volume of the rod as 0.628 cm^3. The final parameter required is the energy of the photon required at the ULL. We know that the YAG laser has a wavelength of 1064 nm, which corresponds to a photon energy of 1.87×10^{-19} J, but we must add this energy to that of the LLL above ground. A good physics reference lists the energy of the LLL as 2.4×10^{-20} J above the ground state of the Nd^{3+} ion, so that $h\nu_{mp}$ is 2.11×10^{-19} J. This may also have been found be calculating the energy of the 946-nm transition in YAG (see Figure 5.3.2) and adding the energy of the LLL above ground for that transition (which, as we had seen in an earlier example, is only 0.1 eV).

The pump energy required is then found using equation (5.10.1) to be

$$P_{minimum} = \frac{dN_{ULL}}{dt} V h\nu_{mp}$$

$$= (2.73 \times 10^{21} \text{ cm}^{-3}/\text{s})(0.628 \text{ cm}^3)(2.11 \times 10^{-19} \text{ J})$$

$$= 362 \text{ J/s (or watts)}$$

This may seem like a small pump power, especially to anyone who has worked with YAG lasers and has seen the large lamps and power supplies used to drive these systems, but we must now depart from our ideal laser analysis and realize (as discussed in Section 5.1) that only a small amount of electrical energy delivered to the lamp is actually converted to pump energy in the YAG system. A typical figure for an optimized optically pumped YAG laser would be around 3%, so the required electrical power supplied to the lamp would be 12.1 kW of power! When pumped by a flashlamp, this represents 12.1 J of energy discharged within 1 ms, obtained easily with a small capacitor and flashlamp. The approximate figure calculated here for the pump power required has been verified in the laboratory to be within 40% of the actual energy required for a flashlamp-pumped YAG laser with these exact parameters. The reasons for deviation from the value predicted are outlined in this section.

In a real laser, 100% of the pump light does not translate into pump energy for the laser (otherwise, the laser described above would indeed operate with 362 W of input). Most gas lasers have overall efficiencies (defined as optical power output divided by electrical power input) of well under 1%. Reasons for low efficiency include the efficiency of the pump source itself at converting electrical energy into optical output, the efficiency with which pump light is coupled into the lasing medium, the absorption efficiency of the lasing material, and the quantum efficiency of the atomic system.

The efficiency of conversion of electrical energy into optical output (denoted $\eta_{optical}$) depends on the lamp technology employed (for an optically pumped laser), which can range from a flashlamp to another laser. Use of a laser diode as an optical pump can lead to conversion efficiencies of up to 50%, while the use of halogen lamps (which may be used with YAG) results in poor efficiency since the lamp itself converts at most 10% of the electrical input power to visible light, with the majority of input energy being converted into heat and infrared radiation. As we have seen in Chapter 1, blackbody-type sources have the worst efficiencies at conversion, while gas discharges and laser diodes have progressively higher efficiencies.

Coupling of the pump light to the laser medium (denoted $\eta_{coupling}$) is highly dependent on the geometry of the laser medium. Many small solid-state lasers pumped by laser diodes are pumped from one end, where the pump beam passes through the HR itself to enter the crystal as depicted in Figure 5.10.1. The HR is designed to be transparent at the pump wavelength and fully reflective at the wavelength of the laser rod. If all the pump light is not absorbed in the rod, the OC can also be designed to reflect leftover pump light back into the rod for higher efficiency. Larger crystals, as employed in high-power diode-pumped lasers, are often side-pumped with arrays of small laser diodes (mounted on bars of copper for good heat conduction) which surround the rod on all sides. Where lamps (either continuous arc or flash types) are used as a pumping source, reflectors are used to focus as much light as possible from the lamp(s) to the rod. In a single-lamp configuration, the reflector is usually elliptical with both the rod and the lamp at the foci for the ellipse, as shown in Figure 5.10.2.

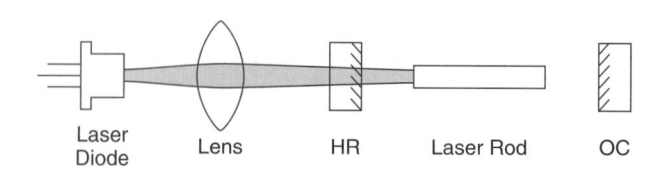

Figure 5.10.1 End-on laser diode pumping.

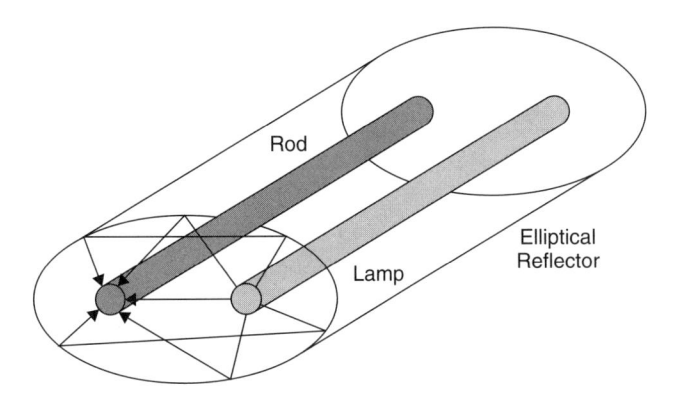

Figure 5.10.2 Elliptical reflector for lamp pumping.

Absorption efficiency (denoted $\eta_{\text{absorption}}$) of the medium varies with the wavelength of the pump source, as have seen in Section 5.1. A laser medium with broad absorption bands is desirable since it can absorb more pump energy from broadband sources such as lamps. Careful choice of wavelength of a laser diode to match an absorption peak in the laser material (e.g., the use of an 808-nm laser diode to pump Nd:YVO$_4$, a material called vanadate) can result in absorption efficiencies as high as 80%.

Finally, the intrinsic quantum efficiency of the atomic system affects the overall efficiency of the laser. In a four-level laser this factor is defined as

$$\eta_{\text{quantum}} = \frac{E_{\text{ULL}} - E_{\text{ground}}}{E_{\text{pump}} - E_{\text{ground}}} \tag{5.10.4}$$

It really represents the energy difference between the pump and ULL and is completely dependent on the particular atomic system. So the overall pump efficiency of the laser (a measure of how well electrical energy is converted to optical output) is a product of all four efficiencies:

$$\eta_{\text{pump}} = \eta_{\text{optical}} + \eta_{\text{coupling}} + \eta_{\text{absorption}} + \eta_{\text{quantum}} \tag{5.10.5}$$

Example 5.10.2 Flashlamp-Pumped YAG Laser Consider a flashlamp-pumped YAG laser (using a nonoptimal xenon flashlamp as discussed in Section 5.1) with the following efficiencies: $\eta_{\text{optical}} = 0.05$ (typical for a flashlamps converting to visible and near-IR light), $\eta_{\text{coupling}} = 0.85$ (the elliptical reflector arrangement is quite efficient), $\eta_{\text{absorption}} = 0.3$ (poor since the flashlamp is a broadband emission source and the YAG rod absorbs only at a few select bands), $\eta_{\text{quantum}} = 0.68$ (an average since the pump energy is spread out across four absorption bands). The overall pump efficiency is just under 1%.

154 LASING TRANSITIONS AND GAIN

Now contrast this to a laser-diode-pumped laser. The quantum efficiency in this case is much more precisely calculated (since it is at a single wavelength) to be

$$\eta_{\text{quantum}} = \frac{E_{\text{ULL}} - E_{\text{ground}}}{E_{\text{pump}} - E_{\text{ground}}}$$

$$= \frac{E_{\text{1064-nm lasing photon}} + E_{\text{LLL}}}{E_{\text{808-nm pump photon}}}$$

$$= \frac{1.87 \times 10^{-19}\,\text{J} + 2.4 \times 10^{-20}}{2.46 \times 10^{-19}\,\text{J}}$$

$$= 0.86$$

The other efficiencies required are $\eta_{\text{optical}} = 0.33$ (typical for a laser diode), $\eta_{\text{coupling}} = 0.95$, and $\eta_{\text{absorption}} = 0.8$. The overall pump efficiency of this laser is about 22%.

It is obvious why in recent years the trend in solid-state laser technology has been toward diode-pumped solid-state (DPSS) lasers. Flashlamp pumping is still used, however, where high CW power levels (over 5 or 10 W) or especially high peak powers (e.g., Q-switched giant-pulse lasers as covered in Chapter 7) are required. Diode pumping also lends itself to compact arrangements, while lamp pumping (either continuous arc or flashlamp) invariably involves a large power supply and often requires water cooling.

Aside from the practical efficiencies already discussed, other factors that affect the efficiency (as well as the minimum pump power required) for real lasers include utilization of the lasing volume by the intracavity laser beam. It is rare that an intracavity laser beam should utilize the entire lasing volume (as we have assumed in our discussion so far), for as we shall see in Chapter 6, photons inside the cavity usually assume patterns that have a much smaller beam waist in the center of the amplifying medium.

5.11 OUTPUT POWER

Using the results of previous sections in which we have derived expressions for inversion, gain, and saturation, we may now formulate an expression to predict the amount of power that we can expect from a particular laser system. In Section 5.8 we stated that at threshold conditions, gain must equal loss in the laser. In terms of a round trip through the laser (i.e., two passes through the gain medium),

round-trip gain = round-trip loss

$$2g_{\text{th}} = 2\gamma l + \ln \frac{1}{R_1 R_2} \quad (5.11.1)$$

where g_{th} is the threshold gain, γ the absorption coefficient for the lasing medium (in m^{-1}), and l the length of the gain medium. If however, gain exceeds the threshold value, it must saturate down to reach an equilibrium point once again. The intensity of light inside the cavity grows in the process, and a usable output beam appears as a fraction of that intracavity intensity (that fraction being $1 - R_{OC}$, where R_{OC} is the reflectivity of the output coupler; the other cavity mirror is assumed to be 100% reflecting in this case). We must substitute the value for saturated gain from equation (5.9.3) (with intensity used in place of photon flux) for g_{th} in equation (5.11.1), to obtain

$$2g_{sat} = 2g_{th}$$

$$\frac{2g_0}{1 + 2I/I_{sat}} = 2\gamma l + \ln \frac{1}{R_1 R_2} \quad (5.11.2)$$

where the 2 in front of the intensity in the denominator of the saturated gain equation denotes that intensity in the cavity originates from a flux of photons moving toward one mirror and a second flux moving toward the opposite mirror (so that I toward one mirror is actually $I_{cavity}/2$). This may now be solved for I:

$$I = \frac{\{2g_0/[2\gamma l + \ln(1/R_1 R_2)] - 1\}I_{sat}}{2} \quad (5.11.3)$$

where g_0 is the unsaturated (single-pass) gain of the laser (proportional to inversion ΔN) and I_{sat} is the saturation intensity as determined in Section 5.9. As expected, the unsaturated gain g_0 is multiplied by 2 since it is usually expressed as a single-pass gain (determined by experimental means using a method similar to that described in Section 4.9, in which $g = \alpha l$, where α, the gain coefficient in units of m^{-1}, is multiplied by the length of the gain medium in a single pass). Multiplying this figure by 2 gives the round-trip gain required for this expression.[8]

Expression (5.11.3) gives us I, the intensity of light inside the laser cavity. We may further express this intensity as output power by multiplying by the area of the laser beam inside the cavity (assuming that t is consistent throughout) since power = intensity × area as well as the transmission of the output coupler $(1 - R)$. Since unsaturated gain may be calculated or measured experimentally, we can calculate the maximum power available from a given laser.

Example 5.11.1 **Red HeNe Laser Tube** Consider a red HeNe laser tube with an experimentally measured (using the method outlined in Section 4.9) unsaturated gain of 0.135 m^{-1}. If the output coupler has 99% reflectivity, the high reflector is 100% reflective, the absorption of the medium is 0.01 m^{-1} (with a tube length

---

[8] The reader is reminded that there is a difference between the gain coefficient, usually denoted α, which has units of m^{-1}, and the gain itself, which is a dimensionless quantity (equal to the gain coefficient times the length of the gain medium. Various references use different notations, so be sure you have the correct quantity.

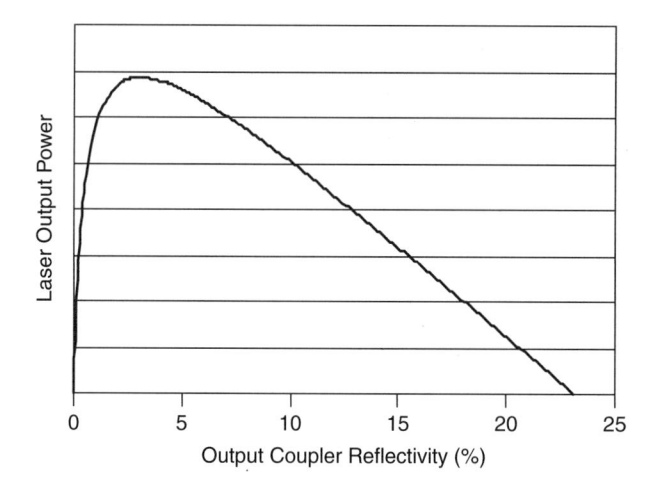

Figure 5.11.1 Power output as a function of OC reflectivity (generalized).

of 27.5 cm), and I_{sat} determined experimentally, the output power of the laser may be calculated as follows:

$$P_{output} = I \times \text{area} \times (1 - R_2)$$

$$= \frac{\{2g_0/[2\gamma l + \ln(1/R_1 R_2)] - 1\} I_{sat}}{2\pi r^2 (1 - R_2)}$$

$$= \frac{\{(2 \times 0.135)/[2(0.01)(0.275) + \ln(1/0.99)] - 1\}(4.49)}{2\pi (0.05)^2 (1 - 0.99)}$$

$$= 2.89 \text{ mW}$$

In the laboratory, an optimized laser of this type has been demonstrated to have an output of approximately 3.0 mW, in good agreement with this calculation.

Finally, we may use the expression for intensity to determine the optimal value of output coupling for a given laser. For any given laser we can plot the expected power output as a function of the reflectivity of the output coupler, which renders a plot like that of Figure 5.11.1. For this particular laser the optimal value for the output coupler is 3% reflectivity. The plot also shows that as expected, there is a minimum value of reflectivity for the output coupler at which the laser ceases oscillation entirely.

PROBLEMS

5.1 To create a population inversion, one must populate the upper lasing level with more atoms than at the lower level. This may be difficult to do if the lower lasing level is very close to the ground level where it is populated thermally. The problem can be alleviated by cooling the laser to depopulate the lower energy levels. Imagine an Nd : YAG laser that is operating at a hot temperature of 1000 K. The laser emits a wavelength of 946 nm, and the lower level of YAG is only 0.1 eV above ground state. If the rod contains a total of 1×10^{19} active lasing ions:

(a) Calculate the number of ions that must be excited at the upper level to ensure that population inversion occurs.

(b) If the rod absorbs 15% of the pump light (in this case 808 nm), calculate the power input required to allow the rod to lase (in joules; assume that pumping occurs as one fast pulse much shorter than the lifetime of the ULL).

(c) To complete this problem, calculate the thermal population of the lower level and equate this to the population of the upper level. Now calculate the energy required by assuming that 15% of incident photons of 808-nm pump light are converted to excited atoms in the upper level. This is a simplified problem but does illustrate the problems involved with thermal populations of levels.

5.2 For each energy-level diagram in Figure P5.2: (a) identify the system as three- or four-level, (b) identify the function of each level, and (c) compute the quantum efficiency of the system.

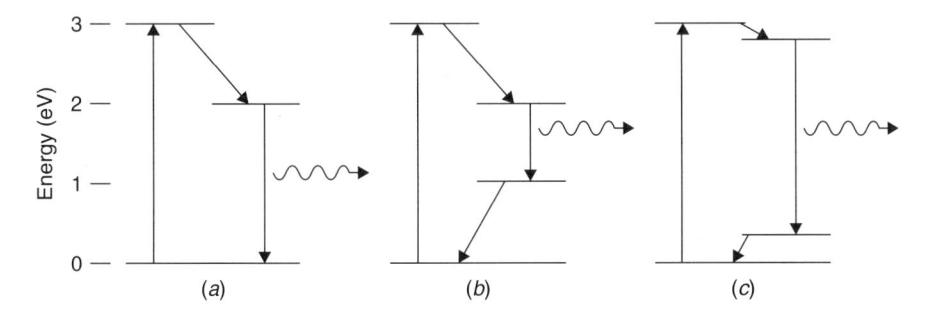

Figure P5.2

5.3 Show how as a CO_2 laser heats up, population inversion is more difficult to achieve. To do this, assume that a population of 5% of the total CO_2 molecules in the laser tube reaches the upper laser level, and calculate the temperature at which the lower level populates to exceed the population of the upper lasing level. This represents the temperature at which lasing will cease completely (due to thermal effects on the LLL). Will the laser cease to oscillate before this temperature is reached? Why?

5.4 A semiconductor laser operating at a peak wavelength of 808 nm (λ_0) has a gain spectrum width (δ) of 5 nm. If the gain of the device is exactly double

158 LASING TRANSITIONS AND GAIN

that of total losses (i.e., lasing begins when the gain reaches a value of one-half the maximum value), calculate the expected spectral width of the output of the laser.

5.5 A semiconductor laser made of GaAs ($n = 3.7$) with a crystal length of 350 μm has losses (primarily, absorption in the crystal itself but not including the cavity mirrors) of 25 cm^{-1}. Mirrors on this laser are formed by cleaving the faces of the crystal parallel to each other and perpendicular to the axis of the laser; each mirror is identical. Calculate the threshold gain required to allow lasing (compare this to the gain of a gas laser such as a HeNe, with gain typically 0.13 m^{-1}).

5.6 Rewrite the three differential equations for the four-level laser (i.e., dN_4/dt, dN_3/dt, and dN_2/dt) to include all possible paths for decay. For example, the equation for dN_4/dt would include possible decay paths to the ULL, LLL, and ground states. Generate steady-state solutions for the ULL and LLL.

5.7 For both a three- and four-level ideal laser, rewrite the solution for the population inversion ΔN in terms of $\Delta N/N$ (this requires considerable algebraic manipulation). For each laser plot (on the same graph), pump energy on the x-axis ($W_{14}\tau_{32}$ for the four-level laser, $W_{13}\tau_{21}$ for the three-level laser) and $\Delta N/N$ on the y-axis.

5.8 Calculate the inversion ΔN required for a HeNe laser operating at 632.8 nm and having an output beam of 1 mW. The HR is 100% reflecting and the OC 98%, and the cross section of the transition is 1×10^{-13} cm^2.

5.9 A 1.550-μm semiconductor laser diode has an active length of 500 μm. The active area of the junction itself is 2.5 μm wide by 0.2 μm high. Assume that the difference between pump level and ULL is negligible, and the difference between LLL and the ground state is negligible (see Figure 5.5.3). The diode is found to lase when current through the device reaches a value of 80 mA, and the device is known to be 50% efficient. Calculate the rate dN_{ULL}/dt. (*Hint*: The voltage across the device may be assumed to be the bandgap energy so that the input power to the device—and hence pump power—are known.)

CHAPTER 6

Cavity Optics

In Chapter 4 we examined the basics of laser action. It became obvious that mirrors were required for most lasers to increase the circulating power within the cavity to the point where gain exceeds losses and to increase the rate of stimulated emission. Laser optics often consist of much more than two simple, flat mirrors though. Aside from the multitude of configurations possibly involving flat and concave reflectors, there are a number of intracavity devices that may be used to enhance specific features of laser output, such as operating mode and spectral width. In this chapter we examine the cavity resonator in detail.

6.1 REQUIREMENTS FOR A RESONATOR

In Chapter 4 we examined the rate equations which govern the laser and determine the criterion for lasing. One important result was equation (4.5.10), which defines the ratio of the rates of stimulated emission to spontaneous emission:

$$\frac{r_{\text{stimulated}}}{r_{\text{spontaneous}}} = \frac{B_{21} N_2 \rho}{A_{21} N_2}$$

The equation was simplified further by substituting for the Einstein coefficients, to yield

$$\frac{r_{\text{stimulated}}}{r_{\text{spontaneous}}} = \frac{c^3 \rho}{8 \pi h \nu^3}$$

The important conclusion was that for this ratio to be large (i.e., for the stimulated emission rate to exceed the spontaneous rate), the energy density of incident photons must be high. Unless the gain of the medium is extremely high in order to create a huge flux of photons as they pass down the tube, or the gain medium is very long, cavity mirrors will be required to contain photons within the cavity, keeping ρ high and creating further amplification. Indeed, the vast majority of lasers require cavity mirrors for oscillation. Furthermore, these mirrors are usually very efficient reflectors

---

Fundamentals of Light Sources and Lasers, by Mark Csele
ISBN 0-471-47660-9 Copyright © 2004 John Wiley & Sons, Inc.

where typical values for a gas laser are 99.99% reflectivity for the HR and 99 to 98% reflectivity for the OC. The resonator must trap photons completely within the cavity since any photons that escape the resonator represent a loss in the laser, which, as we shall see, can drastically affect output power.

You may also recall from Chapter 4 a brief description of superradiant lasers. These lasers have *huge* gains, so high that light is amplified to a usable level in a single pass down the tube. These lasers operate without feedback (i.e., without cavity mirrors). Molecular nitrogen, neutral neon, and molecular hydrogen lasers are usually superradiant. Other lasers, such as copper-vapor lasers, have very high gains requiring minimal or no optical feedback to operate.

6.2 GAIN AND LOSS IN A CAVITY

Assume that a laser operates with an output of 1 mW and has an output coupler with a reflectivity of 99.0% (and hence a transmission of 1.0%). This implies that the power circulating inside the cavity (the *circulating power*) is 100 mW. Of course, such intracavity power level is not usable: It is required to keep the rate of stimulated emission large enough and hence sustain laser action. The output beam (defined by the transmissivity of the output coupler) represents a loss in a laser cavity. Other losses include absorption and scattering in the laser medium itself and losses at windows terminating the lasing medium (e.g., windows on a gas laser plasma tube). Most laser gain media have low gains, so it is essential that losses be kept to a minimum. Since the output coupler represents a loss, the amount of transmissivity depends on the gain of the laser. When total losses in the laser exceed the gain of the medium, the laser will fail to oscillate.

You will recall from Chapter 4 an experiment in which a variable loss (a glass slide) is inserted into the cavity of a HeNe laser. Let us reexamine the experiment by plotting the output power (which is, of course, proportional to circulating power) against loss in the cavity, as shown in Figure 6.2.1. Circulating power is easily measured simply by measuring laser output and multiplying by the transmission of the OC. In this experiment the OC has a transmission of 1.0%, so that for a measured output of 370 µW, the intracavity circulating power is 37 mW. The maximum inserted loss that may be tolerated in this laser is found to be 1.2%. When this loss is summed with other losses in the laser, such as absorption in the medium itself and the loss of the OC (1.0%), the gain of the laser medium may be calculated.

It is obvious from Figure 6.2.1 that even small losses have a large effect on the output power of a laser—losses in a real laser must thus be kept as low as possible. Where possible, such as in a HeNe laser, mirrors are often sealed directly into the ends of the tube so that there are no windows in the optical path to increase loss. In many lasers, though, cavity mirrors must be isolated from the discharge to prevent attack from the plasma. Use of external optics is also required when a laser uses a wavelength selector. In these cases the ends of the plasma tube must be capped with an optical window. To minimize losses, optical windows are angled at

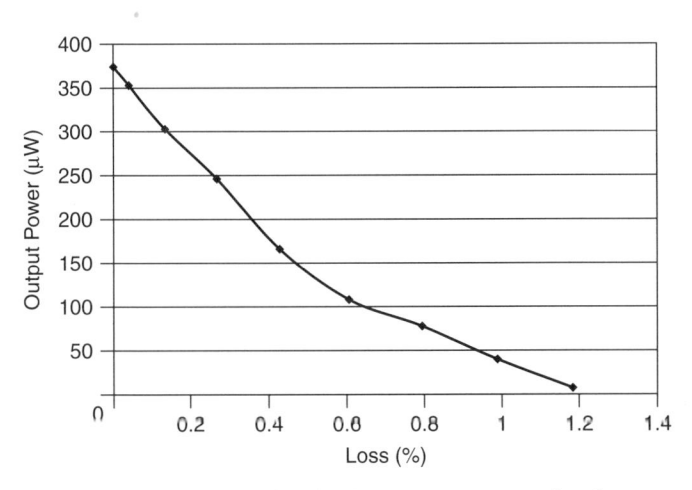

Figure 6.2.1 Effect of cavity loss on output power in a laser.

the Brewster angle, which polarizes the output of the laser since loss is essentially zero in one polarization and very significant in the other polarization.

Figure 6.2.2 shows a Brewster window on a small air-cooled argon laser tube. Protruding from the anode of this tube (with attached cooling fins visible in the

Figure 6.2.2 Brewster window in a gas laser.

Figure 6.2.3 Brewster window in a gas laser.

photograph) is a metal tube that is fused to a quartz tube sealed with a quartz window. The window is angled at Brewster's angle. Most tubes contain two windows, one at each end of the tube, and each must be aligned in the same optical axis. As Figure 6.2.3 shows, where two Brewster windows are used, they should be oriented such that the intracavity beam does not shift overall position as it passes through both windows (as in the top diagram).

The optimal angle can be determined from the Fresnel equation (Section 4.9) as the angle at which reflection (R_p) is reduced to zero. This angle depends on the index of refraction of the window material itself according to

$$\theta = \tan^{-1} \frac{n_2}{n_1} \qquad (6.2.1)$$

where n_2 is the index of refraction of the window and n_1 is the index of refraction of the surrounding media (1.00 for air and most gases). Hence Brewster's angle is independent of the wavelength of the laser except that the index of refraction of the window may change as a function of wavelength, so for a multiple-wavelength laser such as an argon laser, the angle will be optimal for only one wavelength, with slightly higher losses at other wavelengths (only slightly higher since the index of refraction will not change much). The optimal output coupling can be calculated as a function of the gain of the device and total losses in the cavity, as done in Section 5.11. By reducing losses in the cavity such as those at tube windows, power output may be enhanced and weak transitions may be allowed to lase.

6.3 RESONATOR AS AN INTERFEROMETER

In general, a laser cavity acts as an interferometer into which an integral number of waves must fit. The cavity is resonant at wavelengths such that the number of waves inside the cavity is an integer—these are standing waves inside the cavity. At all other wavelengths, destructive interference causes any wave inside the cavity to be extinguished. These resonant wavelengths are called *longitudinal modes* and are spaced apart at regular intervals of frequency. The spacing of modes is called the *free spectral range* (FSR) of the interferometer and is outlined in Figure 6.3.1.

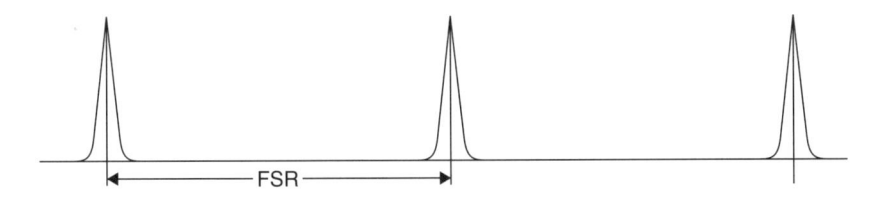

Figure 6.3.1 Response of an interferometer.

Assuming that the distance between the cavity mirrors is L, the condition for a standing wave in the cavity is

$$m\frac{\lambda}{2} = L \tag{6.3.1}$$

where m is an integer, so the spacing (FSR) of resonant modes in the cavity is

$$\Delta\nu = m\frac{c}{2L} \tag{6.3.2}$$

where $\Delta\nu$ is the spacing in hertz. For any interferometer the sharpness of the resonant peaks for an interferometer (measured as a FWHM defined in Section 2.2) is described by

$$\delta = \frac{\nu_f}{F} \tag{6.3.3}$$

where δ is the spectral width (FWHM) in hertz, ν_f is the frequency of the first mode ($m = 1$), and F is the finesse of the interferometer; it is a ratio of the mode separation (FSR) to the spectral width. Finesse is a function of the reflectivity of cavity mirrors:

$$F = \frac{\pi\sqrt{R}}{1 - R} \tag{6.3.4}$$

where R is the total reflectivity of both mirrors. As the reflectivity of cavity mirrors increases, the spectral width of the peaks becomes very narrow, and in most lasers the reflectivity of the cavity mirrors is very large.

Example 6.3.1 Argon Laser Spectral Width Computation An argon laser with a 1-m cavity spacing has an HR with a reflectivity of 99.99% and an OC with a reflectivity of 98.0%. Compute the expected spectral width of a single mode of the cavity.

SOLUTION Begin by computing the FSR as

$$\nu_{FSR} = m\frac{c}{2L} = \frac{3 \times 10^8}{2} = 150\,\text{MHz}$$

This is the spacing between resonant peaks for the interferometer. Now the finesse of the cavity may be computed first by calculating the total reflectivity of the cavity as

$$R_{\text{total}} = R_{HR}R_{OC} = (0.9999)(0.98)$$
$$= 0.9799$$

This reflectivity is then substituting into equation (6.3.4), allowing the calculation of finesse, in this case, 154.7. Knowing that finesse is a ratio of FSR to spectral width, spectral width can be solved by substituting into equation (6.3.3) to yield an answer of 969 kHz. This corresponds (using the 488-nm blue line of the argon laser) to a spectral width of 7.7×10^{-7} nm. This is an extremely narrow spectral width; however, special techniques (outlined in this chapter) are required to isolate a single mode from adjacent modes (which in this case are 150 MHz apart). As we shall see in this chapter, most lasers have a much larger spectral width, originating from the fact that many modes can oscillate simultaneously in most lasers.

6.4 LONGITUDINAL MODES

In Chapter 4 we introduced the notion of linewidth and stated that real laser gain media do not have sharp, defined wavelengths but rather, amplify over a relatively wide range. In a gas laser, Doppler broadening leads to the existence of a gain curve in which the gain of the laser (and hence the output as well) peaks at a center wavelength as shown in Figure 6.4.1. At all points on the gain curve where the gain is sufficient to overcome losses in the laser (and the laser cavity is resonant), the laser may oscillate and have output. We now know that the cavity itself is an interferometer and is resonant only at wavelengths spaced apart by the FSR of the

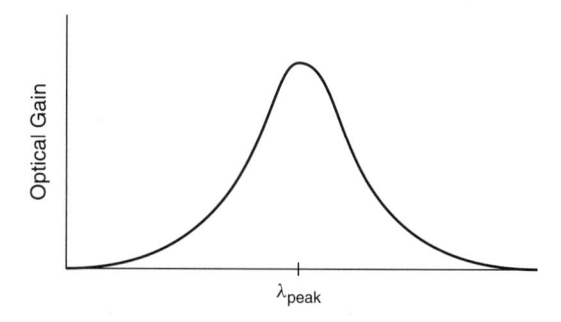

Figure 6.4.1 Gain curve for a practical laser.

arrangement. Refining the model above, we can see that the actual output of the laser will not be a smooth curve as depicted in Figure 6.4.1, but rather, a series of closely spaced wavelengths which follow the general envelope of the curve and exist at points where the gain exceeds losses in the cavity. The development of the output is seen in Figure 6.4.2, where the top diagram shows the gain curve of the laser medium along with the lasing threshold (the point at which gain equals losses). In a simple model one would expect an output spectra resembling the shaded area on the top diagram. The response of the cavity, resonant at frequencies spaced apart by the FSR, is shown in the middle diagram. The result is depicted on the lower diagram, in which the output is seen to be several modes (11 in this case), with the strongest at the center of the curve (where the gain is highest) and the weakest near the edges, where gain just slightly exceeds losses.

Since the laser cavity is an interferometer the resonant modes of the cavity are spaced apart by free spectral range (FSR), which is defined as

$$\text{FSR} = \frac{c}{2nl} \tag{6.4.1}$$

where n is the refractive index of the medium inside the cavity and l is the spacing of the cavity mirrors. The FSR, in this case, is in terms of frequency (Hz). The number of modes that will oscillate simultaneously may be determined by dividing the spectral width of the laser by the FSR.

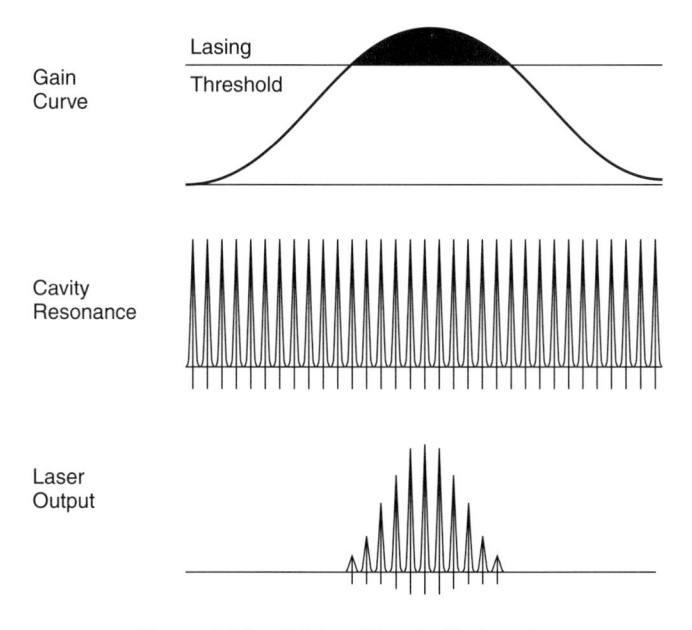

Figure 6.4.2 Origin of longitudinal modes.

166 CAVITY OPTICS

Example 6.4.1 Number of Modes An argon laser with a 90-cm cavity has a spectral width of 5 GHz (see Section 4.7). The FSR of the cavity is

$$\text{FSR} = \frac{c}{2nl} = \frac{3 \times 10^8}{2 \times 0.9} = 167\,\text{MHz}$$

The number of modes is then

$$\text{No. modes} = \frac{5 \times 10^9}{167 \times 10^6} = 30$$

Thirty closely spaced (167 MHz apart) modes will oscillate simultaneously. Modes at the center of the gain curve will have the highest output powers, while modes near the edge will have considerably less. Argon lasers have large linewidths, due to the high operating temperatures of the plasma. Other lasers, such as HeNe lasers, have much smaller widths and hence fewer modes in the output.

6.5 WAVELENGTH SELECTION IN MULTILINE LASERS

Frequently, it is desired to operate a laser on a single wavelength. Multiple-wavelength lasers, such as the argon ion, can oscillate up to 10 discrete lines simultaneously, and dye lasers have broad gain profiles that can span over 100 nm. Single-line operation on these lasers can be achieved by selective feedback in which the optics are designed to reflect only a single wavelength. The first method of wavelength selection is to design the cavity optics themselves to be highly reflective at a single wavelength (or a set of chosen wavelengths) and transmissive (which represents a loss in the cavity) at all others. This method is commonly used with helium–neon lasers, in which it is necessary to suppress the strong infrared transition to boost output on the 632.8-nm red line. In a green helium–neon laser, optics are designed carefully to reflect light centered at 543.5 nm but to transmit wavelengths such as the red. If the optics were broadband and reflected equally well in the red and green regions, the dominant red line would appear—it must be suppressed to allow the green transition to oscillate. Furthermore, to oscillate, even the red helium–neon laser requires optics that suppress an infrared transition, which lases preferentially.

Selective optics are also commonly employed in krypton-ion lasers in which the user may select which lines oscillate simply by replacing the HR with one suited for red (the single powerful Kr$^+$ line at 647.1 nm), red–yellow (allowing both the red 647.1-nm line and the yellow line at 568.2 nm to lase simultaneously), or multiline, which allows various laser lines, spanning from violet to red to oscillate. Of course, using a single-line optic (such as the red line only) will yield a much higher output power on that one line than a multiline optic would.

As an example of the usefulness of selective optics, consider a particular krypton laser in the author's lab which produces white light. The optics are designed precisely to produce multiple wavelengths in the output beam which when added together produce white. The various major lines and relative intensities for a krypton laser are as shown in Table 6.5.1. This is a perfect laser for display purposes and is used extensively for laser light shows since it features output lines in every color of the spectrum, allowing full-color displays. To produce white light, the output powers of various wavelength components must be balanced. While the red line is quite powerful (producing 39% of the total power of a multiline krypton laser), and other lines, such as the yellow line, produce reasonably powerful output, krypton lacks a powerful line in the blue region of the spectrum. Also, the green line at 530.9 nm, although powerful, is a lime-green color that is undesirable for display effects.

The problem now is to design a set of optics for the laser which, for efficiency's sake, allows only a few desired lines to oscillate and in proportions that allow the addition of these to produce white light. The reflectivity of the mirror was analyzed using a spectrophotometer and is depicted in Figure 6.5.1 along with krypton laser lines for reference. Owing to the fact that this is a small air-cooled laser and has low overall gain, the only lines that will lase are those where the reflectivity of the OC is considerably greater than 95% at a particular wavelength. The HR was also analyzed and has a reflectivity of greater than 99.9% throughout the range spanning from 450 to 710 nm. A quick examination of the figure shows that the reflectivity is too low at 406.7, 520.8, 530.9, and 676.4 nm to allow these krypton lines to oscillate, leaving two closely spaced blue wavelengths, one yellow wavelength, and a red wavelength. In practice, the output powers of these four lines balance, so that the beam appears white. In other words, the reflectivity of this optic at the blue wavelengths is slightly higher than at the red wavelength, so that the total output power of the blue lines is equal to that of the red line. These are minor variations in reflectivity and do not appear on the graph, given the scale involved.

Two other methods for selecting a single wavelength involve modifying the HR of the laser cavity. The addition of a prism between the plasma tube and the HR allows selection of a single line, since only one unique wavelength can make the path through the prism, reflecting off the HR and back through the prism in exactly

TABLE 6.5.1 Major Krypton-Ion Laser Lines and Relative Output Powers

Wavelength (nm)	Color	Relative Strength
676.4	Red	0.09
647.1	Red	0.39
568.2	Yellow	0.12
530.9	Green	0.16
520.8	Green	0.06
482.5	Blue	0.02
476.2	Blue	0.04
406.7	Violet	0.12

168   CAVITY OPTICS

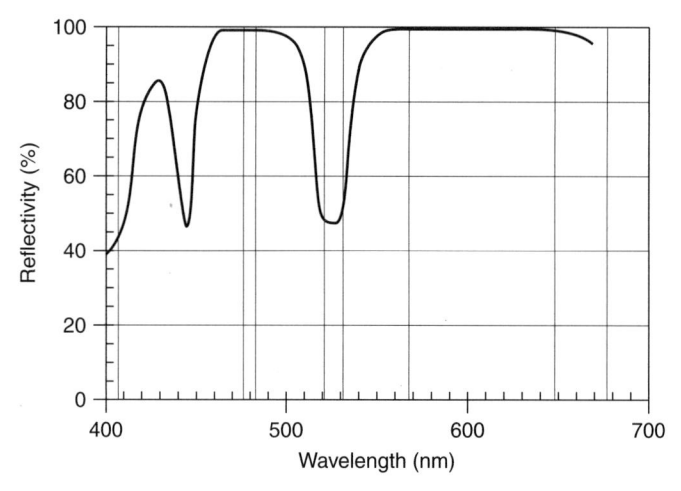

Figure 6.5.1   Reflectivity of the OC for a krypton white light laser.

the same path. By changing the angle of the prism and HR relative to the laser's axis, this arrangement allows tuning through a large range. Alternatively, the HR may be replaced by a diffraction grating which, in a similar manner, reflects only a single wavelength of light back into the plasma tube for amplification. Both methods are outlined in Figure 6.5.2. Practically speaking, prisms have lower losses than diffraction gratings and so are used in lower-gain (e.g., argon) lasers, where even small losses can significantly impair output power (or halt oscillation completely). Diffraction gratings, on the other hand, feature higher angular dispersions of incident wavelengths and so are easier to tune when multiple wavelengths are close together or the medium has a large continuous wavelength range, such as a dye laser.

A wavelength selector employing a prism and used to tune an argon laser is pictured in Figure 6.5.3. Part (*a*) shows the actual wavelength-selective elements, including the prism and HR, mounted together in an assembly that is locked into the rear of the laser cavity, replacing a simple HR. A precision mechanism [part (*b*) of the figure] allows the angle of this assembly to be tilted relative to the laser's axis. In part (*a*) the metal can protecting the optical elements from dust has been removed, while in part (*b*) only the end of the can, with two knurled nuts, is visible in the center of the laser. The mechanism is calibrated in nanometers and allows tuning over the entire visible and near-UV range covered by the laser. While wavelength selection allows oscillation of a single wavelength, that line is relatively broad, owing to Doppler and other broadening mechanisms at play in the laser.

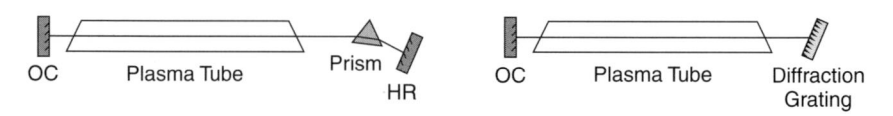

Figure 6.5.2   Wavelength-selection optics.

(a)

(b)

Figure 6.5.3 Actual wavelength selector for an argon laser.

The cavity of the laser, itself an interferometer, divides this broadened line into a number of longitudinal modes each separated slightly in frequency. For true single-wavelength operation (with an extremely narrow spectral width), further refinement is required.

6.6 SINGLE-FREQUENCY OPERATION

Consider an argon laser that has a relatively large bandwidth, allowing a large number of longitudinal modes (with slightly different frequencies) to oscillate simul-

170 CAVITY OPTICS

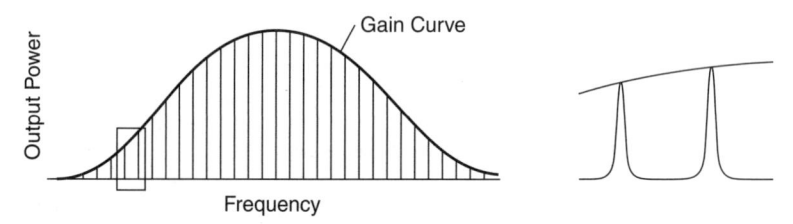

Figure 6.6.1 Longitudinal modes and gain profile in a laser.

taneously as in Figure 6.6.1, in which a small section of the output is expanded to show details of each individual mode. From a problem in Chapter 4, we discovered that the gain bandwidth of the argon laser (broadened primarily by Doppler effects) is about 5 GHz. Inside this 5-GHz range there are many longitudinal modes: The 90-cm laser used in the example had modes spaced 167 MHz apart, for a total of 30 modes. If an etalon (basically, a compact interferometer) is designed such that it is resonant only at wavelengths spaced farther apart than 5 GHz, the laser can be made to operate on a single mode (and hence a single frequency). The spacing of the resonant peaks of an etalon, called the *free spectral range* (FSR) of an etalon, is evident in the middle diagram in Figure 6.6.2. Only when a resonant peak of the etalon and a longitudinal mode of the laser have the same frequency will the laser have enough gain to oscillate.

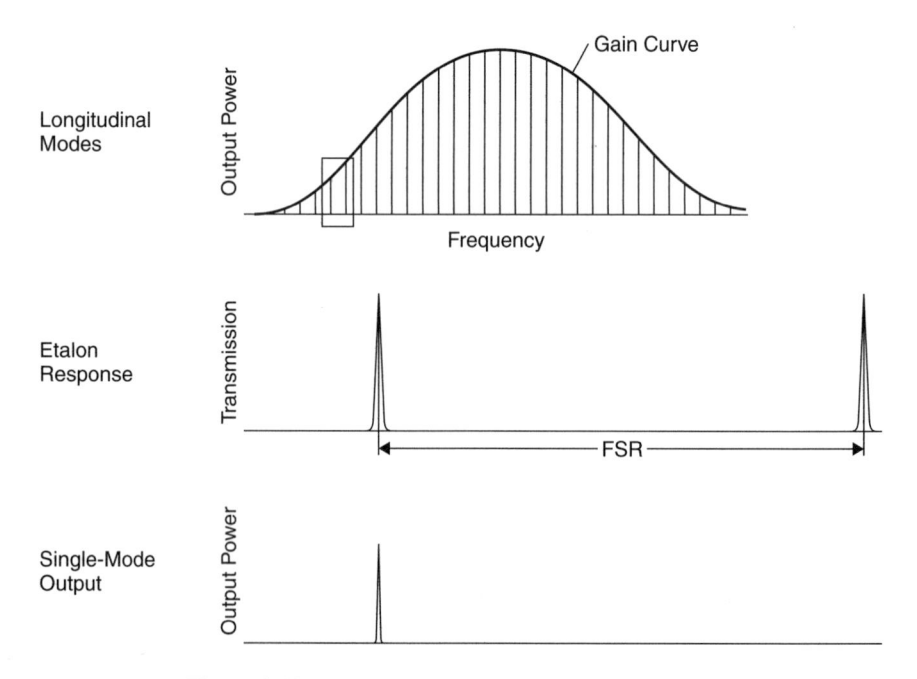

Figure 6.6.2 Selection of a single mode with an etalon.

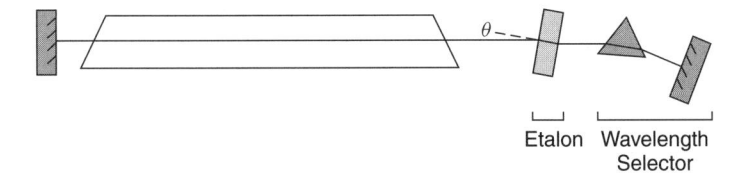

Figure 6.6.3 Intracavity etalon use in a gas laser.

A practical etalon consists of a glass slide (usually quartz, to reduce absorption) coated on either side with a slightly reflective thin-film coating. The etalon acts as an interferometer which is resonant only at certain wavelengths. It is inserted inside the cavity of a laser, which has a wavelength selector in the cavity so that only a single line oscillates, as in Figure 6.6.3. A wavelength selector is required for a multiline laser such as an argon-ion laser, but where the laser optics themselves are resonant at only a single wavelength (as is frequently the case with helium–neon lasers optimized for the red wavelength), the wavelength selector is unnecessary.

The etalon is always tilted with respect to the optical axis of the laser—were it not, it would act as a simple mirror, which would not select wavelength. An examination of the optical path inside the etalon provided in Figure 6.6.4 reveals that an incoming beam (beam 1) enters the etalon and is refracted. It refracts again upon exit and leaves the etalon as beam 2, but a portion of the beam (about 20% for a typical etalon used with a low-gain gas laser) is reflected from the coating on the side of the etalon and is reflected back into the etalon along path a–b. The beam is reflected again, becoming b–c, and is refracted upon exit, becoming beam 3. If the difference in path length between beams 2 and 3 is an integral number of wavelengths, constructive interference occurs and the transmission of the etalon is seen as nearly

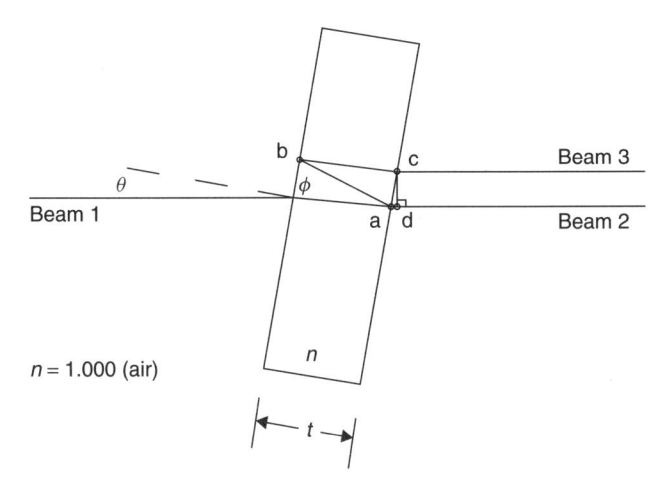

Figure 6.6.4 Optical path inside an etalon.

100%. Conversely, when the paths are not separated by an exact number of wavelengths, beam 3 exits the etalon with a phase different from that of beam 2. Destructive interference occurs in this case, and the transmission is seen as low for the filter.

Mathematically, we can express the condition for constructive interference (i.e., maximum transmission) as the point where the path length for beam 3 is an integer number:

$$N\lambda = n(\text{distance a–b} + \text{distance b–c}) + \text{distance a–d} \qquad (6.6.1)$$

where N is an integer. Geometrically, we can express this (the condition for exit beams to all be in phase) as a function of the angle at which incident light is refracted in the etalon:

$$N\lambda = 2nt \cos \phi \qquad (6.6.2)$$

This is, of course, the familiar formula used to describe the Fabry–Perot interferometer. Small-angle approximations can be applied (for a standard application in which the etalon is just barely tilted in the laser cavity). In this case we can approximate ϕ with θ/n. Finally, the spacing of resonant peaks (the FSR) of this etalon is found to be

$$\text{FSR} = \frac{c}{2nt} \qquad (6.6.3)$$

This is the same result as that developed earlier for the longitudinal modes of a laser cavity. For an etalon made of quartz with a thickness of 10 mm, the FSR is 10.3 GHz, well suited for use in the argon laser above, where the gain bandwidth is 5 GHz (i.e., it would be impossible for two longitudinal modes to overlap and oscillate simultaneously). Note that the faces of the etalon must be very flat and parallel and that thicker etalons will have smaller FSR spacing (which must be chosen accordingly to ensure that FSR is larger than the gain bandwidth).

Equation (6.6.2) indicates that tuning of the etalon can be accomplished by changing the angle of the device with respect to the optical axis of the laser. Although this seems reasonable, one must be cautious, since the interfering beams (beams 2 and 3 in the figure) will separate (distance c–d increases) as the etalon is tilted, resulting in what is termed *walk-off loss*. This complication usually prevents angle tuning of an etalon in a low-gain laser, but an alternative method (the one generally used with etalons of this type) is temperature tuning.

Consider that for essentially all materials, the index of refraction depends on the ambient temperature. In the case of quartz, n will vary by about 1×10^{-5} per degree C. This corresponds to a frequency change of about 5 GHz per degree C of temperature change. By controlling the temperature carefully, one can select any one longitudinal mode within the gain bandwidth of the laser. Obviously, extremely accurate temperature control of the etalon is required for stability (usually, about 0.01 degree C) and the etalon is enclosed in an oven.

SINGLE-FREQUENCY OPERATION 173

Figure 6.6.5 Air-spaced etalon.

Finally, etalons may be constructed using air (or another gas) as a spacer, as shown in Figure 6.6.5. In this etalon two partially reflective optical windows are separated by a mechanical mount designed to keep them perfectly aligned (the two knobs protruding from the device allow alignment of the optics). The entire etalon may be tuned by either changing the angle of the entire etalon in the cavity or changing the gas and gas pressure within the device, which affects the index of refraction. The latter is usually preferred, to prevent walk-off losses. In the case of this particular etalon (which is used with an ion laser), the spacing between the optics is 10 cm, so the FSR is calculated to be 15 GHz when air is used. As we had calculated earlier (in Section 6.3), the linewidth of a single longitudinal mode is predicted to be (for a 488-nm argon laser) under 1 MHz.

Narrow spectral widths result in long *coherence lengths*, coherence length being the distance over which the phases of multiple wavelengths (longitudinal modes) in the output beam stay reasonably in phase with each other. If the output beam consists of many longitudinal modes (each at slightly different frequencies), they will interfere destructively at some point, destroying the coherence of the beam. If, on the other hand, the output beam consists of a single extremely narrow (in frequency spread) mode, the beam will stay coherent for a great distance. Coherence length is hence related to spectral width: The wider the spectral width, the shorter the coherence length. It is defined (in units of meters) as

$$l_c = \frac{c}{\Delta \nu} \qquad (6.6.4)$$

where $\Delta \nu$ is the spectral width in hertz.

Example 6.6.1 Spectral Width versus Coherence Length Consider the spectral width of a regular and a single-frequency HeNe laser. In Section 4.7 we calculated the spectral width of a HeNe laser to be 1.56 GHz and so:

$$l_c = \frac{3 \times 10^8}{1.56 \times 10^9} = 0.19 \,\text{m}$$

By selecting a single mode (with a spectral width of about 1 MHz), the coherence length increases to 300 m! For a laser with a larger inherent spectral width, such as an argon laser, single-frequency operation is quite desirable to allow this laser to be used as a source in making holograms (where a spectrally narrow source leads to better depth in the recorded image).

6.7 CHARACTERIZATION OF A RESONATOR

Resonators (and indeed, interferometers for the most part) may be characterized in terms of frequency or wavelength behavior by a number of parameters, including finesse and free spectral range. We defined these in Sections 6.3 and 6.4. In terms of optical loss, two additional parameters are now introduced which characterize the resonator: the total loss coefficient (γ_r) and the photon lifetime (τ_c). Individual resonator losses occur because of absorption and scattering in the lasing medium itself, at tube windows, as a result of unintended loss at the HR and as an intended loss at the OC which forms the output beam. It is useful to express losses as coefficients with units of m^{-1} (per meter). This may seem counterintuitive since the loss caused by a mirror occurs at a single point in a laser and is not spread out across the entire tube as this approach should do, but the use of this gamma notation shall become evident as we proceed through this and the next chapter.

Losses at each mirror are expressed as loss coefficients (γ_1 for one mirror and γ_2 for the other mirror) as if the loss was distributed throughout the entire laser:

$$\gamma_1 = \frac{\ln \dfrac{1}{R_1}}{2l} \tag{6.7.1}$$

where R_1 is the reflectivity of the mirror. You will note that this is of the same form as the gain threshold formula of equation (4.9.2).

The other primary loss, due to absorption or scattering in the lasing medium, and designated γ_a, is usually given as a property of the medium itself in units of m^{-1} or cm^{-1}. Where the gain medium spans the entire distance between cavity mirrors, such as the case with most helium–neon gas lasers, one may simply use this number directly; however, where the gain medium is shorter than the cavity, e.g., in a solid-state laser where the rod is relatively short in length, we must spread out the loss across the cavity as per the following approximation:

$$\gamma_a = \frac{(2^* \gamma^*_{rod} l_{rod})}{2l} \qquad (6.7.2)$$

In this case the numerator evaluates to the (dimensionless) total loss for a round-trip through the rod which is then spread out across the length of the cavity ($2l$). In a solid-state laser this would not compensate for the index of refraction of the rod which leads to an apparent length (optically speaking) of $n^*_{rod} l_{rod}$, but for most practical lasers the rod is much shorter than the total cavity length and so the cavity is mainly air with $n = 1.00$. So now an overall distributed-loss coefficient (γ_r) can be used to describe the total cavity losses as

$$\gamma_r = \gamma_a + \gamma_1 + \gamma_2 \qquad (6.7.3)$$

Such a coefficient, a *distributed-loss coefficient*, sums all losses in the resonator and has units of m^{-1}. It is useful in further calculations of resonator parameters, as we show below.

Photon lifetime (τ_c) refers to the average time that a photon spends in the cavity of a laser before passing through the OC and becoming part of the output beam or being absorbed in the lasing medium itself. It is best illustrated by an example in which a simple laser cavity consists of one fully reflecting mirror and one mirror of 98% reflectivity (and otherwise is essentially lossless). If the mirrors are separated by 50 cm, the expected photon lifetime is

$$\tau_c = \frac{\text{round-trip distance through the cavity}}{\text{speed of light}} \times \frac{1}{\text{cavity loss factor}}$$

$$= \frac{2(0.5)}{3 \times 10^8} \times \frac{1}{0.02} = 165 \text{ ns}$$

While the photon takes only 3.3 ns to traverse the entire 1-m round trip of the cavity, the probability of passing through the OC is low since it has high reflectivity. On average, then, a photon will make 50 such round trips before exiting the cavity.

Because we have defined the cavity loss factor as a function of length (in m^{-1}), the product of the loss factor with the speed of light (in m/s) defines the loss of photons per unit time (in s^{-1}), so the expression for photon lifetime simplifies to

$$\tau_c = \frac{1}{c\gamma_r} \qquad (6.7.4)$$

This relation is, once again, useful in further calculations involving energy storage in a laser cavity (as we shall do in Chapter 7, where techniques of fast-pulse production are examined). Finally, photon lifetime is also related to the spectral linewidth of

176 CAVITY OPTICS

laser output by

$$\Delta \nu = \frac{1}{2\pi\tau_c} \tag{6.7.5}$$

Mathematically, this is simply the Fourier transform relationship between frequency and time (e.g., the Fourier transform of an infinitely narrow pulse in the time domain is a broadband output covering the entire spectrum of available frequencies). This is also a mechanism by which laser lines are broadened, called *lifetime broadening*.

6.8 GAUSSIAN BEAM

The Gaussian output beam (also called a TEM$_{00}$ beam, signifying that it has the lowest electromagnetic mode structure possible, a concept we cover later in this chapter) is, spatially speaking, the purest laser beam possible and is characterized by the lowest divergence of any mode; it is limited only by diffraction. The beam is characterized in intensity in both axes by

$$I(y) = I_0 \exp\left(-\frac{2y^2}{w^2}\right) \tag{6.8.1}$$

where I_0 is the maximum intensity of the beam, y the distance from the center of the beam, and w the radius of the beam. The intensity profile of the beam is depicted in Figure 6.8.1. Note from the equation that the radius of the beam (also called the *spot size*) expands as we move away from the laser, so the calculation of intensity of equation (6.8.1) applies to a cross section of the beam at an arbitrary distance away from the laser.

Inside a cavity consisting of two concave mirrors with radius of curvature equal to exactly the distance between them (an arrangement called a *symmetric confocal*

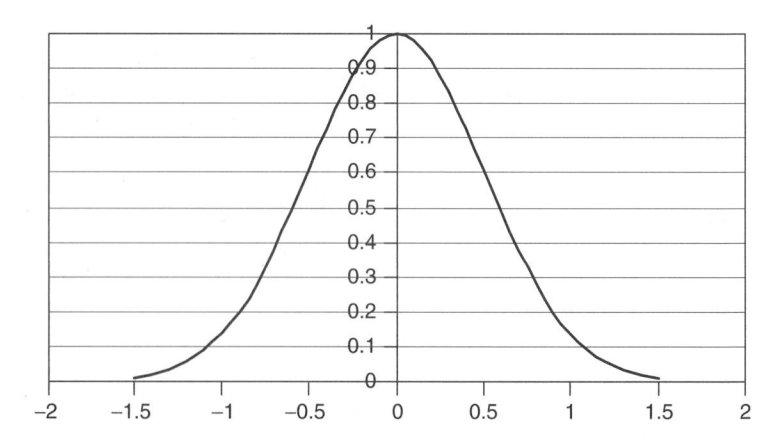

Figure 6.8.1 Gaussian beam spatial profile.

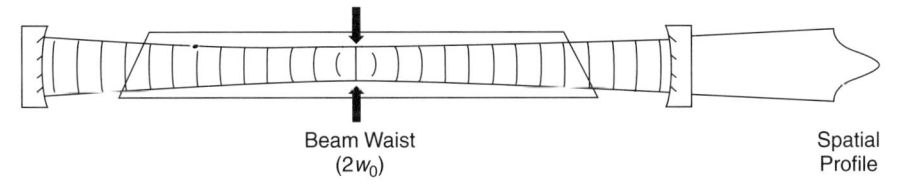

Figure 6.8.2 Gaussian beam inside a laser cavity.

resonator and the most general example, as we shall see later in this chapter) the beam converges at the center of the gain medium in what is called the *beam waist* denoted as w_0 in Figure 6.8.2. This is the smallest spot size of the beam, and the width of the laser beam at this point is governed by the wavelength and the separation of the cavity mirrors according to

$$w_0 = \left(\frac{\lambda L}{2\pi}\right)^{1/2} \quad (6.8.2)$$

where λ is the wavelength of the laser and L is the distance between the cavity mirrors. It should also be noted from Figure 6.8.2 that the volume of the amplifier utilized in the laser process is not the entire volume of the gain medium. A Gaussian beam in a laser cavity does not utilize the entire lasing volume and will hence not render the highest output power for a given laser. Higher-order modes, covered later in this chapter, can often utilize the gain medium more effectively than a Gaussian (TEM_{00}) beam can.

Inside the cavity, the wavefronts of the Gaussian beam do not stay constant but rather, deviate. At the beam waist wavefronts are plane, but as they move toward the cavity mirrors the shape changes to match that of the radius of curvature of the mirrors—essentially that of a spherical wave. Wavefronts exiting the laser through

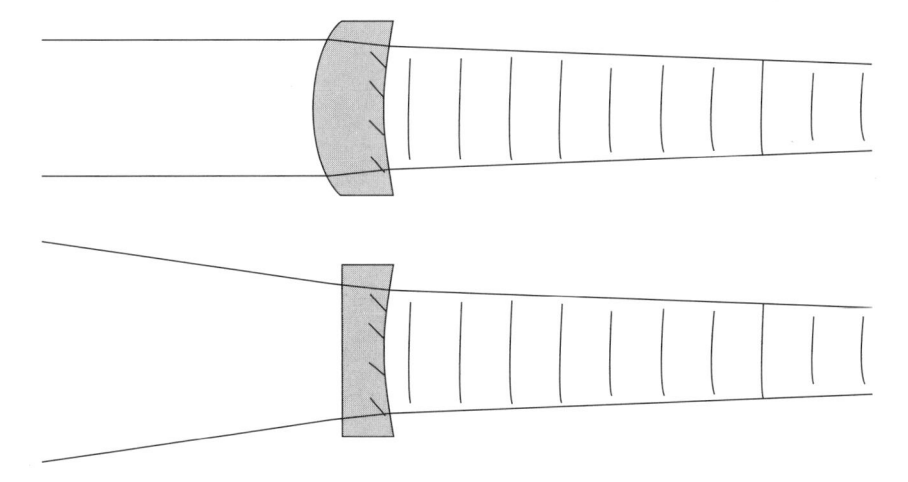

Figure 6.8.3 Collimating lenses used with lasers.

the output coupler also have this characteristic shape, and upon exiting through the OC diverge at an angle of

$$\theta = \frac{\lambda}{\pi w_0} \tag{6.8.3}$$

where θ is the half-angle of the divergence. This formula is derived from the radial expression for a Gaussian beam, which is a complex (i.e., involving a complex term $-j$) expression. The angle 2θ then defines a cone which contains 86% of the power of the laser beam. Note that this approximation does not apply to beams close to the waist since it is a simplified expression with neglected terms. It is entirely valid outside the cavity, however, so defines the divergence of the exit beam. Since divergence is a function of wavelength as well as beam waist, laser beams are inherently more collimated if they are of shorter wavelengths and have large beam waists (e.g., in the case of a large gain medium). This explains why shorter wavelength (e.g., UV) lasers have a distinct advantage in performing delicate cutting and high-resolution photolithography applications requiring small spot sizes (as well as why next-generation CDs and DVDs will use shorter wavelength lasers than the infrared lasers currently in use). Regardless of the wavelength, however, the beam will continue to diverge after exiting the laser, the divergence made even worse by the fact that the output coupler itself will act as a lens. To collimate the beam, most lasers use either an external collimating lens or, more frequently (especially compact lasers such as HeNe tubes with integral tube mirrors), the outer surface of the output coupler itself is shaped as a convex lens, as depicted in Figure 6.8.3.

6.9 RESONATOR STABILITY

The most general form of a laser cavity, as used in Section 6.8, is one consisting of two concave mirrors with coincident foci in the center of the laser medium, as outlined in Figure 6.8.2. This is contrary, perhaps, to popular opinion, which is usually that two plane mirrors are used, although that arrangement, too, is stable (although only marginally) and is employed in some lasers. In the confocal arrangement, a true Gaussian beam traverses between the concave mirrors, and it can be found in which the curvature of the wavefront of the intracavity beam approaching the mirrors matches the curvature of the cavity mirrors themselves. This ensures that the beam reflects perfectly back on itself and is completely trapped within the cavity—the definition of stability. Any ray within the cavity can retrace itself exactly after one round trip through the cavity. If one mirror is partially transmitting, the beam that passes through the mirror (which becomes the output beam) continues to diverge.

We now examine the parameters behind the design of resonators in order that one may be designed to confine radiation properly in the laser. The vast majority of lasers use stable resonator configurations, those in which light between the mirrors

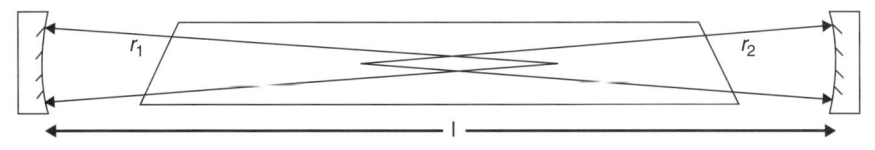

Figure 6.9.1 Parameters for resonator stability.

is trapped completely and the beam is reflected back on itself within the cavity, keeping the power density large, to encourage stimulated emissions. To determine stability (mathematically) we introduce resonator g parameters, one representing each mirror, which define the beam path relative to the entire cavity. Knowing the distance between the cavity mirrors as well as the radius of curvature for a mirror, the expression for a g parameter is

$$g = 1 - \frac{L}{r} \qquad (6.9.1)$$

where L is the distance between the cavity mirrors and r is the radius of curvature. These parameters are defined in Figure 6.9.1, which shows a generalized laser cavity with concave mirrors. *Stability*, then, is defined as the condition where

$$0 \leq g_1 g_2 \leq 1 \qquad (6.9.2)$$

where g_1 and g_2 are the g parameters for each mirror. In the case of a plane mirror, the radius is infinity, so the corresponding g parameter is 1.

Example 6.9.1 Resonator Stability Consider a laser cavity consisting of two spherical mirrors resembling that of Figure 6.9.1. The distance between the two mirrors is 1 m and each mirror has a radius of 60 cm. Using equation (6.9.1), the g parameter for each mirror is calculated to be $1 - 1/0.6 = -0.67$, so that the product of the g parameters is 0.44. This resonator arrangement is hence found to be stable.

Consider a similar second cavity where each mirror has a radius of 45 cm. The g parameter of each mirror is now calculated to be -1.22, so that the product of the g parameters is 1.49. Since this value is greater than 1, the configuration is not stable. Various cavity arrangements are examined in the next section.

It must also be noted that some cavity arrangements utilize the lasing volume better than others. A confocal cavity (see Figure 6.10.1), for example, uses only a fraction of the gain medium, whereas a cavity consisting of two plane mirrors can use much more of the gain medium, presumably allowing higher efficiency and power output. A cavity consisting of two plane mirrors, however, is difficult to align (it presents marginal stability, as the product of the g parameters is exactly 1), so most practical laser cavities consist of one or more concave or spherical mirrors, as we examine in the next section. It is also possible to use intracavity beam expanders (consisting of two or more lenses) to improve utilization of the entire volume of the lasing medium (this is sometimes done in YAG lasers to ensure use of the entire rod).

6.10 COMMON CAVITY CONFIGURATIONS

Consider the case of a resonator with two plane mirrors (Figure 6.10.1a). Parameters g_1 and g_2 are both equal to unity (1), so the arrangement is stable although stability is marginal (i.e., the product of $g_1 g_2$ is 1). In practical terms, *marginally stable* means extreme difficulty in alignment, and a cavity that can become misaligned very easily, usually resulting in the ceasing of lasing oscillation. For this reason, two plane mirrors are rarely used.

For any cavity resonator consisting of two spherical mirrors, the arrangement is stable within limits, as determined by the *g* parameters. A true confocal arrangement (Figure 6.10.1b) in which the radius of both mirrors is exactly equal to the separation between the mirrors (i.e., L) has *g* parameters equal to $g_1 = g_2 = 0$. The product of the *g* parameters is hence zero again, so this arrangement is stable.

The confocal configuration yields the smallest average spot size of any stable resonator with the beam waist being $w_0^2 = L(\lambda/2\pi)$ occurring at the center of the resonator and the largest spot size $w_1^2 = L(\lambda/\pi)$ occurring at each mirror. This defines the diameter of the output beam if collimated (by a lens) at that point.

(a) Plane Mirror Resonator

(b) Confocal Resonator

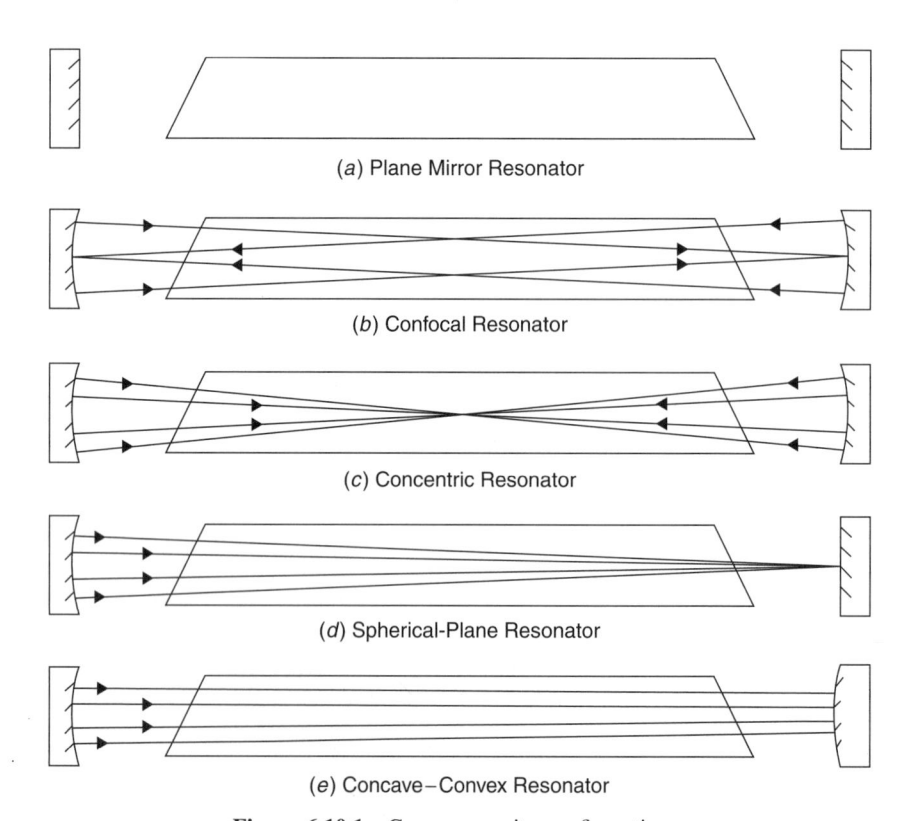

(c) Concentric Resonator

(d) Spherical-Plane Resonator

(e) Concave–Convex Resonator

Figure 6.10.1 Common cavity configurations.

Other resonator configurations may have smaller spot sizes at one mirror or the other, as we shall see.

Example 6.10.1   Spot Sizes   While a small spot size is desirable for many applications, confocal arrangements do not make efficient use of a large gain volume (such as in CO_2 lasers, which frequently feature a large plasma tube bore). As an example, consider a CO_2 laser with a 1-m-long tube. The spot size at the waist and at each mirror would be:

$$\text{Beam waist:} \quad 2w_0 = \left(\frac{L\lambda}{2\pi}\right)^{1/2} = 2.6 \text{ mm}$$

$$\text{Exit beam:} \quad 2w_1 = \left(\frac{L\lambda}{\pi}\right)^{1/2} = 3.7 \text{ mm}$$

Considering that many CO_2 lasers have plasma tube diameters of between 10 and 25 mm, it is easy to see how inefficient this cavity would be at utilizing the large amplifier volume. In contrast to this, consider a 30-cm-long HeNe in which the spot size at the mirrors would be 0.5 mm. This configuration is considerably more reasonable for a laser of this type since many HeNe tubes have plasma tubes with a diameter of 1 mm.

The confocal arrangement is extremely tolerant to misalignment of either mirror. A small angular tilt of either mirror still maintains the center of curvature of one mirror on the surface of the second cavity mirror, as shown in Figure 6.10.2. It is also forgiving of manufacturing tolerances in the radius of the cavity mirrors. Because of this feature, it is an excellent choice for a research laser, where frequent alignment may be required or where alignment cannot be performed by "rocking" or other means (see Section 6.13).

Figure 6.10.1(c) is a concentric configuration in which the radius of each mirror is exactly $L/2$. It represents a confocal resonator, where the radius of curvature is reduced to its lowest limit. From the point of view of stability, this cavity is stable unless the radius of curvature of the mirrors is even slightly under $L/2$, in which case the arrangement becomes unstable (this may occur due to alignment or simply, manufacturing tolerances). This arrangement also suffers from difficulties in alignment since the focus of each mirror is coincident and hence difficult to align precisely.

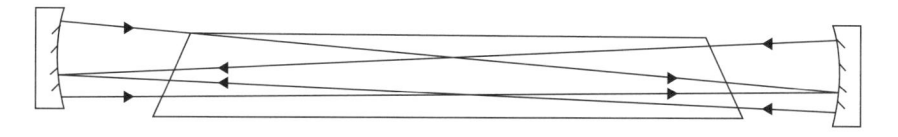

Figure 6.10.2   Misalignment of a confocal cavity.

At the upper limit of the radius of curvature the mirrors become plane (and this arrangement is only marginally stable). However, in many practical lasers a variation of a confocal arrangement is used in which mirrors with radii just slightly over L (i.e., just longer than the cavity length) are employed. A cavity of this type features a larger waist diameter and hence better utilization of large tube bores. Compared to the true confocal arrangement, these cavities are more tedious to align and so, practically speaking, the radii of curvature are often designed to be only slightly greater (perhaps 5%) than the cavity length.

Other arrangements, such as Figure 6.10.1(*d*), use a combination of a long-radius spherical mirror (with the radius at least equal to and sometimes much longer than the cavity length) and a plane mirror. This arrangement is the most popular for low- and medium-power lasers such as HeNe and argon lasers. It is used in many commercial gas lasers (e.g., large-frame argons) where the OC is spherical and the HR plane allows the use of various optical configurations. While the OC stays in place, the rear optic may be changed to a wavelength selector for single-line use or a broadband reflector for multiline use.

If the radius of curvature is exactly L (the cavity length), the spot size at the plane mirror is minimal, since this is indeed the beam waist (but again this leads to stability problems similar to those with a true confocal cavity). The beam occupies a cone-shaped volume inside the gain medium and so utilizes the amplifier volume more efficiently at one end than the other. In a small HeNe laser, the OC is often flat and is placed at the end of the tube away from the discharge. By placing the active gain medium (which in a small HeNe laser is the actual small capillary tube inside the larger laser tube) near the spherical mirror, the gain volume is used more effectively and higher powers are extracted from it, as shown in Figure 6.10.3.

An interesting feature of a spherical-plane cavity configuration is demonstrated in older HeNe tubes, which exploited this feature to assist in alignment of the cavity. During the 1970s and early 1980s, HeNe tubes were made entirely of glass, with mirrors affixed to the ground tube ends by epoxy (called a *soft seal*, this is no longer used in mass-produced HeNe tubes because such seals allow helium to diffuse through them slowly, and hence the resulting tubes had short lives). The use of a concave mirror allows alignment of the laser cavity using translation of the concave mirror alone; no angular adjustment was required. Figure 6.10.4 shows a concave mirror terminating the flat end of a laser which is not perfectly perpendicular to

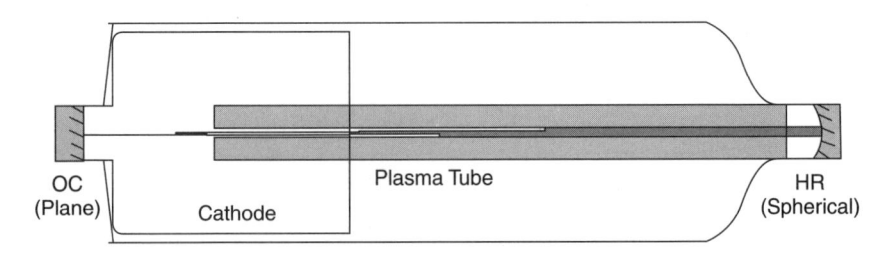

Figure 6.10.3 Spherical-plane resonator for a HeNe laser.

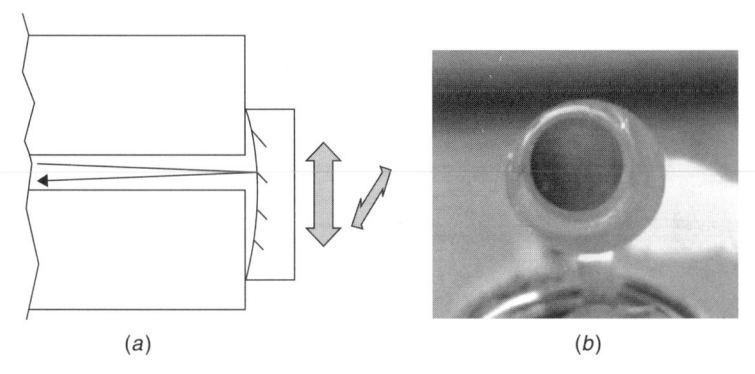

Figure 6.10.4 Spherical reflector alignment on a HeNe tube.

the bore (indeed, it would be quite difficult to grind both glass tube ends perfectly parallel and perpendicular to the bore as would be required for lasing action). This technique of alignment, in which the edges of the mirror were in contact with the flat end of the glass tube, was a great boon to mass manufacturing of these tubes because it required only that the mirror be moved side to side and up/down until alignment was achieved, at which point the mirror was simply fixed in place with epoxy adhesive. Figure 6.10.4(b) shows one of these mirrors, which appears to be haphazardly affixed to the flat end of the glass tube. It is clearly not concentric with the tube bore, but this is exactly how alignment was achieved! The alternative, used on some early HeNe tubes, would be to have each mirror mounted on a three-point mount allowing angular adjustment and attaching each mount to the tube using flexible metal tubing.[1]

In the final resonator arrangement (Figure 6.10.1e), a concave and a convex mirror are used. Again, this arrangement is stable within the confines set out by the g-parameter equation. Concave–convex resonators can utilize much more of the lasing volume since the smallest spot size, at the focus of the concave mirror, is outside the cavity itself! These types of cavities are very sensitive to misalignment, though, so are rarely used in commercial lasers.

Not all lasers use stable resonators, and for certain high-power lasers such as excimer and carbon dioxide TEA lasers, unstable resonators are a popular option. Consider the two unstable resonator configurations depicted in Figure 6.10.5. Part (a), a positive branch confocal resonator, consists of a small convex and large concave mirror. The beam exits around the edges of the smaller mirror and has an annular shape. In part (b), a negative branch confocal resonator, the beam is also annular in shape. Because these resonators are not stable, light is not trapped

---

[1] Old tubes made by Spectra-Physics have exactly this feature, in which the mirror is contained in a cell that can be angularly adjusted. These cells were vacuum-tight, with the mirror sealed to the cell via O-rings, and each was connected to the tube via a short length of stainless steel bellows. As one might imagine, this was a considerably more involved, and costly, tube than simple one-piece tubes, since it required a base to keep everything aligned.

184 CAVITY OPTICS

(a)

(b)

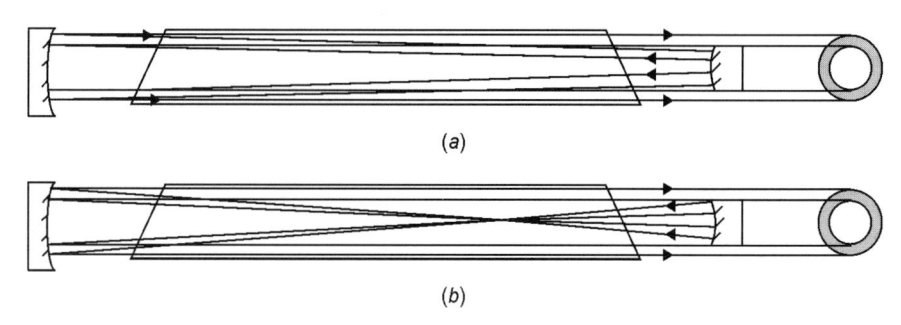

Figure 6.10.5 Unstable resonators.

in the cavity, at least for many round trips, so this arrangement is suitable only for use with high-gain lasers, in which only a few transits through the gain medium are required to amplify the oscillations to a usable power level. The primary benefit of an unstable resonator is allowing the use of total reflectors that would not be damaged by high power levels (in the tens of kilowatts level) as a partially reflecting output coupler might be. The output beam does not pass through the output coupler itself as it does in a stable resonator configuration. In the case of large carbon dioxide lasers, solid metal mirrors (usually, copper with gold plating and often water cooled) may be used, which have very high damage thresholds.

Examining Figure 6.10.5, it is also evident that an unstable resonator utilizes a large volume of the lasing medium, allowing efficient extraction of energy. The biggest apparent problem with this configuration is that the shape of the beam is not Gaussian, so cannot be focused to a sharp point. In reality, it can be focused to almost as sharp a point (certainly better than many stable resonators yield when operating in high-order modes), so is quite suitable for materials-processing applications.

Finally, it may be worth noting that an OC can be avoided altogether in a stable cavity configuration by constructing a cavity in which one mirror has a hole in it. This approach, popular with amateur laser constructors[2] for carbon dioxide lasers, usually utilizes a spherical-plane cavity for ease of alignment. A small hole with a diameter about 10% of the diameter of the plasma tube is drilled into the flat mirror and sealed with a window transparent to infrared radiation. In this manner, inexpensive metal–film mirrors (usually, glass coated with copper or gold, both of which reflect infrared radiation well) may be used in place of an OC made of germanium or zinc selenide, which are comparatively expensive. The approach works but does not allow good mode performance and so is rarely used in commercial lasers.

---

[2]See "The Amateur Scientist" column in *Scientific American*, September 1971, for a description of such a laser, which utilizes inexpensive glass blanks coated with gold or copper film. A hole is drilled through the plane mirror and is covered (to maintain vacuum integrity) with a window of salt or barium fluoride.

6.11 SPATIAL ENERGY DISTRIBUTIONS: TRANSVERSE MODES

In an ideal situation, energy would be stored throughout the entire lasing medium in a consistent manner and the entire lasing volume would be utilized. In reality the nature of the cavity gives rise to electromagnetic modes in which standing waves are set up not only in the longitudinal direction (i.e., the length of the cavity; these modes vary slightly in frequency) but also in the transverse direction. These represent alternative solutions to the wave equation which defines the beam (the complex equation that previously yielded the Gaussian solution). With energy stored in various areas of the lasing medium, the patterns formed are manifested in the output beam as well, which can assume shapes such as those shown in Figure 6.11.1, in which six common modes are photographed. As you peruse these photographs, note that they were shot at the same exposure, so that the relative brightness of the pattern indicates relative power of the output beam. In this case it is evident that some of the high-order modes, such as TEM_{11}, have much higher powers than that of the TEM_{00} mode. Also, the camera was kept at the same position so that the relative size of the various modes can be compared (indicating the divergence, since the beam expands as it exits the laser).

TEM_{00}, the Gaussian mode, is shown in the upper left corner. Two modes are labeled TEM_{01}: the first, commonly called the 01 *donut mode*, appears to have circular symmetry, and the other mode has rectangular (x/y axis) symmetry. The donut mode is often the superposition of TEM_{01} and TEM_{10} modes, both with rectangular symmetry, rather than a circularly symmetrical Laguerre–Gaussian mode. Several higher-order modes, all with rectangular symmetry, are shown in the bottom three photographs. All of the modes in Figure 6.11.1 are Hermite–Gaussian modes. In general, the mode designation (the subscripted numbers) may easily be

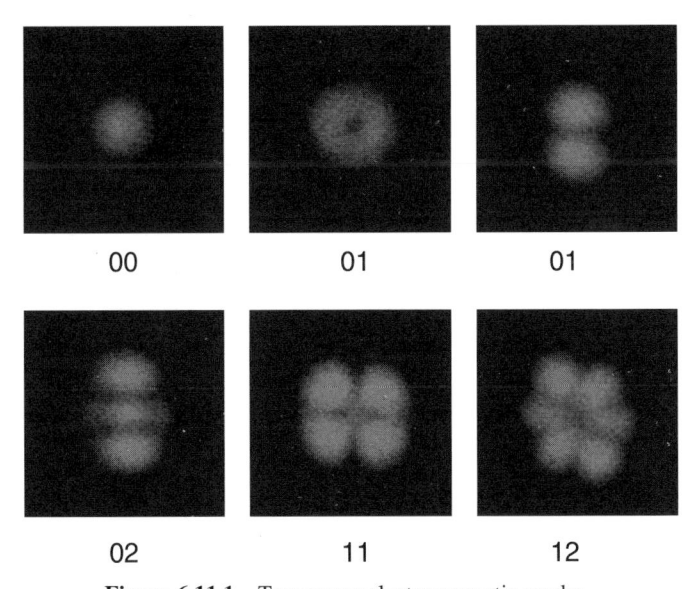

Figure 6.11.1 Transverse electromagnetic modes.

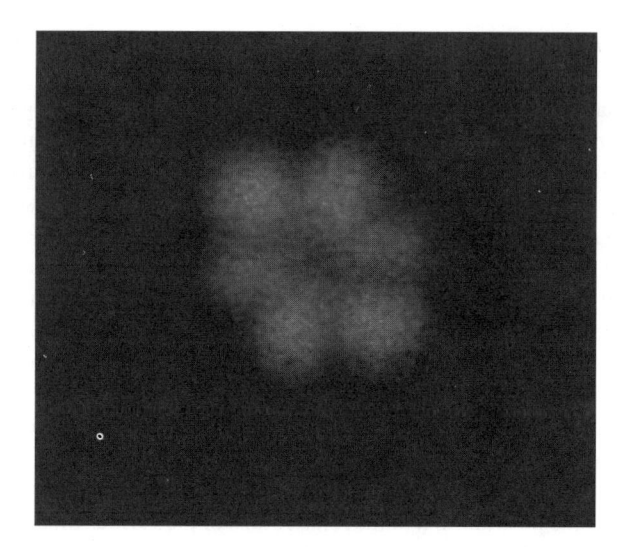

Figure 6.11.2 Typical Laguerre–Gaussian mode (TEM$_{03}$).

determined for a Hermite–Gaussian mode by counting the number of dark lines in each direction so the mode designated as TEM$_{02}$ has no dark lines in the horizontal axis and two in the vertical axis. There is no preference to the order of the subscripted numbers, so TEM$_{02}$ is equivalent to TEM$_{20}$.

Although Hermite–Gaussian modes (with rectangular symmetry) are favored in most operating lasers, other, circularly symmetrical modes are possible. These Laguerre–Gaussian modes are composed of concentric rings which are usually split into pie-shaped wedges like that shown in Figure 6.11.2. The mode in the figure is designated TEM$_{03}$. The first subscripted number for modes of this type is determined by counting the number of dark rings around the center of the pattern (in this example there are none), and the second number is determined by counting the number of lines bisecting the pattern (in this case, three).

In many cases the highest power output from a particular laser can be obtained on a high-order mode in which the volume of the gain medium is utilized more effectively than is possible with the TEM$_{00}$ mode. This is evident by the fact that the TEM$_{11}$ mode and some other higher-order modes are considerably brighter than lower-order modes shown in Figure 6.11.1. This fact can be exploited (in reverse) to limit a laser to operation to TEM$_{00}$ mode.

6.12 LIMITING MODES

Large apeture gain media as well as certain cavity configurations often give rise to high-order modes such as TEM$_{10}$ or higher. Physically speaking, high-order modes consume a larger volume of the gain medium than do low-order modes. For this

(a) Large Gain Medium Aperture Supports High-Order Modes

(b) By Limiting the Aperture of the Gain Medium, Only the TEM₀₀ Mode Is Supported

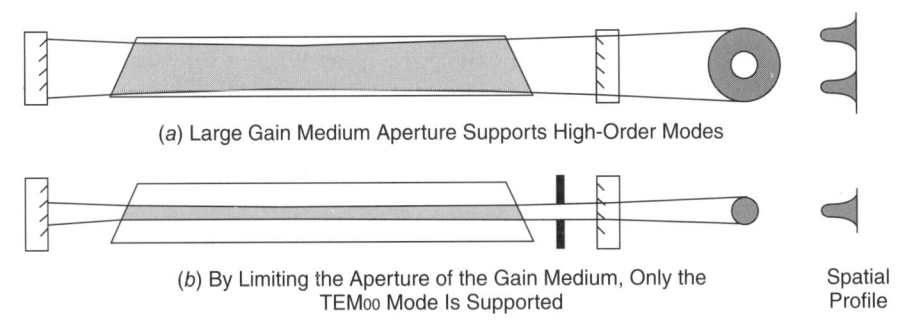

Spatial Profile

Figure 6.12.1 Use of an aperture to limit modes.

reason, many small-bore lasers often operate exclusively in TEM_{00} mode.[3] This effect can be exploited to prevent a laser from oscillating in higher-order modes by placing an aperture of the proper size inside the cavity so that only the TEM_{00} mode will fit through it, as illustrated in Figure 6.12.1. Higher-order modes will be extinguished because the loss imposed on them by the aperture will be greater than the gain provided by the active lasing medium. For example, many large argon lab lasers have an apeture disk or mechanically adjustable iris installed between the laser tube and the output coupler, allowing the user to select the size of hole inserted into the cavity. At a large apeture, the TEM_{10} *donut* mode frequently shows up, but insertion of a smaller apeture forces the laser to operate strictly in the TEM_{00} mode, accompanied by a drastic loss in power! From the figure it is evident that the TEM_{01} mode occupies a larger volume in the gain medium than the TEM_{00} mode does, so the higher-order mode can therefore interact with more of the excited laser medium and hence extract more power from the laser. Lasers oscillating in high-order modes usually produce more power than do similar lasers limited to the TEM_{00} mode. Still, the beam purity of the TEM_{00} mode is often sought after, especially for applications where minimum beam divergence is required.

6.13 RESONATOR ALIGNMENT: A PRACTICAL APPROACH

The alignment of cavity optics can be done using a number of methods, depending on the laser—specifically on the diameter of the laser gain medium itself. When the laser undergoing alignment has a large bore, such as a carbon dioxide laser (which frequently have tube diameters of 10 mm or more) or a YAG laser (with diameters typically ranging from 4 to 8 mm), the simplest method often involves the use of a small visible alignment laser, such as a HeNe or red diode laser. The alignment process begins by removing the OC and aligning the beam from the HeNe laser such that it

---

[3] True in theory, but many large-frame argon lasers (which have small-bore plasma tubes) operate in TEM_{01} donut mode, preferentially with an unrestricted cavity. Still, with large-bore CO_2 lasers, the output is almost always at some high-order mode, which is easily detected by placing a piece of wood in front of the laser and turning the power supply on for only a moment, in which case the mode pattern is burned into the wood for analysis.

enters the front of the laser, passes through the gain medium, reflecting off the HR back through the gain medium, and exiting the laser. When a card with a small hole in it is placed in front of a HeNe laser, the HR is easily aligned such that the beam reflected from the HR and exiting the laser is parallel to the beam entering the laser (literally, "on top" of it). Visually, the mirror is aligned when the reflected and original beams become one. The OC may now be replaced and aligned in a similar manner, but in this case there will be a plethora of reflections since some rays will reflect from the OC, striking the HR, and then exit to be seen on the card as shown in Figure 6.13.1. Still, it is usually easy to identify the first reflection from the OC (it is the second brightest spot) and to align it so that all spots on the card converge to a single spot. At that point, almost by definition, the mirrors are parallel (but not necessarily optimally aligned, as we shall see later in this section).

This technique works only for lasers where the visible alignment beam can pass through the cavity optics. In most cases it works well with YAG lasers, since the mirrors often have high transmission at the HeNe wavelength of 632.8 nm. This is because it is common with high-power lasers such as YAG to use a small HeNe laser as a coaxial targeting laser in which the red HeNe beam emerges from the laser to identify where the target will be hit when the YAG is energized. In many cases, CO_2 laser optics allow alignment by the same method (e.g., while the HR in many CO_2 lasers is made of germanium, which does not pass visible light, the output couplers of many lasers are made of ZnSe, which reflects about 60 to 70% at 10.6 μm and is also partially transmitting at the red wavelength of the HeNe laser).

For small-bore lasers such as HeNe and argon lasers (where the inside diameter of the plasma tube is 1 mm or less), the alignment of an external laser beam down the bore is very difficult. In many cases it is difficult to distinguish total internal reflection (from the almost parallel angle of the alignment laser) from the alignment beam itself, so an autocollimator is often used. The autocollimator produces a beam of collimated light (i.e., light with parallel rays), which is then directed through the laser tube, reflected from the cavity mirror at the far end, and back through the tube to be seen by the observer. An incandescent lamp is used as a light source, and when the mirror is aligned properly, the observer will see an image of the filament. The configuration is shown in Figure 6.13.2, and a photograph of the image of the filament as seen reflected from the mirror being aligned is shown in Figure 6.13.3.

The next step in the alignment procedure is to align the OC. This may be accomplished by using the autocollimator by removing the (just aligned) HR, installing the OC, and aligning it separately in a manner similar to that used for the HR. The

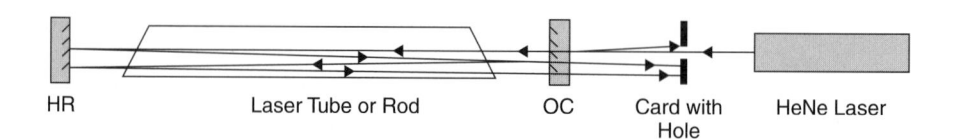

Figure 6.13.1 Alignment using a small HeNe laser.

RESONATOR ALIGNMENT: A PRACTICAL APPROACH   **189**

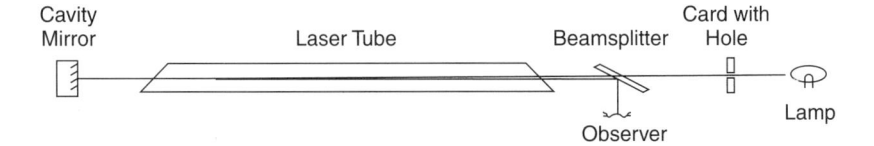

Figure 6.13.2 Alignment using an autocollimator.

location of the autocollimator is simply changed to the rear of the laser. The alternative is a simple search procedure in which the mirror is rocked with the laser energized until aligned. To perform the search pattern, the OC is installed and is aligned so that the reflected beam falls to one side of the tube. The mirror is now rocked in an up–down motion while the mirror is scanned slowly from one side to the other as shown in Figure 6.13.4. As the search pattern progresses with the mirror scanning up and down and side to side, a blink of laser light will be observed. At this point the mirror is aligned in the horizontal (sideways) direction. The adjusting screw for the vertical axis is now adjusted until a continuous beam appears.

Finally, both the OC and HR mirrors are adjusted for maximum output. More specifically, the mirrors are "tweaked" to ensure that they are parallel. It is quite likely that upon first alignment the mirrors are not perfectly perpendicular to the tube, as shown in the top diagram of Figure 6.13.5, in which the path of the intracavity beam is shaded. In this situation the mirrors are parallel to each other, with a clear path through the gain medium allowing oscillation. However, the utilization of the volume of the gain medium is poor, so power output will be much lower than when the mirrors are fully optimized and perpendicular to the tube, so that the entire volume of the laser tube is used as depicted in the lower diagram of the figure. By

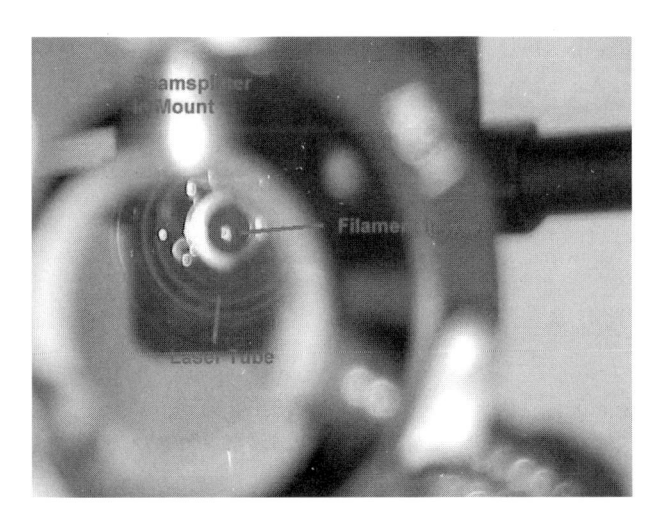

Figure 6.13.3 View of the filament image through the autocollimator.

190 CAVITY OPTICS

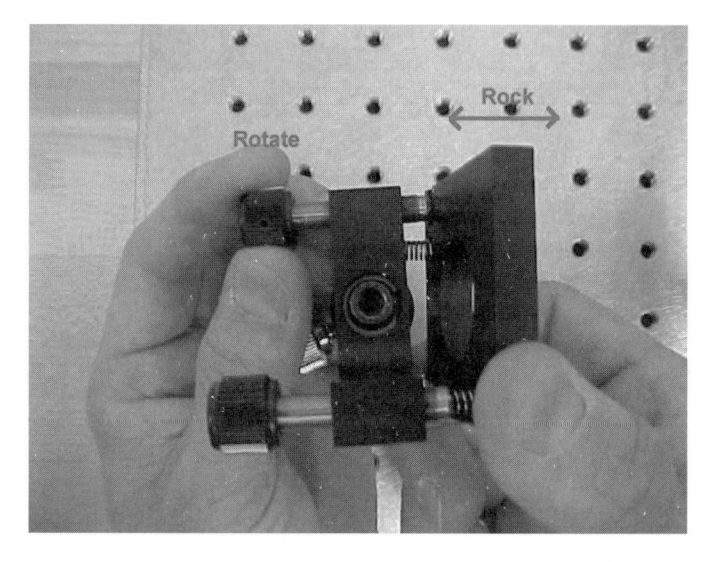

Figure 6.13.4 Rocking procedure to align the OC.

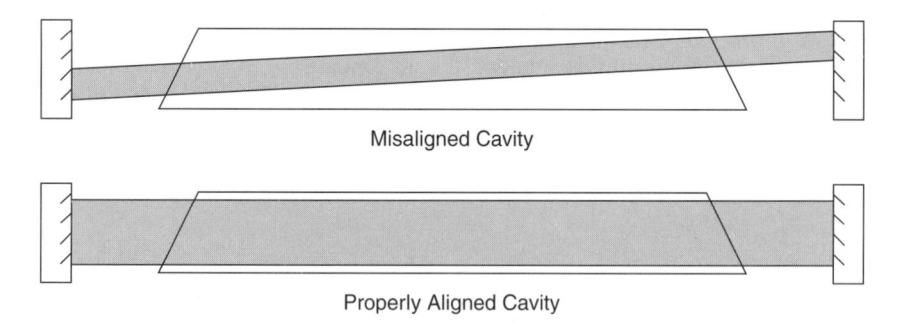

Figure 6.13.5 Possible misalignment of the mirrors.

adjusting the mirrors carefully in both directions by "rocking" the screws in either direction, one may optimize the setup for maximum power output.

PROBLEMS

6.1 A semiconductor laser using a crystal of GaAs material 400 μm in length has $n = 3.4$ and an absorption coefficient of 3000 m^{-1}. Mirrors are fabricated by cleaving the crystal ends and are uncoated.
  (a) Calculate the reflectivity of the cavity mirrors.
  (b) Calculate the required gain of the lasing medium.

6.2 A quartz window is used for an argon laser tube. Although Brewster's angle does not depend on wavelength, the index of refraction of many materials does (which is why a prism works). If a quartz window is optimized for the 488-nm transition of the argon laser, what loss is imposed by this window at the much weaker transition of 454.5 nm? The properties of quartz can be found in a good physics reference [such as the *CRC Handbook of Chemistry and Physics* (Boca Raton, FL: CRC Press)].

6.3 An argon laser has a 90-cm gain tube and operates at a wavelength of 488 nm. First, calculate the Doppler linewidth of the laser (from Chapter 4). Plasma temperatures in an argon laser run at about 5000 K.

(a) Calculate the spacing of longitudinal modes in the laser.

(b) Calculate the number of longitudinal modes that can oscillate in this laser simultaneously.

(c) If this laser is a small air-cooled laser and is desired to operate strictly in single frequency mode, what is the maximum length of the gain tube?

6.4 A dye laser has a very wide spectral width. Cavity optics (separated by 30 cm) consist of a diffraction grating (R approximately 90%) and an OC. To reduce the spectral width of the output (without resorting to the use of an etalon), should the OC be set for a high (90%) or low (50%) reflectivity? Calculate the linewidth predicted in both cases. (The high gain in this laser, which is pumped by another laser, will allow oscillation using either OC.)

6.5 Compare the characteristics of a solid quartz etalon of thickness 8 and 25 mm. Which is most suitable for single-wavelength operation of (a) an argon laser and (b) a HeNe laser?

6.6 Which of the following cavities is considered stable?

(a) A 1-m cavity consisting of two concave ($r = 50$ cm) mirrors

(b) A 0.75-m cavity consisting of one flat mirror and one concave ($r = 50$ cm) mirror

(c) A 0.5-m cavity consisting of one flat mirror and one concave ($r = 50$ cm) mirror

(d) A 1-m cavity consisting of one concave ($r = 0.75$ m) and one convex ($r = -2$ m) mirror

CHAPTER 7
Fast-Pulse Production

Pulsed solid-state lasers are among the most important and widely deployed commercial laser systems. The Nd:YAG laser, for example, is the most popular commercial laser for applications such as marking, cutting, drilling, range finders, and retinal surgery. For many of these applications, the shorter the pulse, the better. Fast, powerful pulses tend to ablate material quickly without heating (and potentially altering) surrounding material or tissue. As an example, consider the precision trimming of thick-film resistors as outlined in Figure 7.0.1. The photo on the left shows a thick-film circuit on a ceramic substrate in which resistors are fabricated using screen-printing techniques, then trimmed to the correct value using a YAG laser. The photomicrograph on the right side shows the actual cut made by the laser. Fast pulses will quickly vaporize resistor material in a controlled manner, increasing the resistance of the structure by a controlled amount. A CW laser beam or one with a long pulse would tend to heat the material surrounding the target area and either damage the substrate or alter the resistor in an uncontrolled way by changing the properties of the resistor film. For this reason a technique called Q-switching is used to produce very short pulses for these types of applications. A number of techniques are used to produce fast pulses, as discussed in this chapter.

7.1 CONCEPT OF Q-SWITCHING

The most simplistic method that can be envisioned to produce a pulsed laser is to switch the gain of the medium on and off. This can be accomplished easily by switching the pump energy to the laser medium. In the case of a gas laser, the current through the discharge may be controlled easily, or in the case of an optically pumped laser such as a YAG, the lamp may be turned on and off as required. By switching the pump energy to the medium on and off, gain inside the laser is also switched on and off. When pump energy is sufficient to allow laser gain to exceed the threshold, an output beam appears. The problem with this scheme is that the output pulses will be quite rounded, since there is a delay as population inversion and hence gain builds in the laser; this also sets limits on the pulse length and repetition rate for the laser.

Fundamentals of Light Sources and Lasers, by Mark Csele
ISBN 0-471-47660-9 Copyright © 2004 John Wiley & Sons, Inc.

194   FAST-PULSE PRODUCTION

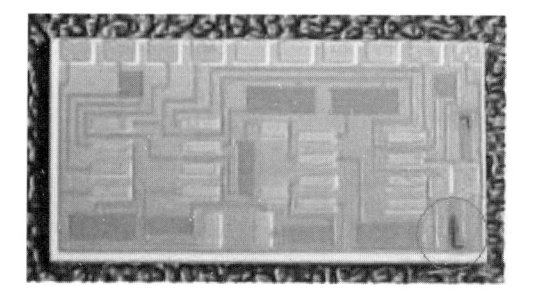

Figure 7.0.1   Precision thick-film resistor trimming by a YAG laser.

This technique is commonly used with semiconductor lasers, in which the laser diode current is modulated to control the laser output.

In a *Q-switching technique*, the laser output is switched by controlling loss within the laser cavity as outlined in Figure 7.1.1. More correctly, Q-switching is *loss switching* in which a loss is inserted into the cavity, thus spoiling it for laser action. In the simplest manner, a Q-switch can be though of as an optical gate blocking the optical path to one cavity mirror and hence causing laser action to cease. In reality, it is not necessary to block the optical path completely. Simply inserting a loss high enough to raise the lasing threshold beyond the maximum gain of the laser is sufficient. When the Q-switch is on, it blocks the intracavity beam. This state is called a *low-Q state*, meaning that the quality factor, or Q, of the cavity (which measures the ability of a laser cavity to act as a resonator) is ruined. With the switch off, losses in the cavity are reduced, and the cavity is resonant—a high-Q state. During the time when the Q factor of the cavity is low, the laser is not oscillating and hence has no usable output beam, but the pump energy continues to drive the laser medium. In this state population inversion continues to build and an inversion much larger than normally possible builds in the laser gain medium. When the Q of the cavity is restored in such a condition, the laser begins to oscillate immediately and the large population inversion releases energy in the laser medium in one enormous pulse. This technique allows energy to be stored in the laser medium during the low-Q state and released in a single massive pulse. The peak power of the pulse is much larger than is possible with gain switching, which for a CW laser, yields a peak power equal to the CW output power of the laser.

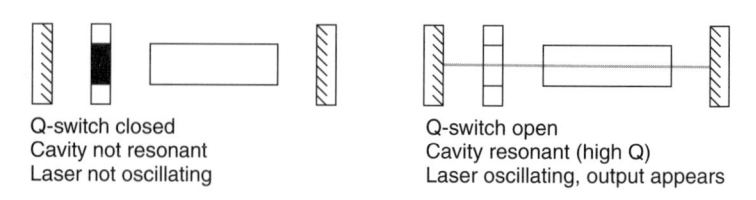

Q-switch closed
Cavity not resonant
Laser not oscillating

Q-switch open
Cavity resonant (high Q)
Laser oscillating, output appears

Figure 7.1.1   Q-switching a laser.

7.2 INTRACAVITY SWITCHES

A Q-switch consists of a mechanism that spoils the resonance of the laser cavity. There are a number of ways to accomplish this, from altering the alignment of a cavity mirror mechanically (e.g., by a rotating mirror), insertion of an optical switch within the cavity of the laser itself (e.g., an EO or AO modulator), or a saturable dye switch within the cavity. These methods are outlined in Figure 7.2.1, in which each type of Q-switch is shown within in a solid-state laser cavity.

A rotating mirror, the simplest method, is rarely used except in a few older military rangefinders that use ruby rods. A high-speed motor drives the mirror, which, once per revolution, aligns with the other cavity mirror. Just prior to this alignment, the optical pump (usually, a flashlamp) is fired to ensure that a massive population inversion has built up before the cavity is aligned. The shortcoming of this method is that the experiment must be synchronized to the laser, which fires at repeated intervals. Control over when the laser fires is not possible. As well, the pulse is not as formed as other Q-switching methods since the Q of the cavity does not suddenly jump to a maximum value but rather, increases as the mirror rotates and the mirror becomes mechanically close to the optimal alignment position.

Electrooptic and acoustooptic modulators work quite literally as optical switches, allowing intracavity light to pass only when the switch is open. These are controllable and allow the laser to be fired at the operator's command, as required by many experiments. These modulators, the most popular for Q-switching applications, are discussed in detail in the next few sections of this chapter.

Saturable dye switches are simply a cell filled with organic dye (similar to the dye used in a dye laser) placed inside the laser cavity. These work by absorbing intracavity radiation, inducing a large loss in the cavity, until the dye becomes saturated or bleached, at which point it cannot absorb more laser energy and the remaining light passes through the cell to be amplified in the laser (i.e., the laser begins to oscillate). The time it takes to saturate a dye is characteristic of the dye itself and depends on the lifetime of the energy levels in the dye molecule itself.

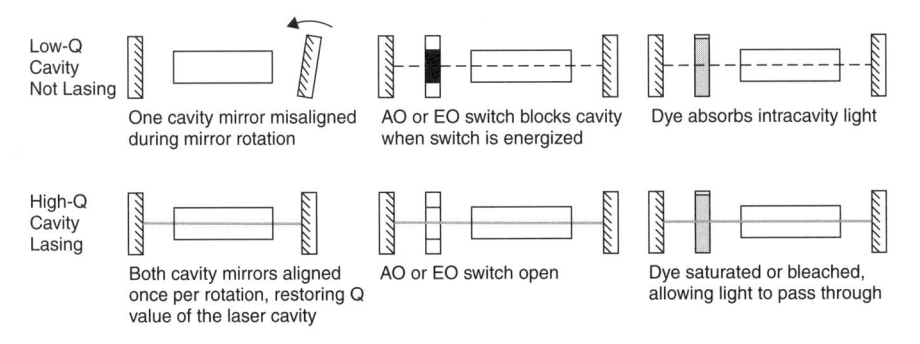

Figure 7.2.1 Q-switching methods.

7.3 ENERGY STORAGE IN LASER MEDIA

The entire concept of Q-switching relies on the fact that energy can be stored in the lasing medium itself in the form of an excited atomic population at the ULL. The very definition of cavity quality factor (or Q factor) basically defines the situation:

$$Q = 2\pi \times \frac{\text{energy stored in the cavity}}{\text{energy lost per cycle}} \qquad (7.3.1)$$

A large Q factor represents a low-loss resonator that can store a large amount of energy. In Q-switching the Q of the cavity is spoiled (the Q factor is purposely made low), so that it is not resonant and hence lasing is not possible. Energy storage takes place, but rather than within the cavity as for optical energy, energy is stored in the atomic population.

In a Q-switched laser the laser medium itself is used as a sort of capacitor, storing energy gradually and releasing it in a single burst. Not surprisingly, the capacity of the medium to store energy depends on the lifetime of the upper lasing level (ULL). A long lifetime implies that the lasing species can absorb energy over a long period without losing it to spontaneous emission; hence it has a large storage capacity and is a good candidate for Q-switching.

Suppose a laser cavity is blocked, so that laser action cannot occur. Unlike previous examples of in this book, CW lasers, where the level of population inversion reaches an equilibrium level, the inversion is now free to build. The primary limit on this inversion is the lifetime of the upper lasing level, which serves to deplete the upper level through spontaneous emission. With continuous pumping the inversion builds until reaching a level equal to the rate of pumping times the lifetime of the upper level. The population rate equation during this interval when the cavity is blocked becomes

$$r_{\text{inversion}} = r_{\text{pumping}} - \frac{\Delta N(t)}{\tau_{\text{ULL}}} \qquad (7.3.2)$$

where $r_{\text{inversion}}$ is the rate at which the inversion builds, r_{pumping} the rate at which the laser is pumped, and $\Delta N(t)$ the population difference at any time t. The solution to this equation at any time t becomes

$$\Delta N(t) = (r_{\text{pumping}} \tau_{\text{ULL}}) \left[1 - \exp\left(-\frac{t}{\tau_{\text{ULL}}} \right) \right] \qquad (7.3.3)$$

The solution to this equation is plotted in Figure 7.3.1, which depicts how the population builds in time. After pumping for one τ_{ULL} cycle, a population inversion of 66% has built up. After two such time periods, the inversion is at 88% of its maximum value. It is clear that in such a Q-switched laser it is not productive to pump the

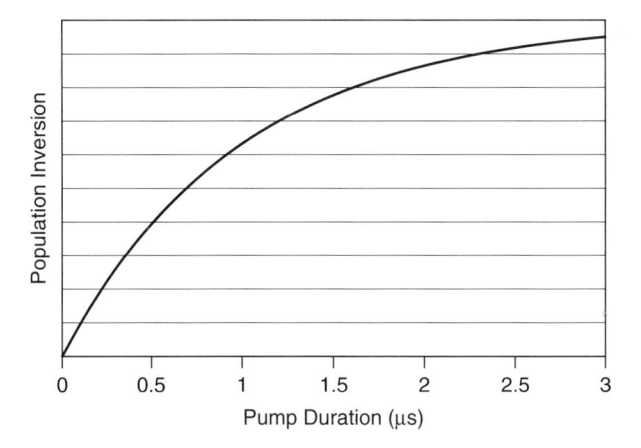

Figure 7.3.1 Energy storage in Q-switched media.

laser medium with energy beyond two or three times the lifetime of the upper lasing level. After this period the population inversion does not really grow much but rather, slowly approaches (and never reaches) the maximum level, as shown in the figure.

Excess pump energy is essentially wasted. Still, in many lasers, especially those driven by continuous sources such as CW arc lamps, the time during which the laser is pumped may be very long. In this case the Q-switch is also used to allow the laser to fire on command and so when not actively firing, the pump source (the lamps) is kept on, so that the laser is always in a ready state. Implications are that any laser medium may be Q-switched, but it should be evident that the lifetime of the upper lasing level (ULL) must be much longer than the opening time for the Q-switch. For most gas lasers with ULL lifetimes of under a 1 ns, Q-switching simply will not work, since no practical switch exists that can open in considerably less than this time. One of the few exceptions is the carbon dioxide gas laser, which has an exceptionally long ULL for a gas laser, although in practice, Q-switching of a CO_2 laser is rarely done.

Solid-state lasers, on the other hand, invariably feature long ULL lifetimes. The ruby laser, for example, has a lifetime of 3 ms, and the YAG laser has a lifetime of 1.2 ms. These are ideal candidates for Q-switching, which is a standard option on most solid-state lasers. An interesting practical demonstration of energy storage in solid-state laser media is in the setup procedure of a double-pulse ruby laser for holography. These lasers can output two short (often around 10 ns) pulses spaced a selectable time interval (usually in microseconds) apart. Since flashlamp pulses are quite long compared to the spacing of the output pulses (and flashlamps have a relatively long recovery time, so that they cannot be fired twice in rapid succession), a laser such as this uses a single flashlamp pulse to produce both output pulses by opening and closing the Q-switch twice, rapidly, during the pumping interval. When the pulses are spaced widely apart in time ("widely" being defined as a few

198 FAST-PULSE PRODUCTION

hundred microseconds) the rod loses stored energy in the first pulse but is re-pumped by absorbing pump light once the Q-switch is closed. Energy in the rod builds once again, so the pulses are somewhat independent of each other. When the two pulses are only microseconds apart, though, pump light is insufficient to repump the rod during the time between the pulses, so the total energy stored in the rod is divided between the two output pulses. If the switch is opened fully for the first pulse, it can deplete the energy stored in the rod, resulting in a small or nonexistent second pulse. To produce two pulses of equal power, the Q-switch must be opened only partially for the first pulse, closed again, and opened fully on the second pulse to utilize all stored energy remaining in the rod. Not all Q-switches are capable of opening partially (with EO types, discussed below, being employed for essentially all lasers of this type). This process of balancing pulses is usually done by setting the Q-switch to open only slightly for the first pulse and fully for the second pulse, test firing the laser, and observing the relative output power of each pulse. By opening the Q-switch progressively more and more for the first pulse and test firing the laser between adjustments, it will be seen that the first pulse robs power from the second pulse, until they eventually match in amplitude and the balancing is complete.

7.4 PULSE POWER AND ENERGY

It becomes evident, then, that at the moment the switch is opened, the initial inversion (ΔN_i) is many times larger than the threshold inversion (ΔN_{th}), which a continuous laser cannot normally exceed. We begin a mathematical analysis of a Q-switched laser by considering the number of coherent photons inside the cavity (per unit volume) at any given time, denoted n. More specifically, we consider the rate at which the number of photons increases or decreases in the cavity as

$$\frac{dn}{dt} = \frac{n}{\tau_c} + \Delta N W_{\text{pump}} \qquad (7.4.1)$$

where τ_c is the lifetime of a photon in the cavity (from Section 6.7), ΔN is the population inversion at any given time t, and W_{pump} is the pumping probability. The left side of the equation (n/τ_c) represents the rate of photon loss in the cavity—simply the number of photons divided by the lifetime of a photon—and the right side of the equation represents the rate of photon gain—simply the inversion times the pumping probability. We may now simplify the equation knowing that the pump rate can be expanded according to

$$W_{\text{pump}} = \frac{n}{\Delta N_{th} \tau_c} \qquad (7.4.2)$$

where ΔN_{th} is the threshold population inversion. Equation (7.4.1) then simplifies to

$$\frac{dn}{dt} = \frac{n}{\tau_c} + \frac{\Delta N n}{\Delta N_{th} \tau_c}$$

$$= \frac{n}{\tau_c}\left(1 + \frac{\Delta N}{\Delta N_{th}}\right) \qquad (7.4.3)$$

which expresses the rate of change of the number of photons in the cavity. Just before the switch is opened, $\Delta N/\Delta N_{th}$ assumes a large value and n is essentially zero.

To solve the problem, two equations are required. For the second we now revisit the rate equations for a laser to generate an expression for the rate of change ΔN. From Chapter 5 you will recall the development of rate equations for the ULL of two-, three-, and four-level atomic systems, but (as mentioned in Section 5.9) in a real laser, stimulated processes are also at work. In a three-level laser, for example, the rate equation for the ULL becomes

$$\frac{dN_2}{dt} = \frac{N_3}{\tau_{32}} - \frac{N_2}{\tau_{21}} - W\Delta N \qquad (7.4.4)$$

which is simply equation (5.7.2) with added terms for stimulated processes where the first term represents input from the pump level; the second term, spontaneous decay. We may develop an exceedingly simply expression for $d\Delta N/dt$ (the rate of change of the inversion) by making a few assumptions: namely, that the Q-switch pulse occurs so quickly that the effects of decay from the pump level (the first term) as well as the effects of spontaneous emission (the second term) are quite negligible during the time the actual pulse is generated. We also note that for a three-level laser a loss of one atom from the ULL results in a gain of one atom in the LLL, so the total change in the difference (ΔN) is 2. The answer is then simply

$$\frac{d\Delta N}{dt} = -2W\Delta N \qquad (7.4.5)$$

Further substitution for W using equation (7.4.2) yields

$$\frac{d\Delta N}{dt} = \frac{-2\Delta N n}{\Delta N_{th} \tau_c} = -2\frac{n}{\tau_c}\frac{\Delta N}{\Delta N_{th}} \qquad (7.4.6)$$

which allows us to simply divide equation (7.4.3) by equation (7.4.6) to form an expression for the change in the number of coherent photons with respect to the change in population inversion:

$$\frac{dn}{d\Delta N} = \frac{\Delta N/\Delta N_{th} - 1}{-2\Delta N/\Delta N_{th}}$$

$$= \tfrac{1}{2}\left(-1 + \frac{\Delta N_{th}}{\Delta N}\right) \tag{7.4.7}$$

This differential equation may be mathematically integrated[1] to yield an answer in terms of the number of coherent photons:

$$n = \tfrac{1}{2}\Delta N_{th} \ln \Delta N - \tfrac{1}{2}\Delta N + k \tag{7.4.8}$$

where k is a constant. To solve for the constant, we apply initial conditions in which there is no output (i.e., $n = 0$) just before the switch opens and the inversion is at its initial, peak value of $\Delta N_{initial}$:

$$\tfrac{1}{2}\Delta N_{th} \ln \Delta N_{initial} - \tfrac{1}{2}\Delta N_{initial} + k = 0$$

$$k = -\tfrac{1}{2}\Delta N_{th} \ln \Delta N_{initial} + \tfrac{1}{2}\Delta N_{initial} \tag{7.4.9}$$

so that equation (7.4.8) assumes the final form

$$n = \tfrac{1}{2}\Delta N_{th} \ln \Delta N - \tfrac{1}{2}\Delta N + k$$

$$= \tfrac{1}{2}\Delta N_{th} \ln \Delta N - \tfrac{1}{2}\Delta N - \tfrac{1}{2}\Delta N_{th} \ln \Delta N_{initial} + \tfrac{1}{2}\Delta N_{initial}$$

$$= \tfrac{1}{2}\Delta N_{th} \ln \frac{\Delta N}{\Delta N_{initial}} - \tfrac{1}{2}(\Delta N - \Delta N_{initial}) \tag{7.4.10}$$

We have now solved equation (7.4.10) for the number of photons per unit volume in the cavity [the reader is reminded, though, that n is actually a function of time, so that, most properly, it should be written as $n(t)$]. We also presumably know the volume of the cavity and the energy of each photon ($h\nu$), so that the output power for the Q-switched pulse can be computed as

$$P_{output} = \text{OC transmission} \times \text{energy per photon}$$
$$\times \text{number of photons} \times \text{cavity loss per unit time}$$

$$= (1 - R_{OC})h\nu n V \frac{1}{\tau_c} \tag{7.4.11}$$

---

[1] A huge help here are calculus substitutions provided in the *CRC Standard Mathematical Tables* (Boca Raton, FL: CRC Press), which provides a number of time-saving calculus and trigonometric identities.

You will recall from Section 6.7 that τ_c represents the total distributed losses (of coherent photons) in the cavity. This expression, then, gives the instantaneous power of the pulse (in watts) at any time t. The total energy of the pulse can be determined by integrating the power over the entire time frame of the pulse:

$$E_{\text{pulse}} = \int \left((1 - R_{\text{OC}})h\nu n V \frac{1}{\tau_c}\right) dt \qquad (7.4.12)$$

Since only n is a function of time, the calculus is straightforward, but the time dependence of ΔN must then be determined. The integration may be made easier by realizing that we may integrate over ΔN since we know that it starts at value $\Delta N_{\text{initial}}$ and ends at some terminal value ΔN_{final} (for which we may, for many practical purposes, substitute $\Delta N_{\text{threshold}}$, especially if pumping is strong enough so that the laser continues to operate in CW mode after the switch is opened and the pulse is produced). The integral then becomes

$$E_{\text{pulse}} = 2 \int_{\Delta N_{\text{initial}}}^{\Delta N_{\text{threshold}}} \left\{ \left[(1 - R_{\text{OC}})h\nu n V \frac{1}{\tau_c}\right] \frac{dt}{d\Delta N} \right\} d\Delta N \qquad (7.4.13)$$

where the "2" in front corrects for the fact that the pulse is only half over when the threshold value is reached (assuming a symmetrical pulse). The term $dt/d\Delta N$ can then be substituted for the inverse of equation (7.4.6) (also in terms of ΔN). Constant terms (such as R_{OC}, $h\nu$, and threshold inversion ΔN_{th}) are pulled out of the integral and several terms cancel, to yield a simple integral:

$$E_{\text{pulse}} = 2 \times \tfrac{1}{2}(1 - R_{\text{OC}})Vh\nu \int_{\Delta N_{\text{initial}}}^{\Delta N_{\text{threshold}}} \left(\frac{\Delta N_{\text{th}}}{\Delta N}\right) d\Delta N$$

$$= (1 - R_{\text{OC}})Vh\nu\Delta N_{\text{th}} \int_{\Delta N_{\text{initial}}}^{\Delta N_{\text{threshold}}} \left(\frac{1}{\Delta N}\right) d\Delta N$$

$$= (1 - R_{\text{OC}})Vh\nu\Delta N_{\text{th}} \ln \frac{\Delta N_{\text{initial}}}{\Delta N_{\text{threshold}}} \qquad (7.4.14)$$

An exceedingly simple result! Bear in mind that assuming pulse termination at the threshold inversion and then multiplying by 2 is indeed a simplification of the problem, since real Q-switched pulses usually have a longer tail. The estimate provided here will yield an answer somewhat lower than the actual output, although results will vary according to the ratio of $\Delta N_{\text{initial}}/\Delta N_{\text{threshold}}$, but generally it is found to be within 20% of experimental results.

Peak power is of particular interest for a Q-switched laser since this is the reason for using the technique in the first place. At peak power, the rate of change of the number of photons will be zero (i.e., at the top of the peak itself, where the curve is essentially horizontal). Setting $dn/dt = 0$ in equation (7.4.3), we find that the peak power occurs when $\Delta N = \Delta N_{th}$ [of course, we knew this already since we used it as a basis for the simplification in equation (7.4.13)]. By substituting this value into equation (7.4.10), we can obtain an exact expression for the number of coherent photons per unit volume at the peak of the pulse. Further simplification is possible if we assume that $\Delta N_{initial}$ is much, much larger than ΔN_{th} (logical since this is required to produce giant pulses), so we can neglect many terms in the result to find that

$$n_{peak} = \tfrac{1}{2}\Delta N_{initial} \qquad (7.4.15)$$

Substitution into equation (7.4.11) is straightforward and yields an answer of

$$P_{peak} = (1 - R_{OC})h\nu n V \frac{1}{\tau_c}$$

$$= \tfrac{1}{2}(1 - R_{OC})h\nu \Delta N_{initial} V \frac{1}{\tau_c} \qquad (7.4.16)$$

Using previous results, it is also possible to characterize the width of the pulse as

$$t_{pulse} = \frac{E_{pulse}}{P_{peak}} \qquad (7.4.17)$$

Once again, this is an approximation. Exact solutions are possible, but these require a good deal of calculus.

7.5 ELECTROOPTIC MODULATORS

Electrooptic modulators rely on the phenomenon of *birefringence*, as induced in a crystal by the application of an external electric field. Birefringence is an effect exhibited in certain crystalline materials in which an incident light ray will separate into two rays that may travel in different directions, the direction traveled being dependent on that ray's polarization. For each of the two rectangular states of polarization (perpendicular and parallel), the light will travel in a different direction as it exits the crystal. This effect is also called *double refraction*, since light entering the crystal is refracted into two different directions. Calcite is a natural crystal that exhibits birefringence. This crystal will separate an incident beam of light into two beams with different polarizations.

Figure 7.5.1 shows the effect of birefringence on a beam of unpolarized light that is incident on the surface of a birefringent crystal at an angle θ. The two different components of polarization (perpendicular and parallel) take two separate paths through the crystal. This effect is caused by the index of refraction of the crystal, which has the property of being dependent on the polarization of the incident beam. For such a crystal, two indexes of refraction exist, one for each polarization of the incident light. Having different indexes of refraction, each polarization of

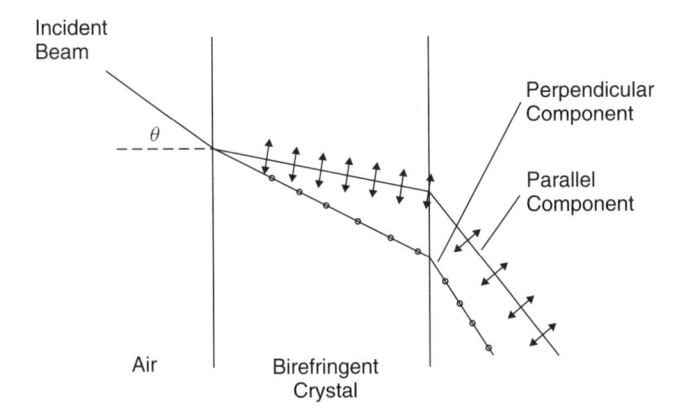

Figure 7.5.1 Birefringence.

light is refracted at a different angle inside the crystal. In the case of calcite, the indexes of refraction are 1.66 for light in the perpendicular plane and 1.49 for light in the parallel plane. Knowing the index of refraction for each component, the behavior of this component in the crystal may be determined using Snell's law.

There are also a number of crystals that exhibit birefringence only when an external electric field is applied. This phenomenon, called the *electrooptic effect*, may be used in the case of a Q-switch to block an intracavity light beam. It may also be used to deflect beams. An electrooptic modulator is shown schematically in Figure 7.5.2. Incident light is polarized and then passed through the electrooptic material (this polarization is not needed if the incident light is already polarized). Light exiting the crystal is passed through a second polarizing filter, called an *analyzer*. Since the crystal causes no change to the light passing through it when not energized, the analyzer, oriented at 90° to the polarizer, prevents any light from being transmitted through the device. In essence, when the crystal is not energized, the device acts as a shutter. When an external electric field is applied to the crystal, the direction of the polarization of light passing through it is rotated by 90°. Then the light will pass through the analyzer; effectively the shutter is open.

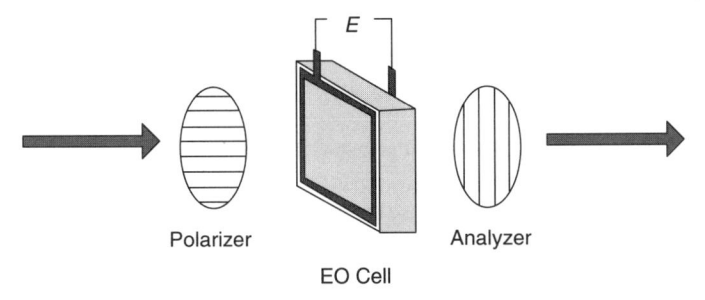

Figure 7.5.2 Electrooptic modulator.

In a modulator, the two components (with different polarizations) travel through the crystal in the same direction; they do not separate from one another. However, the relative phase of each component changes. Since the index of refraction is different for each polarization, they race through the crystal at different velocities and so emerge from the crystal with different phases. As they continue through the crystal, the phase difference increases. This depends on the thickness of the crystal as well as on the electric field applied. When the beams emerge from the crystal, the polarization of the combined single beam depends on the accumulated phase difference between the two individual polarized components. A phase difference of one-half wavelength results in the overall polarization of the beam being rotated by 90° from its original direction. When this polarization of light falls on the analyzer, it passes through.

Two electrooptic effects are possible, classified according to how the refractive index changes relative to the applied electric field: the Pockels effect and the Kerr effect. Which effect occurs depends on the material involved. In the *Pockels effect* the change in refractive index is a linear function of an applied electric field E:

$$\Delta n = a_1 E \qquad (7.5.1)$$

where a_1 is the linear electrooptic coefficient. The coefficient a_1 is a property of a material and is zero in any noncrystalline substance, so Pockels EO modulators are always constructed using a crystal, such as those listed in Table 7.5.1, which also details the range of index of refraction possible for each polarization of light.

The other electrooptic effect, in which the change in refractive index is a function of the square of the applied electric field, is the *Kerr effect*, in which the change of refractive index is described by

$$\Delta n = a_2 E^2 \qquad (7.5.2)$$

where a_2 is a second-order electrooptic coefficient, or, more generally as

$$\Delta n = \lambda K E^2 \qquad (7.5.3)$$

where K is the Kerr coefficient, a property of a particular material. All materials exhibit the Kerr effect, although the effect is very small in many materials, so an extremely high electric field may be required to observe the effect. This may not always be physically possible since dielectric breakdown will occur in any material

TABLE 7.5.1 Common Pockels Cell EO Materials

Material	n_1	n_2
Ammonium dihydrogen phosphate (ADP)	1.51	1.47
Potassium dihydrogen phosphate (KDP)	1.51	1.47
Lithium niobate	2.23	2.16

when too high a voltage is applied. Common materials used for Kerr-type EO modulators include the liquid nitrobenzene (which is toxic) and glass. Kerr-type modulators are uncommon in laser use, and essentially all modern modulators use the Pockels effect. These are often called *Pockels cell modulators*.

In a practical EO modulator an electrooptic crystal (ADP or KDP in this example) is sandwiched between two electrodes. The crystal is then inserted between a pair of crossed polarizing filters in Figure 7.5.2. By applying an external electric field in the form of a direct-current potential of several thousand volts across the crystal (parallel to the optical axis of the crystal), the refraction of the crystal itself changes. The electric field E (in units of V/m) can be found simply by dividing the applied voltage by the thickness of the cell (the distance over which the voltage is applied).

The phase change for light passing through a modulator is proportional to the thickness of the EO material as well as the change in refractive index generated by the electric field according to

$$\Delta\phi = \frac{2\pi \Delta n L}{\lambda} \tag{7.5.4}$$

which allows us to express the transmission of the modulator (T) according to the change in refractive index according to

$$T = T_0 \sin^2\left(\frac{\pi \Delta n L}{\lambda}\right) \tag{7.5.5}$$

where T is the transmission, T_0 the maximum transmission of the assembly when open, Δn the difference in refractive index for the two polarizations, L the length of the crystal, and λ the wavelength of the light. The birefringence is a function of the applied electric field and hence the voltage applied to the crystal. Mathematically, it can be shown that the maximum transmission occurs when the term $\pi \Delta n L/\lambda = \pi/2$ or when

$$\Delta n = \frac{\lambda}{2L} \tag{7.5.6}$$

This occurs at a voltage called the *half-wave voltage*, denoted $V_{1/2}$. The half-wave voltage is a property of the material itself, but this voltage increases with the wavelength according to

$$\frac{V_{1/2-\text{wavelength 1}}}{V_{1/2-\text{wavelength 2}}} = \frac{\lambda_1}{\lambda_2} \tag{7.5.7}$$

EO modulators operating in the infrared thus require much higher voltages than those operating in the visible region (which often operate at voltages of 5 to 10 kV). This factor can limit the use of EO modulators in infrared lasers (such as a CO_2 laser operating at 10.6 μm).

Example 7.5.1 Pockels Cell Voltage Consider a Pockels cell in which the phase shift is given by equation (7.5.4):

$$\Delta\phi = \frac{2\pi\Delta n L}{\lambda}$$

We can substitute for Δn in terms of the half-wave voltage so that in terms of applied voltage the phase shift becomes

$$\Delta\phi = \frac{\pi V}{V_{1/2}}$$

Now, by substituting for $\Delta\phi$ in equation (7.5.5), the relationship of transmission to applied voltage becomes

$$T = T_0 \sin^2\left(\frac{\pi}{2}\frac{V}{V_{1/2}}\right)$$

Hence by applying a variable voltage to the cell, we can modulate the transmission to any value desired. Such techniques have been used to transmit high-speed signals such as video via a laser beam.

Electrooptic modulators are the fastest form of Q-switches for lasers and feature switching times of less than 10 ns. They are very repeatable and have a large *hold-off* or *extinction ratio*. This is the ratio of light passing through the modulator when it is on (open) to when it is off (blocked). For an EO modulator this ratio can be as high as 1000 : 1, which translates into a huge insertion loss when used as a Q-switch inside a laser cavity. In the blocked state the Q of the laser cavity will be extremely slow. The EO modulator is so fast that it is often used as a shutter for ultrafast cameras: for example, those used to photograph atomic bomb blasts. Unfortunately, suitable electrooptic crystals are very expensive, and the electrical driver circuitry, which must produce a high-voltage pulse with a rise time in the nanosecond range, is also critical. Circuitry to drive an EO modulator usually consists of a high-voltage capacitor discharged quickly through the modulator using a thyratron (a high-speed switch). Still, when a Q-switched laser is required with nanosecond pulse times (e.g., a double-pulse ruby laser used for fast holography) an EO modulator (usually, a Pockels cell) is the only option. For less time-critical applications, a less expensive (and simpler) alternative is a modulator based on the interaction of sound with light, although this approach does not yield the performance of an EO switch.

7.6 ACOUSTOOPTIC MODULATORS

Perhaps the simplest (and most common) modulator is the acoustooptic modulator. An acoustic wave, at radio frequencies, originates from a piezoelectric crystal which

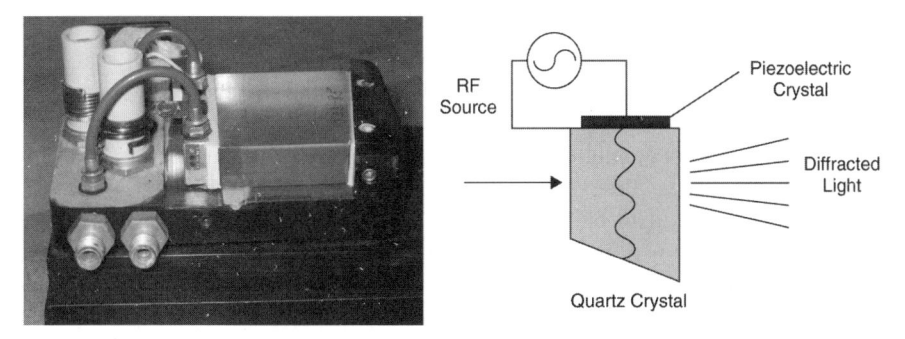

Figure 7.6.1 Acoustooptic modulator.

generates a surface acoustic wave in a crystal as outlined in Figure 7.6.1. The waves generated by the piezoelectric crystal form almost entirely flat plane wavefronts in the crystal. This wave is transmitted through a quartz crystal and induces a strain in the crystal which changes its refractive index on a localized scale. Like a sound wave propogating through air, the acoustic wave creates rarefactions and compressions that change the density of the crystal. This is a photoelastic effect in which the strain distorts bonds between atoms of the crystal. The end of the crystal is either cut at an angle or terminated with an acoustic absorber so that a standing wave is not generated in the crystal.

The acoustic wave, although at radio frequencies, propagates through the crystal at the speed of sound, so the wavelength of the acoustic wave inside the crystal is

$$\Lambda = \frac{v_{\text{acoustic}}}{f} \quad (7.6.1)$$

where Λ is the wavelength inside the crystal and f is the frequency. Thus, we have a relatively long wavelength in the crystal. Regions of high and low index of refraction will similarly exist in the crystal at these points, as outlined on Figure 7.6.2.

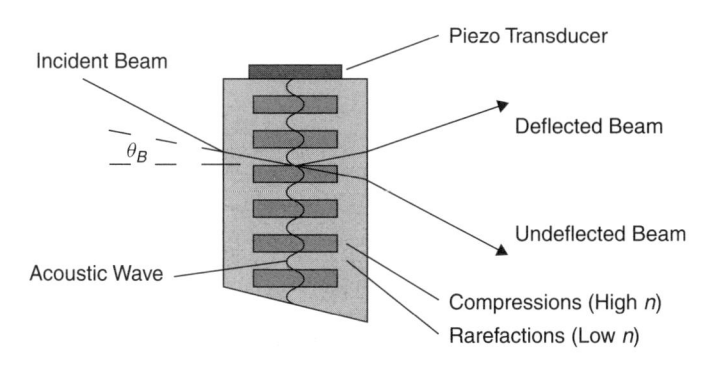

Figure 7.6.2 Acoustic waves in an AO modulator (Bragg diffraction condition).

Because these layers of varying n are stacked together, diffraction will occur and an incoming beam of light will change deflection upon passing through the crystal. Two types of diffraction may occur here, Bragg diffraction and Raman–Nath diffraction, depending on the angle of incidence of the beam to be diffracted.

In the Bragg regime, a light beam is deflected by the alternating layers of differing refractive indices. Bragg diffraction refers to a type of diffraction generated when x-rays are incident on crystals. The very short wavelengths of x-rays rule out using diffraction gratings constructed using parallel lines ruled in, for example, glass. X-ray wavelengths, however, correspond to the spacings of atoms in crystal lattices, so crystals may be used as three-dimensional diffraction gratings. In 1912, W. L. Bragg discovered that the structure of a crystal is repetitive in three dimensions, with atoms aligned in parallel planes on each of which the arrangement of atoms is the same. These parallel planes are called *Bragg planes*. Incident x-rays are reflected from such a crystal (actually diffracted) according to a simple law by Bragg.

In the case of an AO modulator, the acoustic wave inside a crystal sets up parallel planes of varying indexes of refraction, which like the parallel planes of atoms in the crystal, generate Bragg diffraction of an incident light beam. The wavelength of the acoustic wave in the crystal corresponds to the wavelength of the incident light beam just as the spacing of the atoms in the crystal corresponds to x-ray wavelengths. An incident light beam on an AO modulator crystal is deflected only when an acoustic wave is present in the crystal (i.e., it is driven by a radio-frequency (RF) source). Mathematically, this deflection can be expressed in a manner similar to that for an optical diffraction grating:

$$2\Lambda \sin \theta_B = \frac{\lambda}{n} \qquad (7.6.2)$$

where θ_B is the Bragg angle and λ is the wavelength in air. When this equation is satisfied, reflected waves from each parallel plane will be separated by a phase shift of exactly 2π, so that constructive interference takes place. The exit beam is deflected out of the crystal at the same angle as it enters, shown in Figure 7.6.2 as the *deflected beam*. As well as being useful as a modulator (by deflecting the intracavity beam out of the cavity, hence preventing amplification when the modulator is on), this type of diffraction may be used to deflect a beam and hence may be used as a beam scanner.

Example 7.6.1 Calculation of the Angle of Deflection Consider an AO modulator driven with a 28-MHz signal and made of quartz used to deflect the beam of a HeNe laser. The speed of sound in quartz is 3750 m/s and quartz has an index of refraction of 1.46. Calculate the angle of deflection caused by this modulator when used in the Bragg regime.

SOLUTION The wavelength of the acoustic wave in the quartz crystal is

$$\Lambda = \frac{v_{acoustic}}{f} = \frac{3750 \text{ m/s}}{28 \times 10^6 \text{ Hz}} = 134 \text{ μm}$$

The angle of deflection may now be found using equation (7.4.2) as

$$\sin\theta = \frac{\lambda}{2n\Lambda} = \frac{632.8 \times 10^{-9} \text{ m}}{(2)(1.46)(134 \times 10^{-6} \text{ m})}$$

The angle is hence found to be 0.093 degree. The modulator is placed within the laser cavity so that radiation within the laser cavity is incident on the plane waves at this angle. When the modulator is off, light passes through unimpeded; when on, light is deflected out of the cavity, spoiling the resonance (Q) of the laser cavity.

It must be noted that the angle of diffraction does not depend on the intensity of the acoustic wave: Increasing this parameter (by increasing the RF drive signal) simply changes the intensity of the diffracted light beam. Of course, there are physical limits on how much energy a given switch will tolerate.[2]

The second type of diffraction, *Raman–Nath diffraction*, occurs when the incoming beam is aligned perpendicular to the alternating layers, which now act like parallel slits in a transmission diffraction grating. In this regime the diffraction is modeled to occur from a thin acoustic wave beam inside the crystal. The angle of diffraction is quite different from the Bragg angle and the incoming light beam is deflected in both directions (both upward and downward) away from the plane of

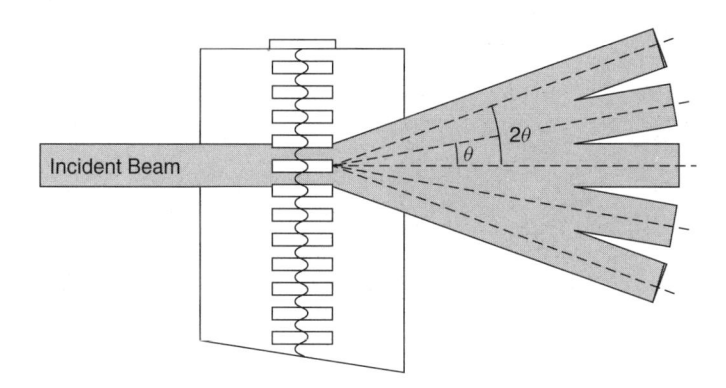

Figure 7.6.3 Raman–Nath diffraction from an AO modulator.

---

[2] Limits on the maximum drive signal that may be applied involve power dissipation in the piezoelectric crystal as well as electrical circuitry. In one test an overdriven Q-switch was observed to fail when arcing (bright sparks) were evident around the piezo crystal, resulting in smoke!

210 FAST-PULSE PRODUCTION

(a)

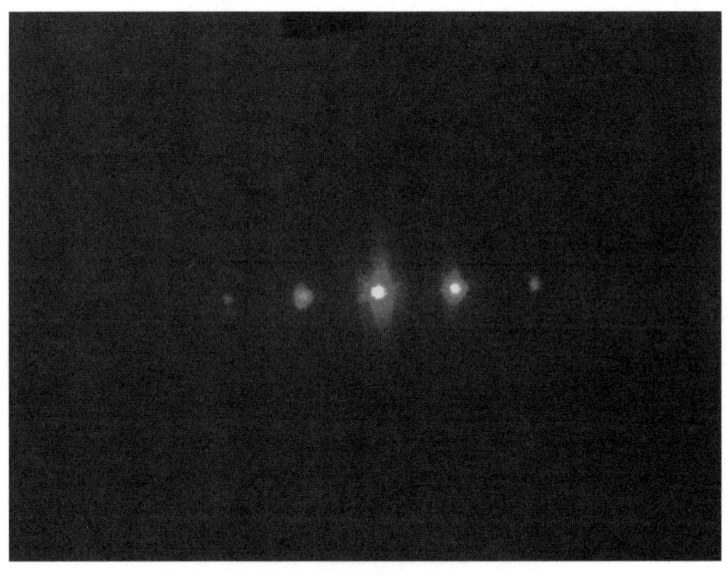

(b)

Figure 7.6.4 Diffraction with an AO modulator.

the incident light beam. Also, multiple high-order diffractions may occur as in Figure 7.6.3.

It is diffraction that allows the AO modulator in this regime to act as a switch. When inserted inside a laser cavity, the diffraction caused by the AO Q-switch causes a portion (but not all) of the intracavity beam to be deflected away from the inside of the cavity so that it is not reflected into the laser medium for further amplification, thereby spoiling laser action. The Q of the cavity falls from its normal value to a much lower value as losses increase in the cavity due to the diffraction. A practical modulator (as employed in a YAG laser) is depicted in Figure 7.6.1. In a test setup a laser beam is passed through this modulator. The modulator requires an RF drive signal at a frequency of 27 to 28 MHz and a minimum power of 10 W. Most commercial Q-switches operate at frequencies of 24, 27, or 80 MHz with drive power levels of 5 to 50 W. In this test, whose results are shown in Figure 7.6.4, the output beam (having passed through the device) is observed with the RF drive signal turned on and off. The diffraction pattern showing a strong first-order and weak second-order components is clearly visible when the RF source is energized.

As evident in Figure 7.6.4, AO modulators used in this mode have low hold-off ratios. In the case of this modulator, the intensity of the central beam is of concern since it is this component that is amplified in the laser. In the test above, the modulator was observed to cause a 10% loss in the power of the central beam when energized. In Figure 7.6.4(*b*), the brightness of the central spot with the switch energized is still significant, so this switch is by no means absolute: The blocked state (with the RF supply on) still allows a reasonably large amount of light to pass unimpeded (EO modulators have much better extinction of light when energized). This is still quite suitable for use in low-gain lasers (with less than 10% net gain per pass) such as CW lamp-pumped YAG lasers, but not in higher-gain lasers, where a 10% loss will still allow lasing to occur. When operated in the Bragg regime, the same modulator shows a higher hold-off ratio (>90%) but still not nearly as high as that of an EO modulator.

AO modulators are by far the most common for laser Q-switches. They are inexpensive (the crystal for most modulators is made of quartz), and the drive circuitry is a simple RF oscillator and power amplifier. With a suitable antireflective coating on the faces of the crystal (which is wavelength specific), the insertion loss is almost negligible, allowing it to be used with low-gain lasers.

7.7 CAVITY DUMPING

Imagine a laser in which there was no output coupler but rather, two high reflectors in the cavity. Although there would be no usable output beam, the intracavity power would be very high, much higher than in a laser that has an output coupler (which represents a loss in the cavity). Now, if one of those mirrors was suddenly removed, a massive output pulse would result. All of the energy circulating in the cavity would

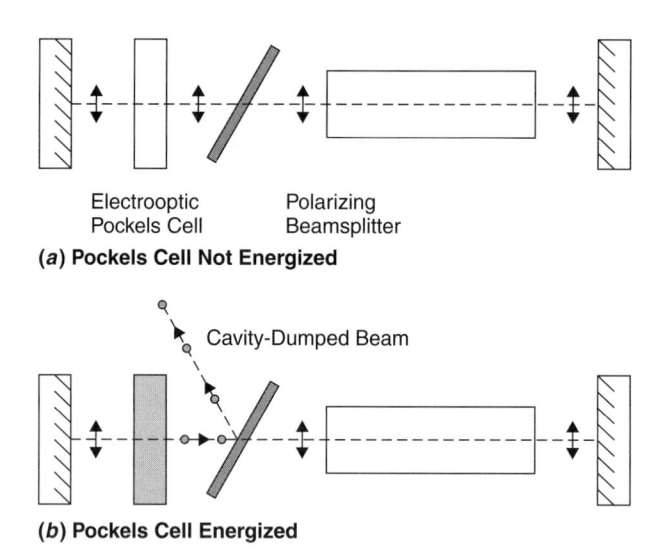

Figure 7.7.1 Cavity dumping using an EO cell.

exit in a single pulse with a pulse length equal to the round-trip time for radiation in the cavity. This is the premise for cavity dumping.

Theoretically, a mechanism to remove a mirror instantly could be used as a cavity dumper, but such an arrangement would not be practical given the large time it would take to move the mirror physically. Practically, an EO modulator may be used as a cavity dumper by changing the polarization of intracavity radiation, which is then deflected from an intracavity polarizing beamsplitter as shown in Figure 7.7.1. With the Pockels cell unenergized, intracavity light passes through it without a phase shift. Light inside the cavity is polarized such that it passes through a polarizing beamsplitter without loss, as shown in Figure 7.7.1(a). When the cell is energized, it effects a polarization change of 90 degrees, and light of this polarization is reflected from the beamsplitter in a single massive pulse. The losses imposed by the cell and beamsplitter are almost 100%, more than enough to cause lasing to cease. To repeat the process, the EO cell is unenergized, lasing begins again, and an intense intracavity light builds rapidly, ready for the next pulse to be generated. Cavity dumping techniques will work with any laser, including solid-state and gas lasers, to produce fast pulses, up to about 20 times more powerful than CW laser power. Such a technique can be done repetitively to produce a train of pulses.

7.8 MODELOCKING

Modelocking techniques are used to generate the shortest pulses of light ever produced. There are two ways to look at this technique: in the time domain and the frequency domain. Although the latter approach is required to fully understand the technique, the time domain is the simplest view and will be dealt with first to give us a basic understanding of the technique. In a modelocked laser, energy in the laser itself is compressed into a single packet of light that traverses the laser, reflecting from cavity mirrors and through the gain medium. The pulse inside the cavity is much shorter than the cavity itself: If the cavity was 1 m long, the pulse might typically be 10 cm within this cavity. If the cavity contains a partially reflecting mirror as an output coupler, a short output pulse is transmitted each time the pulse is reflected from that mirror.

As depicted in Figure 7.8.1, the modelocked laser consists of a normal Q-switched laser in which the Q-switch is opened at regular intervals corresponding to the transit time of the pulse within the cavity ($c/2L$). Once per round trip through the laser cavity the Q-switch is opened (Figure 7.8.1c) to allow the pulse to pass; at all other times the switch is closed to prevent any other light from oscillating in the cavity except for this modelocked pulse. With the modulator in the center of the cavity, it is required to open twice for each round trip of the pulse. If it is placed near one mirror in the cavity (as in the figure), only one opening of the switch is required per round trip.

The output of a modelocked laser with the configuration described is a continuous series of short pulses. In the case of a laser with mirrors 1 m apart, the pulses will appear at a frequency of $c/2L$ or 150 MHz. Pulse duration depends, among other factors, on the time for which the Q-switch is open as well as the gain bandwidth of the lasing species (which we shall examine when considering the modelocker in the frequency domain). Q-switches for a modelocked laser must open and close in a very short time period. Regular Q-switches, such as the AO modulators used for a Q-switched laser, are generally not fast enough for these purposes. Consider a typical AO modulator that can open and close in 100 ns. The total optical length

(a) a pulse travels through the amplifier

(b) the pulse approaches the closed Q-switch

(c) which opens to let the pulse through

(d) the Q-switch closes

(e) the pulse traverses the amplifier again

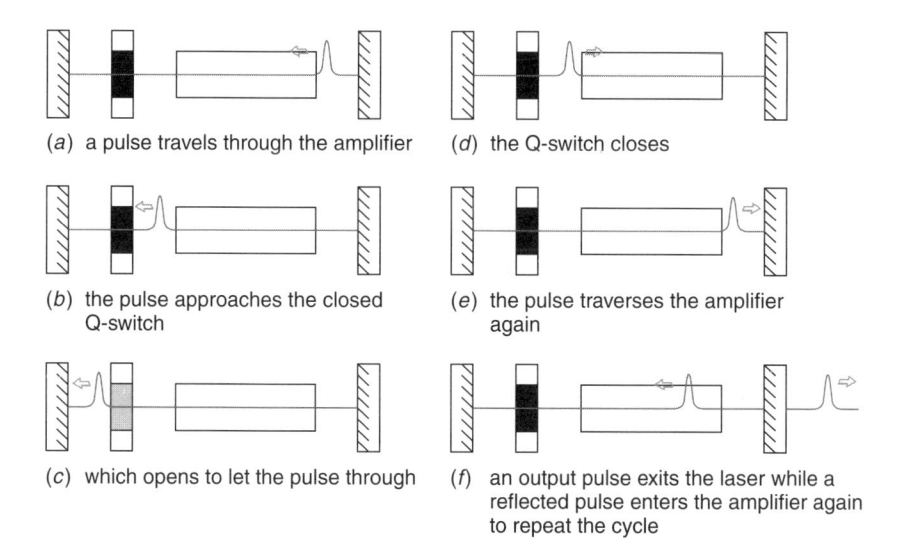

(f) an output pulse exits the laser while a reflected pulse enters the amplifier again to repeat the cycle

Figure 7.8.1 Modelock pulse development in the time domain.

of a 1-m laser is 2 m, so light makes a round trip through the entire laser in 6 ns. This modulator will clearly not work for modelocking; the modulator must open and close in much less time than the transit time for the pulse in the laser cavity. One possibility for an AO modulator, however, is to set up a standing wave in the modulator. An acoustic wave can be generated that bounces back and forth through the crystal. At two points in the period of the wave there is a point where light is not diffracted (i.e., a node in the standing wave where the electric field is zero), and hence the switch is open at those points. In our example of a 1-m cavity, such a modulator opening at a rate of 150 MHz will modelock the laser. Note that the frequency required of the modelocker increases as the cavity size decreases.

Compared to AO modulators, EO modulators have much faster opening times of 1 to 2 ns and so can be used directly as modelockers (i.e., no standing-wave scheme is required). A further advantage is that EO modulators can be inserted directly into the cavity *without* using a polarizer/analyzer filter combination as usually required when used as an optical switch. This is desirable since the inserted device does not absorb large quantities of intracavity light (which the polarizer and analyzer do absorb in a EO switch as used for Q-switching applications). In such an application, the EO modulator is in the *open state* when light passes through without phase change and in the *closed state* when it changes the phase of light passing through it. By changing the phase of light passing through the modulator (in the closed state), such light is really shifted in time with respect to unchanged light. Since only certain resonant frequencies can exist in the cavity (i.e., only those with an integral number of wavelengths that fit into the cavity), light waves that are shifted in phase cannot exist and are extinguished by destructive interference. This behavior is outlined in Figure 7.8.2, in which two waves inside the laser cavity are shown. In the figure, wave 1 passes through the EO cell without a change in phase. This wave satisfies the basic lasing criteria in that an integral number of waves fits inside the laser cavity. Wave 2 is phase-shifted by the modulator (which has varied its index of refraction in response to an external drive signal). This wave no longer satisfies the criteria for a standing wave inside the laser cavity.

The final method to modelock a laser is to use a saturable dye absorber (also used as a Q-switch). These absorbers also serve to Q-switch the laser at the same time; the output is a pulse train with a Q-switched amplitude envelope. As with a simple Q-switch, an incident pulse inside the laser cavity bleaches (saturates) the dye, which opens the switch, allowing the pulse through it.

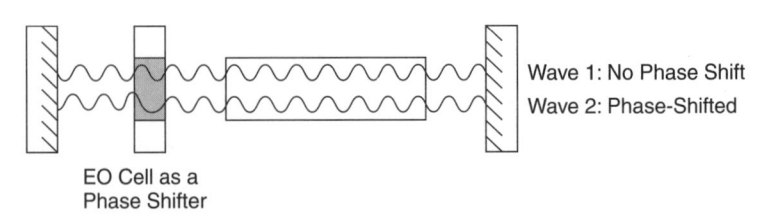

Figure 7.8.2 EO phase shifter as a modelocker.

7.9 MODELOCKING IN THE FREQUENCY DOMAIN

The time-domain view of modelocking is somewhat naive in that it does not explain how this technique literally locks the phases of various longitudinal modes in the laser together. We now take a more mathematical approach to see how this can occur. When considering a practical laser it becomes apparent that many longitudinal modes (with slightly different frequencies) can oscillate simultaneously in a cavity. If these mode components are examined at any point in the tube, you'd expect to find that each of these has a different phase relative to each of the others. In a modelocked laser, all of these components are locked together in phase with each other (so, really, modelocking is *phase locking*). Of course, the modelocked pulse is moving down the laser cavity, each component at the same speed. Where these components interfere constructively, this is the center of the modelocked pulse with a high peak power.

This occurs at all because all modes in the laser are given equal amplification (at the same time) when the switch is open, so that energy is distributed among these modes equally. Where the modes interfere constructively (i.e., they are all in phase), the modelocked pulse peaks; where they interfere destructively, the pulse is extinguished. To see how a modelocked pulse is manifested, consider the Fourier series of the light pulse in the cavity, which consists of various (longitudinal) modes. In most practical lasers the bandwidth of the gain is wide enough to allow a number of longitudinal modes, each of slightly different frequency, to oscillate simultaneously in the laser output. A previous question from Chapter 4 showed that an argon laser 90 cm long may have 31 modes in the output! This is an extreme case, but a solid-state laser such as a ruby or YAG will certainly allow many simultaneous longitudinal modes to oscillate. Modeling each wave in the laser as a sine wave, with slightly different frequencies, the amplitude of the output wave of the laser is the sum of amplitudes of each component:

$$E(t) = \cdots + E_{n-2} \sin \omega_{n-2} t + E_{n-1} \sin \omega_{n-1} t + E_n \sin \omega_n t \\ + E_{n+1} \sin \omega_{n+1} t + E_{n+2} \sin \omega_{n+2} t + \cdots \quad (7.9.1)$$

where the terms of this equation represent each component mode. For example, the component $E_n \sin \omega_n t$ represents a mode with frequency ω_n, where, say 1×10^6 standing waves exist in this particular laser cavity mode (wavelength divided by cavity length). In this case, E_n is the amplitude of the component which follows the gain bandwidth of the lasing species. Each component of the series in the equation represents a component that is a standing wave with an integral number of waves inside the laser cavity: the $(n + 1)$ component having 1,000,001 standing waves within the cavity, the $(n - 1)$ component having 999,999 waves, and so on.

The amplitudes of each of these component waves is summed (as happens where multiple waves are present) by interference. When all component waves are in phase, constructive interference occurs, producing a large amplitude. At some point these amplitudes will sum to zero, and the resulting summed amplitude will

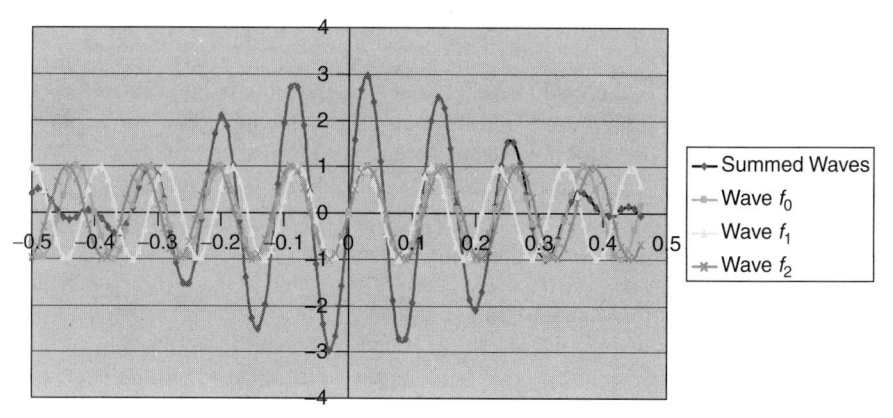

Figure 7.9.1 Modelocked pulse synthesis.

also be zero. This situation is outlined in Figure 7.9.1, where three individual modes, each with a slightly different frequency, are plotted in time. The amplitude of all three modes is also summed, revealing a pulse of large amplitude that results where the phases are the same (i.e., at the center of the figure).

By opening the Q-switch at periodic intervals, all modes in the laser are given equal amplification, which momentarily locks their phases together as shown by the central maxima in Figure 7.9.1. A fast, large pulse results: a modelocked pulse. If the switch is opened at regular intervals at a frequency of $c/2L$, the pulse continues traversing through the laser. In addition, energy builds in the upper lasing level when the pulse is not present in that area of the laser medium (in the same manner that Q-switching builds energy in the lasing medium), resulting in a high gain as the pulse passes through the laser. As the pulse is reflected from the output coupler, an output pulse results: The output from the modelocked laser thus consists of a train of extremely short pulses. A single pulse may be extracted by adding a second optical switch (such as an EO modulator) in front of the laser.

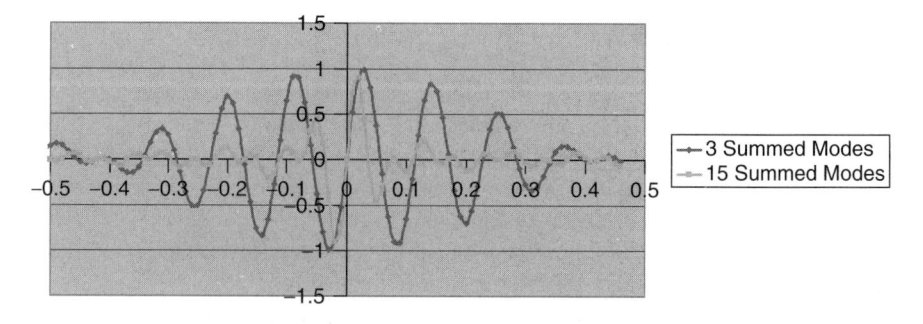

Figure 7.9.2 Effect of number of modes on pulse shape.

Considering that numerous longitudinal modes are required in the laser for this technique to work, the more modes that oscillate, the shorter the resulting pulse will be. A laser species with a wide spectral range will allow more modes to oscillate, and hence the pulse will be shorter. This situation is demonstrated in Figure 7.9.2, where the pulse shape is plotted when three and 15 component modes are summed. With three modes active, the envelope is obvious, but as more modes are summed, the envelope becomes quite narrow. In this case, the pulse appears to consist of a single wave! As more modes are allowed to exist, the pulse shape becomes shorter, which explains why the shortest pulses ever produced have been from modelocked lasers with extraordinarily wide spectral width.

For an Nd:YAG laser, the spectral width is reasonably narrow at about 0.5 nm, but Nd:YLF or Nd:glass lasers have larger spectral widths of about 5 nm, so the minimum pulse duration that may be produced is approximately 1 ps. When a laser with a larger bandwidth yet, such as a dye laser with a spectral width spanning of perhaps 100 nm, is modelocked, pulses as short as 6 fs (6×10^{-15} s) may be produced. Consider that the duration of this pulse is so short that it contains literally only a few cycles. In free space this pulse is only 9 μm long! Earlier, the impression may have been given that a narrow linewidth is always desirable; in the case of modelocking, however, a large bandwidth is required.

PROBLEMS

7.1 An AO Q-switch is set up to modulate a HeNe laser beam. RF energy at 28.0 MHz is used to drive the switch, which is made of quartz, and when energized a diffraction pattern is produced similar to that of Figure 7.6.4. It is observed that at a distance of 2.36 m from the screen, the bright dots in the pattern are 12 mm apart. Calculate the speed of the acoustic wave in the quartz crystal.

7.2 An AO modulator using tellurium dioxide is used to deflect a HeNe laser beam. RF energy is delivered to the crystal (which has an index of refraction of 2.26 at 632.8 nm and an acoustic velocity of 4200 m/s) at 80 MHz. Compute the Bragg angle for the device.

7.3 An EO modulator has an intrinsic insertion loss of 5% (hence a maximum transmission of 95%). The crystal itself is 4 mm long. A HeNe laser beam is directed through it and a voltage is applied so that the index of refraction changes by 0.0001. Calculate the transmission of the modulator.

7.4 A Q-switched YAG laser has a rod 4 mm in diameter and 50 mm in length. The cavity is 50 cm long and has mirrors of 100% and 90% reflectivity. The peak power of the laser (operating at 1064 nm) is measured to be 100 kW with a pulse width of 100 ns.

218 FAST-PULSE PRODUCTION

(a) Calculate the peak inversion of the system.

(b) If the rod has 1% doping, calculate the percentage of Nd^{3+} ions that are excited during the pumping process. (*Hint*: consider the number of ions in the rod.)

7.5 Assume that a Kerr cell modulator is constructed of a glass slide. The cell has a length of 10 mm (the optical path) and is 0.2 mm thick (across which the voltage is applied). Glass has a Kerr coefficient of 3×10^{-15} m/V^2. Calculate the applied voltage required to turn the modulator on fully.

7.6 Calculate the number of modes in a dye laser that has a 30-cm cavity length and a gain bandwidth that spans from 600 to 550 nm. Contrast this to a carbon dioxide laser of 1-m cavity length. Which will potentially produce the shortest pulse when modelocked?

CHAPTER 8

Nonlinear Optics

In 1961 it was discovered that certain materials could literally double the frequency of laser light passing through them.[1] In the demonstration outlined in Figure 8.0.1, a powerful laser beam from a ruby laser at a wavelength of 694.3 nm in the red was focused on a quartz crystal. The beam exiting the crystal was analyzed by a spectroscope that recorded spectral components of the exiting beam on a photographic plate. Most of the source light passed through the crystal unchanged, but a small portion (1 part in 10^8) was converted to light at exactly double the frequency (half the wavelength) or 347.1 nm in the ultraviolet. Since this discovery, much study has been made of nonlinear materials and techniques have been devised to exploit this phenomenon in an efficient way. Today, laser pointers are commonly available that use this technique to generate green light by doubling the infrared output from a tiny diode-pumped neodymium solid-state laser (which normally has an output in the infrared). In this chapter we examine the physics behind these nonlinear optical effects and applications.

8.1 LINEAR AND NONLINEAR PHENOMENA

Most effects in the physical world are linear. Basic physics shows us that effects such as the deflection of a steel ruler on a table under the influence of a weight (as in Figure 8.1.1) is directly proportional to the weight applied to it. If a given weight deflects the ruler 1 cm, doubling the applied weight will cause a deflection of 2 cm. The ruler is modeled as a simple spring in which the deflection is governed by the applied force according to

$$F = kx \qquad (8.1.1)$$

where F is the applied force, k the spring constant of the material, and x the deflection or displacement. Like most physical relationships, the deflection of the ruler is

---

[1] An excellent description of the original experiment by Franken et al. can be found in *Scientific American*, "The interaction of light with light," by J. A. Giordmaine, April 1964, p. 38.

Fundamentals of Light Sources and Lasers, by Mark Csele
ISBN 0-471-47660-9   Copyright © 2004 John Wiley & Sons, Inc.

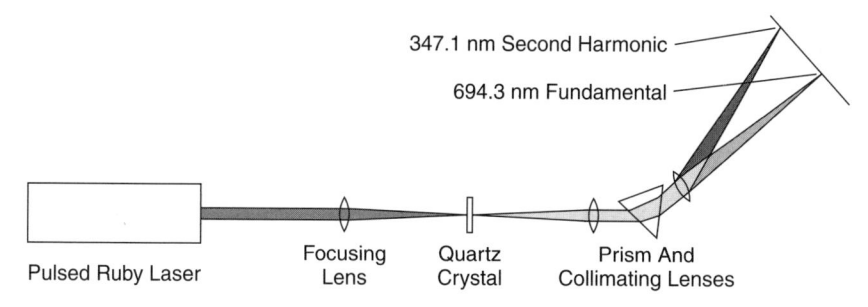

Figure 8.0.1 Generation of second harmonic light.

linear with the force applied. If, however, enough force is applied to the ruler, the deflection is found to be nonlinear and is related to the force applied by a nonlinear relationship such as

$$F = k_1 x + k_2 x^2 + k_3 x^3 + \cdots \tag{8.1.2}$$

High-order terms such as x^2, x^3, and beyond enter into the equation, which is no longer linear. It is important to note that this nonlinear behavior occurs only when large forces are applied to the material. When forces are small (i.e., the deflection of the ruler is small compared to the length of the ruler), the relationship is quite linear and governed by equation (8.1.1). As force is increased, though, stresses will be induced in the material, which will cause permanent damage (in this case, a bent ruler), but before this happens, nonlinear behavior will be seen. In a nutshell, for small forces the behavior of the system is linear, for large forces it is nonlinear, and for enormous forces the system is damaged.

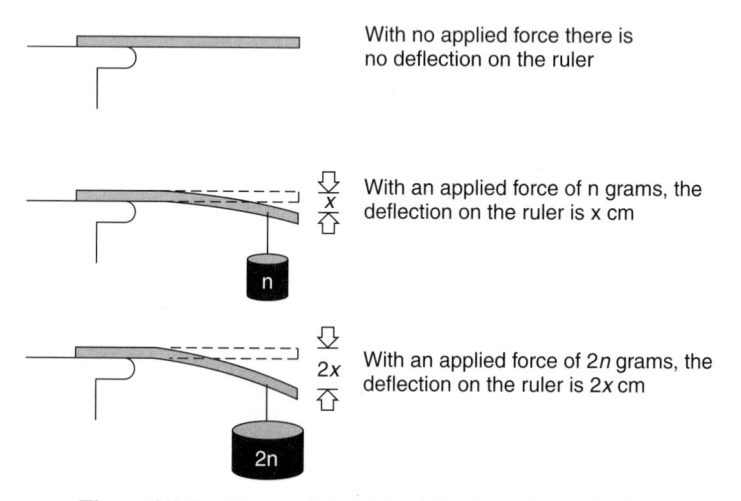

Figure 8.1.1 Linear relationship: deflection of a steel ruler.

Many optical materials behave in a similar manner; when the force applied (i.e., light intensity) is small, they are quite linear, but when a large force is applied, nonlinear behavior is exhibited. Consider the structure of a quartz crystal: a lattice with atoms placed at regular patterns. Electrons in the crystal are held to the nucleus of each atom by a force similar to a spring. When low and moderate light intensities are incident on such a crystal, electrons behave in a linear manner. Bonds stretch in response to an electric field in a linear manner.

Now that the behavior of a simple system can be defined as linear or nonlinear, we can apply the same principles to the behavior of certain materials, such as quartz, when intense laser beams travel through them and predict what happens when an electric field such as the field of an electromagnetic wave such as light interacts with atoms in a crystal. We know from an earlier chapter that the structure of the atoms in crystals consists of a nucleus with shells of electrons surrounding it (the Bohr model is sufficient to describe the atom for our purposes). The outer (valence) electrons in these atoms are more loosely bound to the nucleus than inner electrons and may even be removed (i.e., the atom ionized) when enough energy is injected into the atom.

Using the Bohr analogy, electrons are held in orbit by linear forces, and the radius of the orbit itself is governed by the same type of linear force equation that describes the deflection of the ruler in our example. When light impinges on the atom, it can cause the valence electrons to relocate in the space around the atom in step with the electric field of the incident light. Only valence electrons are affected in such an interaction because the nucleus is too heavy to be affected by an electric field, and electrons in the inner shell are bound very tightly to the nucleus. Valence electrons are held to the atom by comparatively weak forces. When an intense electric field strikes an atom, valence electrons in the shell move to one side of the atom, and the electron shell becomes eccentric and offset. Instead of a spherical shell (in which the shell is centered around the positive nucleus and hence there is no electric dipole), electrons literally move toward one end of the atom and the entire atom becomes polarized. A positive and a negative end develop on the atom, and an electric dipole with a dipole moment develops in the direction of the electric field applied.

In the presence of an applied electric field, atoms in the crystal become polarized in this manner, and the entire material becomes polarized on a localized level (around the intense beam). This is a *macroscopic charge polarization* of the material, and the amount of charge polarization depends on the intensity of the applied electric field according to

$$P = aE \quad (8.1.3)$$

where a is the coefficient of polarizability for the material and E is the applied electric field in V/cm. This is a linear relationship, which, like the previous relationship between force and deflection [equation (8.1.1)] applies for small applied forces such as those encountered everyday: the intensities of light sources such as sunlight and most artificial lights. Because of this, nonlinear effects are not seen when, for example, red light passes through quartz. In the presence of a weak electric field

222 NONLINEAR OPTICS

a polarization is produced that is *in step* with the applied electric field. This polarization and reradiation of photons results in a slowing of the velocity of light through the crystal (this in the origin of the index of refraction of the material), but no change occurs in the nature of the light passing through the crystal.

Now, when the applied force is very large, nonlinear effects are exhibited for the same reason that the ruler exhibits nonlinear behavior. The charge polarization becomes nonlinear according to a geometric series:

$$P = a_1 E + a_2 E^2 + a_3 E^3 + \cdots \tag{8.1.4}$$

The nonlinearity of the charge polarization becomes apparent, and high-order terms such as a_2 and a_3 contribute to the polarization.

To accomplish this nonlinear polarization, an intense electric field is required. The electron is held to the positive nucleus by an internal force of about 10^9 V/cm. Normal light sources with electric fields below 100 V/cm exhibit only linear optical effects. Few sources exist with a high enough electric field (comparable to the internal force) to generate nonlinear effects. A focused laser beam, for example, can impress an electric field of 10^6 to 10^7 V/cm on a crystal. Even higher intensities, with even higher electric fields, can be found inside the cavities of most lasers (where the intracavity intensities are often 100 times the external intensity).

To demonstrate mathematically how second harmonic light is produced under such conditions (e.g., how the ruby laser at 694.3 nm was doubled to 347.2 nm in the example at the beginning of the chapter) consider equation (8.1.4), in which we substitute for the applied electric field E:

$$E = E_0 \cos \omega t \tag{8.1.5}$$

where ω is the angular frequency of the light ($2\pi f$). Equation (8.1.4) then becomes

$$P = a_1 E_0 \cos \omega t + a_2 E_0^2 \cos^2 \omega t + \cdots \tag{8.1.6}$$

Trigonometric substitution can be made for the $\cos^2$ term as

$$\cos^2 \omega t = \tfrac{1}{2} + \tfrac{1}{2} \cos 2\omega t \tag{8.1.7}$$

so that equation (8.1.6) becomes

$$P = a_1 E_0 \cos \omega t + \tfrac{1}{2} a_2 E_0^2 - \tfrac{1}{2} a_2 E_0^2 \cos 2\omega t + \cdots \tag{8.1.8}$$

This leads to an interesting conclusion since the third term in equation (8.1.8) contains a term with order 2ω; one component of the charge polarization equation is at *twice* the fundamental frequency of the original laser beam. As the incident beam falls on the crystal, oscillating electric dipoles absorb light at a frequency ω and reradiate it at the same frequency ω and at twice the original frequency 2ω.

The total energy of the beam is not altered much, but is now split between two components. This is the second harmonic of the original beam and the effect is called *second-harmonic generation* (SHG).

Equation (8.1.8) reveals not two, but three output terms: in addition to the fundamental (ω) and second-harmonic (2ω) term, a constant term $\frac{1}{2}a_2 E_0^2$, which represents a DC offset or *steady polarization*. It also represents an optical rectification analogous to that of a diode rectifying an AC voltage. Indeed, this DC signal (although a minute quantity) can be detected across a nonlinear crystal by placing electrodes on either side of the crystal, and this signal is proportional to the intensity of the light beam passing through the crystal.

8.2 PHASE MATCHING

The three terms that result from equation (8.1.8) are plotted in Figure 8.2.1 and summed together. The ω and 2ω terms are in-phase and add together to form a sine wave, but when all three terms are added together, the result (the *total polarization*) is a warped wave, shown in Figure 8.2.1. The peaks and valleys of the summed wave still coincide with those of the original incident wave (i.e., it is in phase with the original ω term), but the wave is shifted in one direction—in this case, upward. This is possible only if the nonlinear crystal has no internal symmetry, so that electron shells are relocated more effectively in one direction than in the other by the electric field applied. About 10% of all crystals have such a lack of symmetry and hence can be used as nonlinear media.

Now, considering that there are two frequencies of light in the crystal, it is quite likely that each will have a slightly different index of refraction. This effect, of course, is responsible for dispersion of light of various wavelengths by a prism. In most crystals the index of refraction will be higher for the higher-frequency component (2ω) than for the lower-frequency component. A higher index of refraction means that light travels more slowly. Figure 8.2.2 illustrates the situation in which the second harmonic light becomes out of phase with the fundamental component as

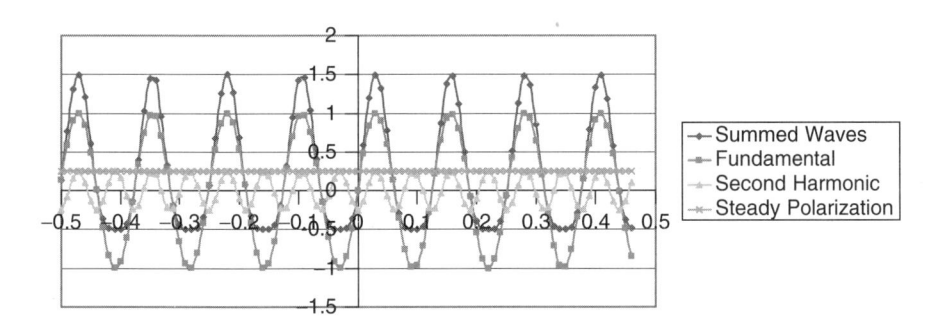

Figure 8.2.1 Output terms from equation (8.1.8).

224 NONLINEAR OPTICS

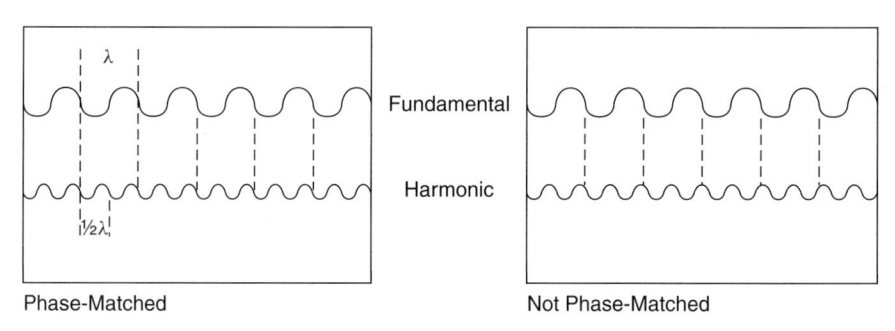

Figure 8.2.2 Phase shift of components in a nonlinear crystal.

it passes through the crystal, causing destructive interference, which leads to extinction of the second harmonic component. The distance over which the waves travel before becoming 180° out of phase is called the *coherence length*. If the crystal is thinner than this length (typically, 10^{-3} cm or less), the components remain in phase. This is somewhat impractical, though, and severely limits the efficiency of the harmonic light generation. If, the crystal is thicker and happens to have a thickness that is an odd multiple of the coherence length, the generated second harmonic light is entirely extinguished, due to destructive interference.

The solution to the problem of phase mismatch between the components is to exploit the phenomenon of birefringence. You will recall from Section 7.5 that this is an effect in which the index of refraction depends on the polarization of light passing through the crystal, or in this case, the individual light component passing through the crystal. By tilting the crystal such that the index of refraction for both the fundamental (with one specific polarization) and the harmonic component (with a different polarization) are the same, phase matching can be accomplished. When the direction of propagation of light in the crystal is chosen so that the indexes of refraction for both the fundamental and second harmonic are exactly the same, the crystal is said to be *phase matched*. The second harmonic component produced in the crystal now exits in phase with the fundamental component, so that constructive interference occurs. This has implications for efficiency of generation of second-harmonic light.

Nonlinear interaction is a general term describing the mixing of two light beams to produce a sum or difference of frequencies. The case of second-harmonic generation, the most common application of nonlinear phenomena, is a specific case in which the input components are identical (with the same frequency) and interact to produce a resulting beam with a doubled frequency. Other interactions are possible, of course, such as the mixing of a fundamental and second-harmonic beam to produce third-harmonic output, the most common application being the mixing of a 1064-nm fundamental from a YAG laser with a 532-nm frequency-doubled beam to produce a third-harmonic output at 355 nm.

Nonlinear crystals are classified by the type of phase matching required, each type requiring a different polarization of each input component. Type I phase-matched crystals utilize input beams with polarizations parallel to each other, while type II phase-matched crystals utilize input beams with polarizations perpendicular to each other. Both types of crystals are outlined in Figure 8.2.3, in which input components ω_1 and ω_2 (identical for a second-harmonic generator) are mixed to produce an output at frequency ω_3.

As an example of phase matching, consider a small green DPSS laser in which vanadate (a common solid-state laser crystal normally with an output at 1064 nm in the infrared) is frequency-doubled to produce green light at 532 nm using a type II phase-matched crystal of KTP (a common nonlinear crystal). Vanadate produces a polarized output at 1064 nm that is oriented at 45 degrees to the optical axis of the KTP crystal. This output is hence split into two polarization components (the o and e components) perpendicular to each other, as required for type II phase matching. A second-harmonic beam is generated with a rotation of 45 degrees relative to the 1064-nm output from the vanadate, which exits through the output coupler. The optical train of such a laser is described in Figure 8.2.4 along with relative polarizations of each component.

Many commercial frequency-doubling crystals, especially those intended for use in compact DPSS lasers, are cut specifically with the faces perpendicular to the optical axis of the laser and such that the crystal may be oriented in the same direction as the lasing crystal. In this manner, the user is generally not aware of the requirement for phase matching, and the crystal is simply installed in the same orientation as the solid-state laser crystal. For a crystal with unknown axes, phase matching can be accomplished by mounting the crystal in a mount allowing angular adjustment, and passing the beam from a Q-switched laser through the crystal (a Q-switched laser is used since the peak intensities are extremely high, as required for high second-harmonic generation efficiency). By rotating the crystal while the beam from the laser passes through it, second-harmonic light will be generated when the phase-matching angle is encountered. The angle may then be optimized for peak second-harmonic output, at which point the phase-matching direction has been found. The crystal may also be inserted into the cavity of a laser and adjusted as well.

The scheme outlined in Figure 8.2.4 may seem somewhat cumbersome. After all, why not simply frequency-double the diode output directly by placing the nonlinear

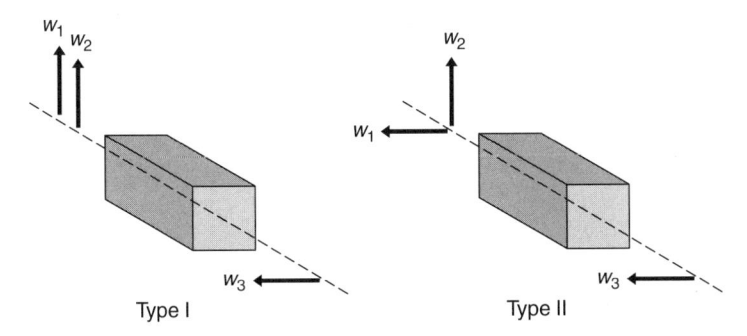

Figure 8.2.3 Types of phase matching.

226 NONLINEAR OPTICS

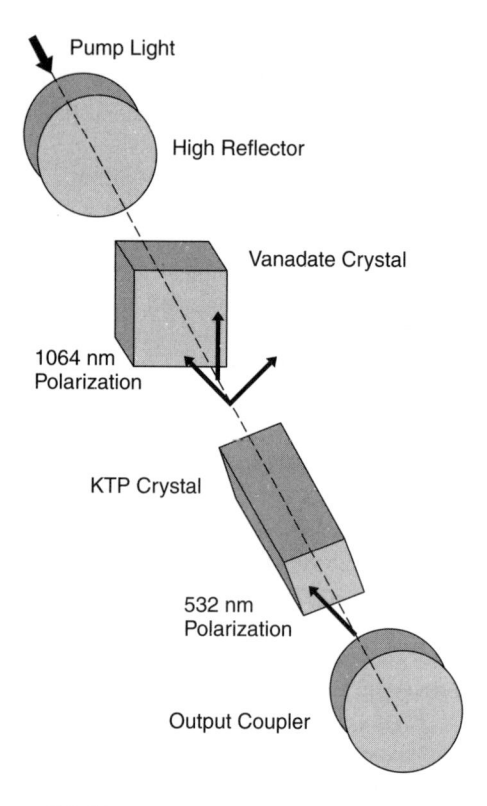

Figure 8.2.4 Type II phase matching in a DPSS laser.

crystal inside the cavity of the laser? With a normal laser diode, this is not possible because the relatively wide gain bandwidth of most diodes prevents proper phase matching at all but one specific wavelength. However, inclusion of a wavelength-selective element such as a Bragg reflector in the cavity narrows the spectral width, allowing efficient harmonic generation.[2] Other approaches include the *optical* pumping of a semiconductor laser (as opposed to a normal solid-state laser crystal) with a nonlinear doubling crystal inside the cavity. Solid-state lasers such as the YAG, on the other hand, feature much narrower gain bandwidths, so are already ideally suited for doubling.

In addition to angular adjustment, phase matching of some nonlinear crystals may be achieved by changing the temperature of the crystal itself. Such crystals exhibit indexes of refraction that change with temperature. Niobates such as lithium niobate

---

[2]Of course, it isn't quite that simple. As an example, the Protera laser from Novalux Inc. uses a three-mirror cavity in which the diode laser has two Bragg reflectors surrounding it as well as an extended cavity, inside which the doubling crystal is situated. This arrangement is claimed to allow better control of longitudinal modes. The diode operates at 976 nm, so that the resulting second-harmonic beam is at 488 nm. This laser is designed as a replacement for small argon-ion lasers.

(LiNbO$_3$) exhibit such behavior (which was discovered by accident when it was evident that the phase-matching direction of this material was different for various temperatures). For these materials the crystal is placed inside an oven, and the temperature varied until phase matching occurs in a manner similar to that employed for angular phase matching. This method is used to tune most parametric oscillators, described later in the chapter.

It is obvious that phase matching is essential for high conversion efficiency, which also depends on the nonlinear coefficient of the material and the intensity of the laser beam (hence, as we mentioned earlier, why harmonic generators are often placed inside the cavity of a laser where the intensity is very high). Efficiency of conversion can be as high as 70% but normally hovers around 30% for a material such as KDP.

An alternative method to ensure phase matching is quasi-phase matching using a periodically poled nonlinear material. In such a material, thin layers of nonlinear material are stacked with polarizations in opposite directions. As the fundamental and harmonic waves both propagate as they pass through the crystal, phase shifts occur due to the differing refractive indexes at each wavelength. By reversing the phase mismatch periodically, phase matching can be accomplished without the use of angular rotation (which exploits birefringence). Such crystals offer wider wavelength ranges since they operate independently of the wavelength range over which birefringence can be exploited. Production of such crystals is a complex process, however, so the cost is high compared with that of standard birefringent crystals.

8.3 NONLINEAR MATERIALS

What makes a material a good candidate as a harmonic light generator? It is evident from equation (8.1.4) that there are a number of coefficients governing the amount of harmonic light generated in relation to the fundamental. These nonlinear coefficients (a_2 and a_3), as well as the coefficient for the linear term (a_1), are properties of the medium itself but can be determined and depend on the applied electric field. At small electric field values, the linear term dominates (i.e., the material behaves in a linear manner), but as this value increases, the nonlinear coefficients increase as well, so the material behaves in a nonlinear manner.

The linear term is related to the index of refraction by

$$a_1 = \varepsilon_0(n^2 - 1) \qquad (8.3.1)$$

where ε_0 is the permittivity of free space (8.85×10^{-12} F/m). This makes this term on the numerical order of 10^{-12} whereas the second-order coefficient a_2 for many nonlinear crystals is on the order of 10^{-24}. Table 8.3.1 lists the constants for actual crystals (the coefficients listed are in conventional MKS units, in which a_2 is one-eighth of the coefficient listed since the series in equation (8.1.4) is actually a Taylor geometric series). Although this coefficient seems extremely small, consider that it is multiplied by E_0^2 in the equation. For a small electric field such as $E_0 = 1000$ W/m (this is actually a reasonably bright light by everyday

TABLE 8.3.1 Nonlinear Optical Coefficients for Various Materials

Material	Nonlinear Coefficient[a]
Quartz	2.6×10^{-26} to 3.0×10^{-24}
ADP (NH$_4$H$_2$PO$_4$)	6.8×10^{-24}
KDP (KH$_2$PO$_4$)	3.8×10^{-24} to 4.1×10^{-24}
KTP (KTiOPO$_4$)	4.4×10^{-23} to 1.2×10^{-22}
Lithium niobate (LiNbO$_3$)	2.3×10^{-23} to 3.9×10^{-22}

[a]The variations given depend on the specific symmetry of the crystal.

standards), the coefficient for the 2ω term is on the order of 10^{-18}. This is extremely small compared to the a_1 term, which is numerically on the order of 10^{-12} and hence 1 million times larger. The linear ω term dominates and nonlinear activity is not seen. Now consider the situation when the applied electric field is large, as in that of a focused laser beam (say 10^6 W/m). The coefficient for the 2ω term is now on the order of 10^{-12}, about the same size as that of the ω term. When applied electric fields are this large, the output of second-harmonic light from the crystal is also large.

Common crystals for use as frequency doublers have large nonlinear coefficients, but not all crystals in nature have this feature. Only crystals that have no symmetry have a nonzero a_2 coefficient. These crystals are similar to those used in Pockels cell EO modulators and are piezoelectric as well. An applied mechanical pressure will cause asymmetry and unequal distribution of charges in the crystal, and a DC voltage appears across the crystal. Piezoelectric crystals are commonly used to generate high voltages for sparks to ignite gas burners for barbecues and lighters. Quartz and other glasses have relatively small coefficients and hence make poor generators. Other crystals commonly used for harmonic generators that have large nonlinear coefficients include phosphates such as ADP (ammonium dihydrogen phosphate), KDP (potassium dihydrogen phosphate), KTP (potassium titanyl phosphate), and niobates such as lithium niobate (LiNbO$_3$).

ADP and KDP were among the first nonlinear materials discovered. They are transparent from about 200 to 1500 nm but have a relatively small nonlinear coefficient. A better and more efficient material is KTP, which is more commonly used nowadays to double in Nd:YAG and Nd:YVO$_4$ lasers. Lithium niobate is another common material useful in the range 400 nm to 5 mm, making it more useful than ADP or KDP in the infrared. It has a large nonlinear coefficient, making it about 100 times more efficient than KDP as a second-harmonic generator. Table 8.3.1 lists some common nonlinear materials and their corresponding coefficients for comparison.

Commercially available nonlinear crystals are often specified by the type of phase-matching required, phase-match angle, and coatings on the crystal faces. In general, crystal faces must be coated with antireflective coatings suitable for both the fundamental and harmonic wavelength involved. An uncoated crystal would lead to a large intracavity loss, which may well prevent oscillation of the laser.

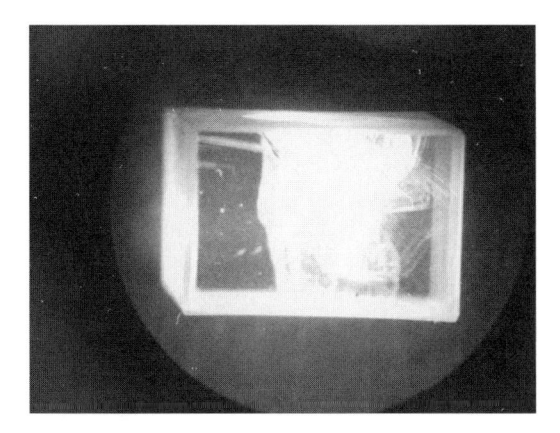

Figure 8.3.1 Damage to a nonlinear crystal.

Finally, a good nonlinear crystal must be able to take the enormous power densities incident on the crystal without permanent damage. As shown in Figure 8.3.1, a nonlinear crystal may be damaged when the incident power reaches a high enough level. In this particular case, heat generated by the incident laser beam caused the crystal to fracture. Each type of crystal then has a characteristic damage threshold, which varies from 20 MW/cm^2 for a lithium niobate crystal (which has a particularly low damage threshold) to 500 MW/cm^2 for an ADP crystal (which has one of the highest damage thresholds of any nonlinear material). Damage is, of course, a concern since the intracavity beam of a Q-switched laser can possess as high a power density as the focused beam of any laser.

8.4 SHG EFFICIENCY

The second-harmonic conversion efficiency, defined as the ratio of the power of the second-harmonic output component to the power of the fundamental component, is directly proportional to the intensity of the electric field, $I = P_{\text{fundamental}}/A$. This explains why many harmonic generator crystals are placed inside the cavity of the laser where the intensity is highest. Incident beams may also be focused inside the crystal to increase the localized intensity in the crystal as the beam passes through it. To prevent damage in this case, one must ensure that the intensity does not exceed the damage threshold of the crystal.

Efficiency is also proportional to the square of the length of the second harmonic crystal so that doubling the length of the crystal quadruples the second-harmonic output. Revisiting equation (8.1.8), we find that the relationship between incident power and second-harmonic output power is

$$P_{\text{second harmonic}} = \frac{Kl^2 P_{\text{incident}}^2}{A} \tag{8.4.1}$$

where K is a proportionality constant, l the length of the crystal, and A the beam area in the crystal.

8.5 SUM AND DIFFERENCE OPTICAL MIXING

So far we have examined primarily the phenomenon of second-harmonic generation, in which two identical incident waves (ω_1 and ω_2 in our case) are mixed to produce a shorter wavelength (doubled frequency) result. In general, though, nonlinear processes extend much further, to the production of both sum and difference frequencies between source beams of different frequencies. It is well known that when two electromagnetic waves of different frequencies are mixed together, additional sum and difference frequencies are generated. In the world of electronics, this process of mixing is used in all radio receivers to generate intermediate frequencies, which are then detected. In AM radio receivers, incoming radio signals are mixed with a variable oscillator to produce a difference signal centered at 455 kHz. In FM receivers an intermediate frequency of 10.7 MHz is used. This scheme, called *superheterodyning*, allows the tuning of the receiver to any frequency desired by tuning an oscillator. The block diagram of an AM radio in Figure 8.5.1 shows how mixing works.

The same mixing effect can be accomplished with light by mixing it in a nonlinear crystal; in fact, most nonlinear optical effects have similar analogies in the world of electronics. In the effect described here, when two beams with frequencies ω_1 and ω_2 interact in a nonlinear crystal, they induce a polarization oscillating at $\omega_1 + \omega_2$ and $\omega_1 - \omega_2$: in other words, a sum and a difference signal. To understand the mixing process, consider a model in which two beams with crests and valleys interact in a nonlinear crystal. Where crests for the two incident beams meet, the electric field strength is maximum and corresponds to the crests of two new

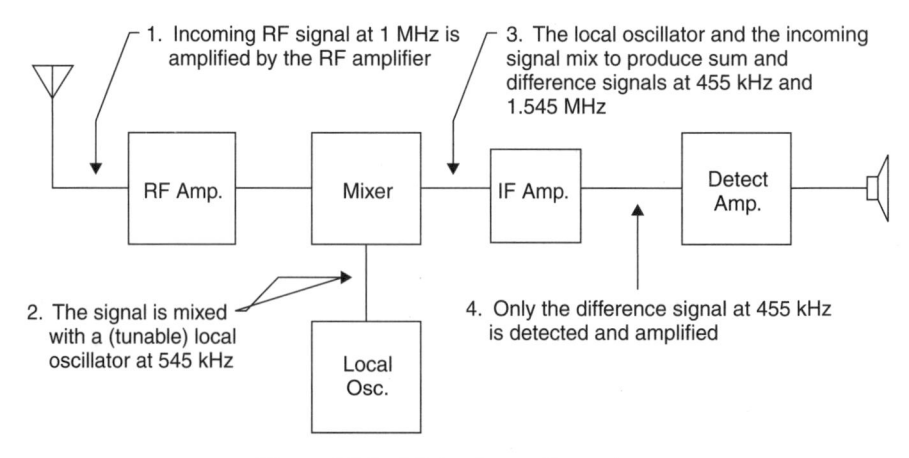

Figure 8.5.1 Mixing in a radio receiver.

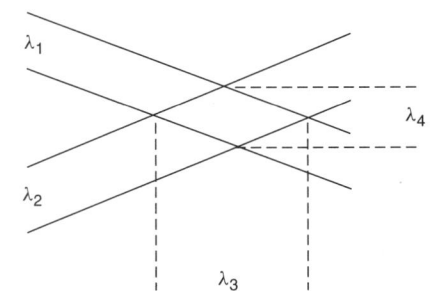

Figure 8.5.2 Model for mixing in a nonlinear crystal.

beams produced in the interaction. This interaction is depicted in Figure 8.5.2. Where λ_1 and λ_2 interact they can sum together to produce λ_4 or a difference in λ_3. Remember that energies of these waves add and subtract so that the shorter wavelength sum (λ_4) contains the sum of energies in photons λ_1 and λ_2. Although laser beams are generally used for nonlinear mixing, intense incoherent sources can be used as well: Lasers simply offer a convenient source of intense light.

In terms of quantum mechanics, each incident photon is annihilated to produce a new photon that contains the total energy of the two individual incident photons. This is simple conservation of energy. Since frequency is proportional to energy, the frequency of the new photon is the sum of the frequencies of each incident photon.

Example 8.5.1 Up-Converted Frequency The process of mixing optical waves by a process such as that described in this section can be used to up-convert frequencies. One application involves the mixing of 10.6-μm radiation from a carbon dioxide laser with the 1.06-μm radiation from a YAG laser in a proustite crystal (one of the few nonlinear materials that is transparent at both wavelengths). The wavelength of the resulting radiation may be found by summing the frequency of each laser (2.83×10^{13} and 2.83×10^{14} Hz, respectively) to yield the summed component at 3.11×10^{14} Hz or 963 nm.

In addition to their value in mixing two frequencies, nonlinear processes may be used to mix three or more laser frequencies in a process called *n-wave mixing*. Such processes usually exhibit low efficiencies, however, and so are often restricted to the laboratory.

8.6 HIGHER-ORDER NONLINEAR EFFECTS

As we have mentioned earlier, not all materials have a nonlinear a_2 coefficient—only 10% of all natural crystals do—but all materials, including glass, have a third-order coefficient. However, as this coefficient is extremely small compared

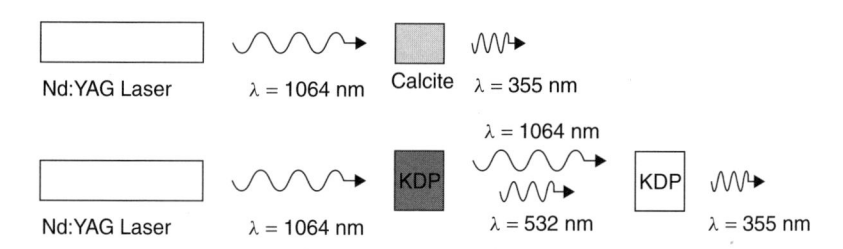

Figure 8.6.1 Generation of high-order harmonics.

to a second-order coefficient, much less efficiency is expected for third-harmonic generation. Calcite, for example, can be used as a third-harmonic generator with efficiencies of less than 10^{-5}. Alternatively, series nonlinear crystals can be used to generate third-harmonic light: the first crystal producing second-harmonic light at 532 nm and the second crystal mixing this light with the original 1064-nm fundamental wavelength, the second crystal often being type I phase-matched. Both methods are outlined in Figure 8.6.1.

8.7 OPTICAL PARAMETRIC OSCILLATORS

In the process of harmonic generation two photons combine their energy into a single photon. We have also seen that the reverse process, in which a single photon splits its energy into two photons which will have lower energy than the incident photon, is also possible. By conservation of energy, the energy of the resulting photons must sum to the energy of the incident photon. In an optical parametric oscillator (OPO) configuration a pump beam incident on a nonlinear crystal produces two resultant photons at different wavelengths. One wavelength, called the *signal*, exits the device as the output beam; the second wavelength, essentially useless, called the *idler beam*, stays within the cavity of the device, as depicted in Figure 8.7.1. The frequencies of the idler and output beam sum to the frequency of the pump beam (as required for conservation of energy). An OPO is not a laser since it does not amplify; rather, it is an oscillator only. It is, however, a coherent oscillator producing laser light.

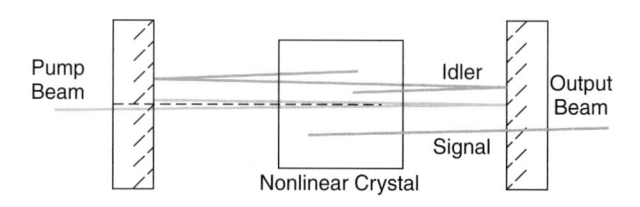

Figure 8.7.1 Simple OPO device.

The question now arises: If two output beams are produced by the crystal, why is only one used as an output? The answer lies in phase matching. Only one wavelength is phase-matched at a time. This is also how the OPO is tuned: by using the same methods as those used to phase-match a second-harmonic generator crystal. The temperature of the crystal or the angle of the crystal within the cavity can be changed to tune the laser. In the case of angular tuning, often two crystals are used that rotate in opposite directions. This scheme eliminates displacement of the beam in the cavity, which leads to walk-off loss. A common arrangement is to use an Nd:YAG laser to pump an OPO employing either lithium niobate (LiNbO$_3$) or beta-barium borate (BBO). Using lithium niobate, for example, an OPO is tunable from about 1.4 μm to almost 4 μm by varying its orientation in the cavity from about 44.5 degrees to 51 degrees. Temperature tuning is also frequently used to simplify the mechanical arrangement of the optics.

PROBLEMS

8.1 Compare the approximate conversion efficiency of quartz (with a nonlinear coefficient of 2.6×10^{-26}) to KTP with a coefficient of 4.4×10^{-23}. The indexes of refraction are 1.46 and 1.51, respectively. Substitute each value into equation (8.1.8) and extract the terms for ω and 2ω. Calculate efficiency at intensities of both 10^3 and 10^6 W/m, as this will allow determination of how intensity affects conversion efficiency.

8.2 Whereas AM radio receivers employ a 455-kHz IF signal, FM receivers employ a 10.7-MHz signal. Explain why a 455-kHz IF will not work for FM signals. (*Hint*: For a given local oscillator frequency, two signals will mix to produce the IF signal, which is detected and amplified.)

CHAPTER 9

Visible Gas Lasers

Early in the development of laser technology it was found that a mixture of helium and neon gases would lase in the infrared. Soon afterward, the familiar red transition of the helium–neon laser was found and so began the life of the most dominant laser for the next 20 years. In this chapter we cover two of the most common and useful lasers: the helium–neon and the argon. Helium–neon lasers have become the de facto standard laboratory laser.

By 1963 it appeared that almost anything would lase when it was discovered that mercury vapor, in a buffer of helium, could be used to produce a pulsed high-gain laser with simultaneous red and green output. A few companies began immediately to produce commercial versions, and R&D departments kept trying to understand and optimize the laser, some hoping to produce a CW mercury ion laser. Quite by accident, argon was found to lase on six visible lines simultaneously when it was tried as a buffer gas for mercury. Argon showed promise of high-power CW operation and quickly became a dominant lasing species. Mercury vanished from the scene and argon-ion lasers have been produced since.

9.1 HELIUM–NEONeon LASERS

The first gas laser, the helium–neon (HeNe) laser, is still an important source of coherent red light with uses ranging from bar-code scanning to alignment. At one time the most popular laser, the dominance of this laser in the market is decaying rapidly, due to inexpensive, reliable, and small semiconductor lasers operating in the same range of wavelengths and power levels as the HeNe. Although the semiconductor cousin of the HeNe lacks the beam quality of the HeNe, this is not important for many applications.

HeNe lasers normally operate with the familiar red (632.8 nm) beam, but multiple transitions are possible, allowing the laser to operate (with suitable optics) at wavelengths in the infrared, orange, yellow, and green. Power output for commercially available HeNe tubes ranges from under 1 mW for a small HeNe tube to just over 100 mW for a large behemoth unit. These lasers typically feature excellent

---

Fundamentals of Light Sources and Lasers, by Mark Csele
ISBN 0-471-47660-9 Copyright © 2004 John Wiley & Sons, Inc.

spectral and coherence characteristics, so are a popular choice for uses in holography, where they are still the laser of choice, preferred over semiconductor types. They are also the de facto standard laboratory laser, used for a myriad of odd jobs around the lab, including alignment and testing of optical components.

9.2 Lasing Medium

The HeNe laser has been used extensively as an example throughout this book, but a basic recap is provided here for the sake of completeness. The HeNe laser uses a mixture of very pure helium and neon gases in the approximate ratio of 10:1. Helium, the gas used to furnish the pump level in this four-level system, is in greater abundance. Tube pressures are typically between 1 and 3 torr, being dependent primarily on the diameter of the plasma tube.

Being a four-level laser with favorable dynamics, HeNe lasers have low thresholds and operate in CW mode. The pump level is supplied from the helium atom, which absorbs energy from the discharge through electron collisions and becomes excited to a level 20.61 eV above the ground state (for all visible transitions; IR transitions at 1.15 and 3.3 μm use different levels). The ULLs in neon (denoted as the $1s^2 2s^2 2p^5 5s^1$ electronic state) are at almost exactly the same energy (20.66 eV) as the excited helium energy level, so that a collision with an excited helium atom will result in the transfer of energy to neon atoms, raising them to an excited state (Section 4.3), with the small difference between the two energies being made up by thermal and kinetic energy in the system. From the ULL they can emit a photon of coherent light (via stimulated emission) to decay to the LLLs (the $1s^2 2s^2 2p^5 3p^1$ electronic state).

You will recall from Chapter 3 that quantum mechanics dictates that this electronic state is actually composed of numerous, closely spaced levels, hence multiple transitions (and wavelengths) are possible in this system. Neon atoms at the LLLs then fall to ground through a depopulation process involving both radiative decay (and the emission of an incoherent 600-nm photon) and collisions with the tube walls (Section 5.5). This also explains why HeNe lasers invariably use small-diameter (<2-mm) plasma tubes. The entire process, including excitation of the helium atoms, transfer to the ULL in neon, lasing transitions, and decay from the LLL is shown in Figure 9.2.1.

This simplified diagram does not show all possible energy levels, so numerous other transitions are possible. The 1.15-μm transition, for example, originates from the $1s^2 2s^2 2p^5 4s^1$ electronic state and terminates at the same lower levels as those used for the visible transitions. Of particular interest in an extremely powerful line at 3.39 μm originating from the same upper level as the visible transitions but terminating at the $1s^2 2s^2 2p^5 4p^1$ electronic state. This powerful transition must be suppressed in all HeNe lasers operating at other wavelengths since it serves to deplete the upper lasing level rapidly and hence would halt oscillation of any visible line. This transition has such high gain that in a long HeNe laser it can produce superradiant output. The second most powerful transition is the 1.15-μm IR line

OPTICS AND CAVITIES **237**

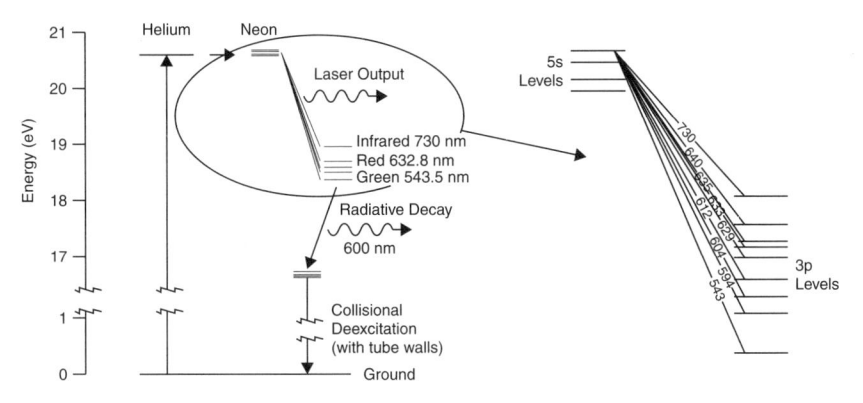

Figure 9.2.1 Helium–neon laser energy levels.

TABLE 9.2.1 Commercially Available HeNe Lasers

Wavelength (nm)	Relative Gain (Compared to 632.8-nm Output)
543.5 (green)	0.06
594.1 (yellow)	0.07
611.9 (orange)	0.2
632.8 (red)	1

(which originates from a different ULL), followed by the ubiquitous 632.8-nm red line. In the visible spectrum there are eight lines, ranging from green to red. Of these, several are too close to more powerful lines to be useful (e.g., the output at 635 nm is too close to 633 nm to separate with a normal wavelength selector). Commercially, four visible wavelengths of HeNe laser are commonly available, as outlined in Table 9.2.1, along with gain relative to the 632.8-nm red line. Because of the low gain of the green HeNe, special low-loss optics are required which are much more critical (and expensive) than red HeNe optics. Where a small red HeNe tube might have an output of up to 20 mW, a comparable green tube would be limited to under 3 mW.

9.3 OPTICS AND CAVITIES

Most commercial HeNe laser tubes are of a one-piece design featuring integral mirrors resembling that of Figure 9.4.1. This compact design has prealigned optics providing trouble-free operation for most users. Optics never require cleaning or realignment since they are sealed into the tube. Whereas in older tubes, mirrors affixed to the glass tube by epoxy were aligned with a simple lateral motion (see Section 6.10), such mirrors have been abandoned in favor of hard-sealed optics in which adjustment is provided during manufacture by flexing a metal tube between the optic and the tube. Most one-piece tubes are randomly polarized, although some tubes have integral Brewster windows within the tube itself, designed solely to polarize the output beam.

Some lasers, primarily research and lab lasers, feature external optics. Tubes often have Brewster windows to reduce loss, but an alternative is a plane window with antireflective coatings. These coatings are required since losses of up to 8% at an uncoated window would certainly prevent oscillation. Some long (up to 1 m) research lasers also feature wavelength selectors allowing tuning of the laser over multiple wavelengths. These selectors are prisms with an affixed HR and allow tuning by changing the angle of the assembly with respect to the tube axis (see Section 6.5 for details). Shorter lasers lack the necessary gain to overcome losses imposed by the wavelength selector, so usually lack these features.

Regardless of the type of optics employed (either internal or external), mirrors for use in a visible HeNe laser must be designed to absorb (or transmit) strongly at 3.39 μm to induce a large loss at that wavelength and hence suppress the strong IR transition there. This transition has extremely high gain which will, if allowed to oscillate, rob power from all other transitions and stop most from oscillating. For this reason, many HeNe lasers that have external optics use borosilicate glass plasma tube windows instead of quartz since borosilicate absorbs a good deal of IR energy. In long tubes (e.g., 1 m) even the use of borosilicate windows may not prevent oscillation of this IR line, in which case magnets are placed along the length of the tube to suppress the transition (by inducing Zeeman splitting of the transition). In Figure 9.3.1, magnets of this type are located under the plasma tube. Also evident

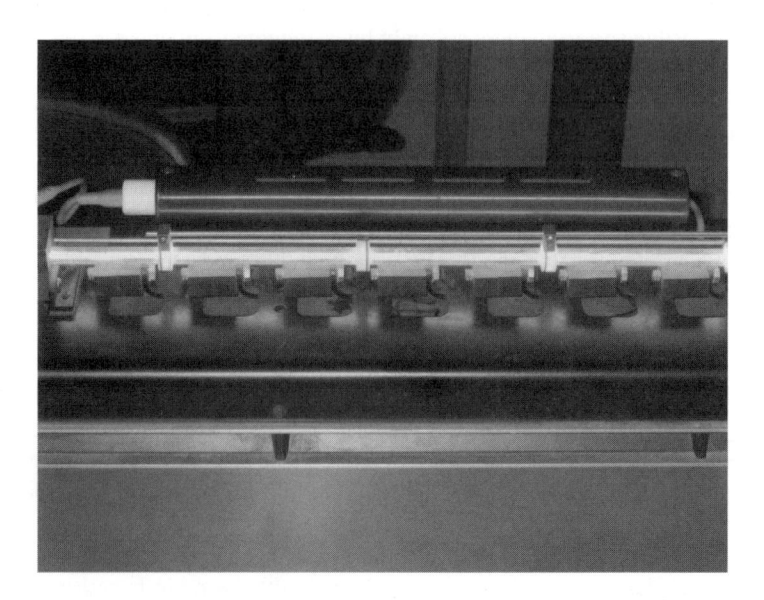

Figure 9.3.1 Magnets to suppress the 3.39-μm IR transition.

are two bare metal rods traversing the length of the laser, which are used to help ignite the plasma on this very long (>1 m) laser. Even in a relatively short HeNe laser, suppression of the IR line (by cavity optics) may lead to significant power increases in the visible transitions.

9.4 LASER STRUCTURE

Figure 9.4.1 shows a typical HeNe tube and a diagram of its principal features. This tube, typical of any modern HeNe tube built by any major manufacturer, features integral cavity optics sealed right into the tube. Other than the initial alignment at the factory, no other adjustments are required for the life of the tube. The anode

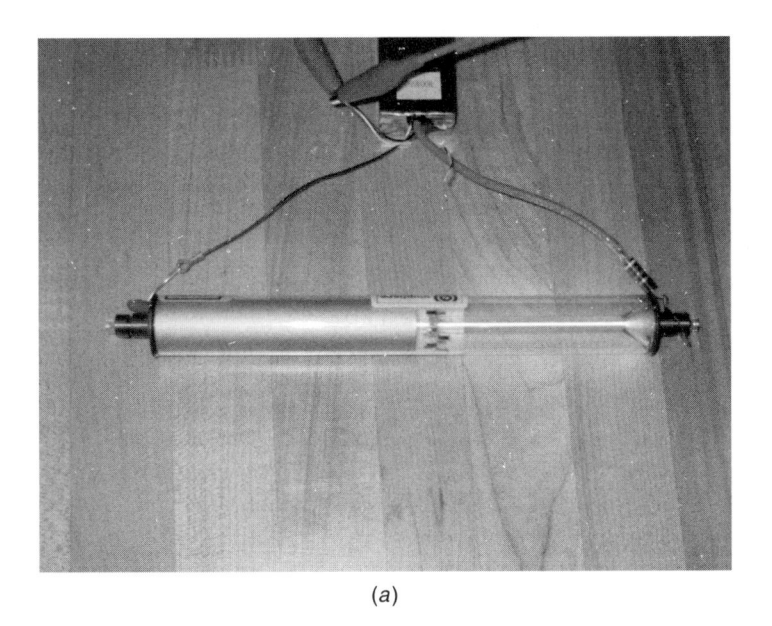

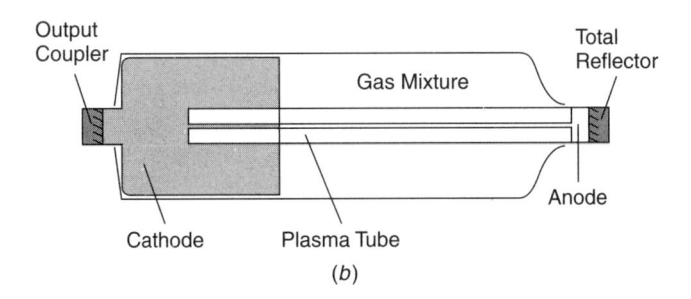

Figure 9.4.1 Typical HeNe tube and structures.

of the tube is small (in older HeNe tubes it was a metal pin), while the cathode is quite large and is contained inside a large gas ballast volume. As the tube operates, gas molecules are pumped toward the anode, increasing the pressure at the anode. Without the large ballast volume the gas pressure at the cathode becomes very low and the discharge becomes unstable. In some lasers, such as argons, a gas return path is provided to equalize gas pressures in the tube; however, HeNe tubes simply use a large ballast volume.

The cathode itself is usually made of high-purity aluminum, and in fact the necessity for purity in tube materials as well as gas mixture is quite extreme. Gas purity is crucial since even small amounts (<1%) of impurities, such as water vapor or hydrogen, will drastically decrease power in the tube and cause lasing to cease entirely. Older HeNe tubes used mirrors that were epoxied to the glass tube. Such tubes, called *soft-sealed tubes*, typically had short lives, since helium diffuses through epoxy at a slow, albeit significant rate. Over the course of a few years the gas mix in the tube became short on helium and the lasers would cease to operate. This process could be exploited in reverse to resurrect outgassed tubes by placing the tube into a chamber of pure helium, which would eventually diffuse back into the tube. By the 1980s, tubes were manufactured with hard glass-to-metal seals, essentially eradicating the problem.[1] Processing is still an example of where high purity is required, though. Extreme care must be taken to rid tubes of impurities (usually, through multiple bake-out cycles) before finally filling them with a high-purity gas mixture and sealing the tube.

HeNe tubes can vary in length from under 12 cm to over 1 m. The smallest tubes easily fit, along with a compact power supply, into handheld bar-code scanners and produce around 1 mW (although for this application semiconductor lasers are now often used). The largest units, like that pictured in Figure 9.4.2, sport tubes over 1 m long and are used for research and holography purposes, producing about 120 mW of red output. Long tubes such as these also have higher gains, allowing the use of wavelength selectors to lase one of several possible wavelengths (although the gain is still generally too low to allow multiline operation as is commonly done with argon lasers).

While most HeNe tubes have integral mirrors, a few (usually, research lasers like the one used in many experiments throughout this book) have external mirrors. In this case the ends of the plasma tube are either terminated with Brewster windows or with plane antireflection (AR)-coated windows. Such AR coatings are required since light passing through an uncoated plane window will incur a loss of about 4% per surface (or 8% in total). Most small HeNe tubes sport a gain of about 0.12 to 0.15 m^{-1}, so a tube 30 cm in length has a gain of only about 4 to 5%. Losses of 16% (two windows) would be completely intolerable! External optics are mounted on a heavy frame or optical bench with optomechanical mounts allowing adjustment in either (horizontal or vertical) direction. Often, dust covers are

---

[1]The author has several hard-sealed tubes manufactured in the 1980s which still have an output within 5% of that rated from the factory. Early, soft-sealed tubes had lives of only a few years. Many manufacturers advocated running older tubes at least once a week to extend the lifetime.

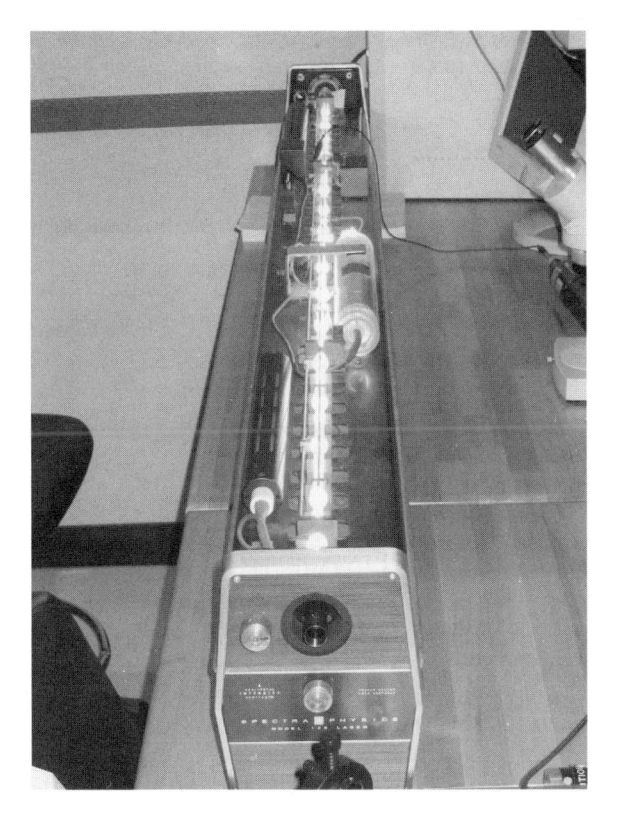

Figure 9.4.2 Large HeNe lab laser.

installed between windows and mirrors to keep the necessity for cleaning the laser optics to a minimum (in a low-gain laser, dust on a window may be enough to extinguish oscillation). If used on a tube with external optics, Brewster windows also serve to polarize the output beam. Some tubes feature internal Brewster windows placed within the tube (usually, within one of the metal stems protruding from the tube), solely for the purpose of polarization.

9.5 HeNe POWER SUPPLIES

Given the modest current requirements of 5 to 6 mA for an average HeNe tube, power supplies are, generally, relatively simple, supplying a run voltage of about 1500 to 2000 V and a start pulse of up to 10 kV. The start pulse ionizes gas inside the tube to start the discharge. Upon ionization, the resistance of the tube drops, current flows, and the voltage across the tube stabilizes to the run voltage. The schematic for a typical AC-powered supply is shown in Figure 9.5.1. Incoming line voltage (110 V AC in North America) is stepped up to about 650 V by transformer

242 VISIBLE GAS LASERS

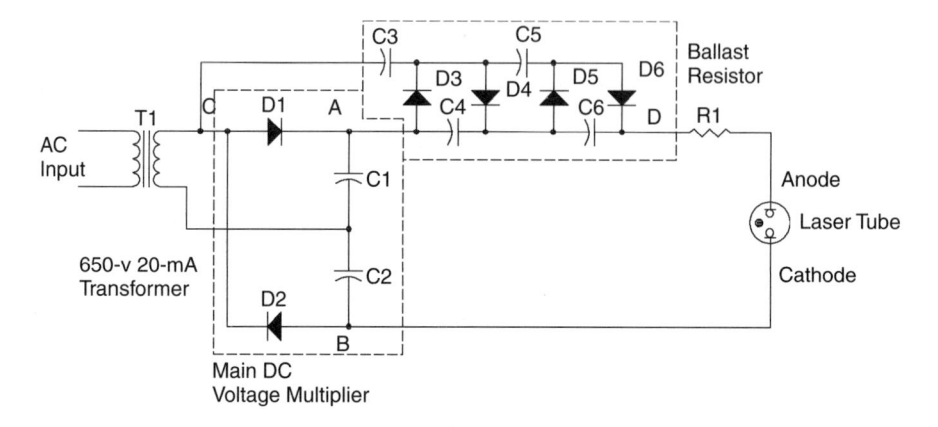

Figure 9.5.1 Helium–neon laser power supply.

T1. When rectified, this becomes 919 V, which is not high enough to run the tube, so it is doubled by a voltage multiplier, consisting of capacitors C1/C2 and diodes D1/D2. On a positive cycle of the AC line, capacitor C1 charges to 919 V DC (=650 V AC × $2^{1/2}$). On a negative cycle, capacitor C2 charges, also to 919 V. Since C1 and C2 are in series, a total voltage of 1838 V is found across terminals A and B on the schematic. Although 1800 V will run most HeNe tubes, it will not initiate the discharge. To do this, a voltage of over 5 kV is required, provided by the multiplier circuit at the top of the diagram consisting of capacitors C3 to C6 and diodes D3 to D6. When the tube is not conducting, this circuit functions as a voltage multiplier, taking AC from the transformer (at point C) and producing a high voltage (of well over 5 kV) at point D (referenced to point B). This high voltage is presented across the tube, which ionizes and begins to conduct current. At that point, current from the main DC voltage multiplier (at terminals A and B) flows through diodes D3 to D6, effectively shorting out the starting circuit. When operating, tube current is limited by ballast resistance R1.

Readers are cautioned that the power supply depicted in Figure 9.5.1 is a simplified one for the sake of explanation; in reality, the circuitry for a power supply would contain safety features such as bleeder resistors to discharge capacitors when the unit is switched off and may well be designed to use readily available components. For example, capacitors rated at 1000 V are expensive, so each capacitor C1 and C2 would actually be composed of two series capacitors rated at 450 V each, which are much more economical. Similarly, high-voltage diodes in the diagram would probably be constructed of three inexpensive 1-kV diodes in series.

In very long HeNe lasers (some up to 1 m in length) the start pulse provided by a circuit like that described above may still be insufficient to initiate a discharge, in which case a high-voltage pulse can be applied to a tube externally to start it. Generated by a pulse transformer inside the laser head (most lasers of this length have external power supplies) and distributed at several points down the length of the plasma tube (see Figure 9.3.1), the pulse assists in ionization of the entire length of the tube, which is necessary to initiate a plasma discharge. It is also possible to use RF energy to assist in the generation of the plasma in the same manner (in fact, RF energy can be used as a pump source for the laser discharge instead of a DC current. This approach was popular in the 1960s and 1970s; however, it is rarely used today).

Another trick used with long lasers is to employ a single cathode with two anodes, as evident in the long laboratory laser of Figure 9.4.2. This approach is used to reduce the electrical path to a manageable length which does not require excessively high voltages to operate or ignite. It is also used in some carbon dioxide laser tubes for the same reasons. In this case, the single cathode connects directly to the power supply, and each anode connects to the supply through a separate ballast resistance. Bear in mind, though, that the majority of HeNe lasers are under 30 cm in length, so do not require special ignition circuitry or techniques.

Many modern HeNe supplies use switching technology for a compact unit. An input voltage between 12 and 120 V is rectified and chopped into AC at high frequencies. A step-up transformer is still employed, but given the high frequencies employed, this transformer is quite miniscule and lightweight compared to one operating at 60 Hz. High frequencies also mean that the capacitors used can be much smaller than those used in a lower-frequency design. The result is a compact power supply that is encapsulated in epoxy similar to the unit pictured in Figure 9.5.2. These units, the standard for a modern HeNe tube supply, have only a few leads protruding, including a high-voltage lead (which connects to the anode of the tube via an external ballast resistance), a ground for the cathode, leads for the input supply, and usually a few leads that configure unit operation, including delayed start for regulatory compliance and interlock inputs. The supply

Figure 9.5.2 Modern HeNe power supply.

is often encased in copper foil, which serves to prevent the emission of high-frequency RF radiation from the supply, which would interfere with nearby equipment and circuitry.

Like any other gas discharge tube, HeNe laser tubes exhibit a feature known as *negative resistance*. Consider a common resistor that obeys Ohm's law according to

$$E = IR$$

where E is the voltage across the resistance in volts, I the current through the resistance in amperes, and R the resistance in ohms. As current through the resistor is increased, voltage across the device also increases since resistance is fixed. In a gas discharge the effective resistance of the tube decreases as current is increased, so that as the current increases through the tube, voltage across the tube decreases—in effect a negative resistance. Unfortunately, this means that a tube connected directly to a power supply will consume current to a point where the resistance of the tube decreases, where, in turn, the tube consumes more current and a runaway condition ensues. The tube then either consumes an extremely high current (usually, catastrophically) or oscillates in an unstable manner.

To correct the problem a (positive) resistance called a *ballast resistance* is added in series with the tube. This serves to limit the current through the tube to an optimal value and stabilize the discharge. The optimal value of ballast resistance is determined by plotting the voltage-to-current characteristics of the tube and determining the slope of the curve at the desired operating current. The slope of the graph being rise (voltage) over run (current) yields, by definition, resistance. If the optimal operating current is known, the ballast resistance may also be found by measuring the tube voltage at that current.

Example 9.5.1 Ballast Resistance A small HeNe tube is operated from a power supply similar to that described in Figure 9.4.3, which has an output of 1838 V. This particular tube has an operating current of 5 mA and a voltage of 1320 V is measured across the tube at that current. The required ballast resistance would be

$$R_{ballast} = \frac{(V_{supply} - V_{tube})}{I_{tube}}$$

$$= \frac{1838 \text{ V} - 1320 \text{ V}}{0.005 \text{ A}}$$

$$= 104 \text{ k}\Omega$$

A ballast resistance of 104 kΩ is required for the tube. Note that when operating, this resistance has 518 V across it and hence dissipates 2.6 W of heat (power = VI).

Ballast resistances typically vary between 25 and 200 kΩ, depending on the tube and power supply and should be mounted as close to the tube as possible (if the tube

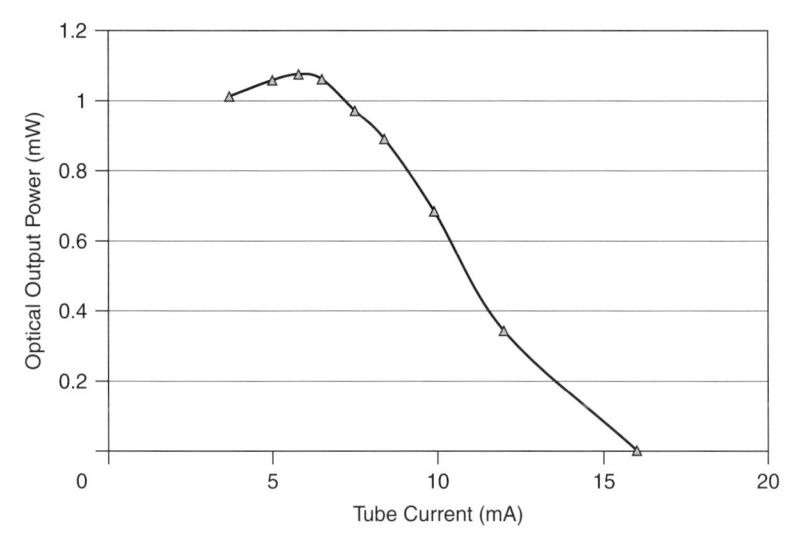

Figure 9.5.3 Power output versus tube current for a HeNe laser.

is separate from the power supply, it is mounted in the housing with the tube itself) to minimize capacitance, which might tend to cause the tube to oscillate. The operating current desired for a given tube can be determined experimentally by plotting optical output power versus current (the operating current chosen for maximum output power or just below that point) or, in most cases, is specified by the manufacturer for a particular tube. HeNe tubes cannot operate at indefinitely high current, and more current does *not* translate into more optical power output, since as depicted in Figure 9.5.3, tube currents above an optimal point actually serve to reduce output power to a point where laser output extinguishes completely. The maximum optical output power is found to occur at a current of 5.8 mA, after which an increase in current leads to a rapid decrease in output power. By 16 mA the laser ceases operation completely (although it becomes quite warm, given the large powers dissipated).

9.6 OUTPUT CHARACTERISTICS

The output of a HeNe laser is not immediately stable, as Figure 9.6.1 shows. At zero a cold HeNe laser tube with a typical output power of 3 mW (and a maximum rated output of 4 mW) was switched on. Lasing output appears almost immediately and rises quickly to a reasonable level, reaching 80% of the final output power in less than 10 s. As the laser warms up, the output power asymptotically approaches the final value of almost 3 mW, but oscillatory behavior is seen in the output power, which varies almost 2% at a period that gradually becomes longer (the period is seen to be 34 s after the laser has been operating for 15 minutes). This behavior eventually disappears as the laser is operated, and for utmost stability it was found that two days was not an

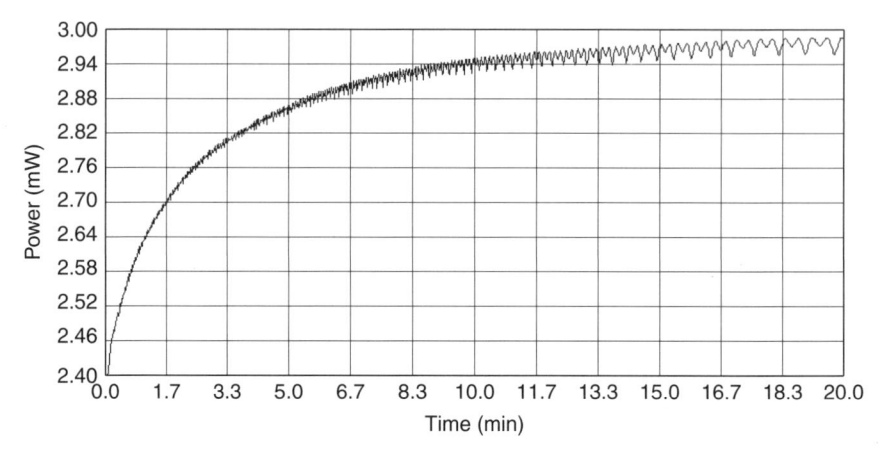

Figure 9.6.1 Time-dependent behavior of HeNe output power.

unreasonable period. As the laser warms, the entire assembly expands, as does the spacing between the cavity mirrors. As the mirrors slowly move apart, the laser hops from (longitudinal) mode to mode, with a corresponding fluctuation in output power. Commercially, stabilized HeNe lasers exist that use a temperature-controlled heater with a number of optical feedback devices and control loops. These lasers do not require long periods of warm-up to becomes stable.[2]

Aside from stability concerns, a HeNe laser produces an excellent-quality output beam, so is the laser of choice for holography applications. With a spectral width of 1.5 GHz, common HeNe lasers exhibit a coherence length of around 20 cm. This is much better than the average semiconductor laser (which is why semiconductor lasers have been slow to move into holographic applications). When configured for single-mode operation, coherence lengths of over 300 m can be achieved (although this is no longer a common "garden variety" HeNe laser). Most HeNe lasers operate in TEM_{00} mode and yield excellent beam quality. Although most standard HeNe lasers are randomly polarized, some have integral Brewster windows (in many cases within the tube itself), giving a polarized output beam.

9.7 APPLICATIONS

Until the 1990s, the HeNe laser was the only option for many pointing, alignment, and scanning applications. Not long ago, many supermarket bar-code scanners and even some high-end handheld scanners used HeNe lasers, as did the original (early 1980s) CD and laser disk players. Today, most applications have migrated toward more compact (and inexpensive) semiconductor laser diodes. However, in some

---

[2]It is important to realize that many applications are not particularly sensitive to output power, in which case a standard unstabilized laser is quite sufficient.

applications where beam quality, beam collimation, and coherence length are important, HeNe lasers are still used extensively. For holography, the HeNe laser is still the laser of choice because of its good coherence length compared to other types of lasers. The same is true for many precision measurement applications that utilize interferometry to measure tiny distances. Access to the gain medium and the ability to use unique cavity configurations (e.g., a ring) allows the HeNe laser to be used as the basis for instruments such as laser gyroscopes for aircraft. Finally, the HeNe laser is indeed the de facto standard laboratory laser, with uses ranging from alignment of optics to testing of optical components.

9.8 ION LASERS

The term *ion laser* refers to a laser in which the lasing energy levels exist in the ionized atom of the species. Although ions are the active species in many solid-state lasers (e.g., Cr^{3+} ions in ruby), the term *ion laser* is accepted to apply to gas lasers. Ion lasers are generally high-powered lasers (much higher powered than a HeNe laser) emitting in the green-blue region of the spectrum (for argon) or on many lines across the entire spectrum (for krypton), and even in the UV. Popular for entertainment purposes such as laser light shows, these lasers have a host of other applications, including medical, graphics arts, and general laboratory uses. While power outputs are often impressive (with a large argon laser capable of tens of watts of visible output), power input is equally impressive, with most large-frame ion lasers requiring three-phase power and a supply of cooling water to operate. Even the power supply components of most large ion lasers are water cooled.

Unlike a HeNe, an ion laser is a complex beast with plasma tubes made of exotic ceramic materials. Many large ion lasers have active gas replenishment systems to compensate for gas use as the tube operates. Tubes also feature heated cathodes and the electrical characteristics are radically different from those of a HeNe: Whereas the HeNe operates at a relatively high voltage and low current, ion lasers operate at relatively low voltages but enormous currents (with some ion lasers having continuous discharge currents of over 70 A!).

9.9 LASING MEDIUM

The lasing medium in an ion laser is a rare gas such as argon or krypton that has been ionized; that is, it has one or more electrons removed from the outer shell. Argon and krypton are the most common ion lasers, although neon, xenon, and a few other gases will lase in an ion form as well. Ionized species exhibit different energy levels than neutral species do and the degree of ionization (the number of electrons removed) affects these levels.

As an example of the mechanics of a typical ion laser, consider singly ionized argon (denoted Ar^+), the active species of the most common ion laser by far. Ions are created by discharging a current of up to 40 A through low-pressure (<1 torr) argon gas. Discharges may be pulsed, as the earliest lasers were, but most ion lasers are CW, so a continuous current of 40 A is required (which leads to complex tube and power supply designs, as we shall see). Ar^+ has numerous energy levels in two bands—nine ULLs centered around 36 eV and two LLLs around 33 eV. Many transitions share an upper or lower energy level. Overall, 10 laser transitions exist in the violet-to-green region of the spectrum.

Ions are pumped to the ULL by a variety of methods, some by decay from a higher level (the expected route for a four-level laser) or directly to the ULL by electron impact in a process resembling that of a metal-vapor laser. In the case of argon, the neutral (nonionized) configuration of the atom is $1s^2\,2s^2\,2p^6\,3s^2\,3p^6$ and when ionized (which requires 15.76 eV of energy) the ground state for the ion (Ar^+) becomes $1s^2\,2s^2\,2p^6\,3s^2\,3p^5$. The argon ion can now be promoted to a higher-energy state in which one electron assumes a 4s state. There are two such states possible with different spins ($\frac{1}{2}$ and $\frac{3}{2}$), and these serve as lower lasing levels. Even-higher-energy states are possible, in which an electron enters a 4p state—there are numerous states here that serve collectively as upper lasing levels (in the visible argon laser spectrum seven tightly clustered ULLs are involved).[3] Four distinct pathways for pump energy have been identified which serve to excite ions to the ULL, including the direct pumping of ions from the neutral ground state to the ULL, a two-step process in which argon atoms are ionized to the ion ground state and later pumped to the ULL from there, and two processes in which the ULL is excited via decay from levels above the ULL.

Having made a transition from one of the ULLs to a LLL, the ion decays almost immediately to the ion ground state (still 15.76 eV above the ground state of the neutral argon atom) by a radiative process in which a 74-nm extreme UV photon is emitted. This decay process is fast—a requirement to maintain a large population inversion—but the extreme UV photon created in the process poses problems since it can damage optical windows and degrade tube materials. Windows must be fabricated from special crystalline quartz to withstand the constant bombardment of extreme-UV radiation. The energy levels involved, transitions, and depopulation process are shown in Figure 9.9.1.

Krypton gas may be used in an ion laser as well with various wavelengths, covering the entire visible spectrum from violet to red. Physically, krypton laser tubes are similar to argon tubes with the exception that krypton lasers require a large ballast volume (covered in Section 9.10). Krypton, however, is less efficient than argon, so output powers are lower than those for a comparable argon laser (the most powerful lines of the krypton laser are only one-fifth the power of the most powerful argon lines). The wider range of visible wavelengths available, however (including a

---

[3] A complete energy-level diagram including full notations for each level and approximate gains in pulsed mode can be found in the first published report of the argon ion laser: "Laser oscillation in singly ionized argon in the visible spectrum" by W. B. Bridges in *Applied Physics Letters*, Vol. 4, No. 7 (1964). The diagram of possible transitions in Figure 9.9.1 is adapted from this paper. A few energy-level assignments have been changed since this first report.

LASING MEDIUM   **249**

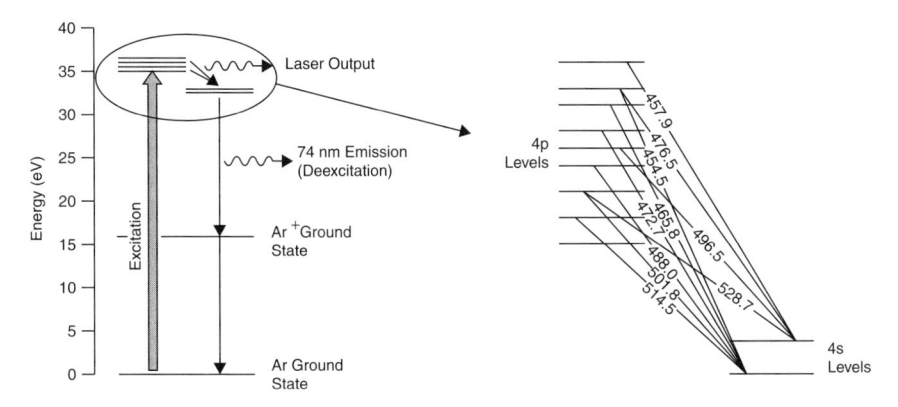

Figure 9.9.1 Argon-ion laser energy levels and transitions.

powerful red line and a yellow line, both lacking in the argon laser spectrum), make this laser a popular choice for entertainment purposes.

Table 9.9.1 lists the common visible wavelengths of argon and krypton ion lasers and typical output power for a comparably sized single-line (wavelength-selected) laser using each gas. With broadband optics it is possible to have several lines oscillate simultaneously, but in many cases these levels share energy levels, so competition between lines can occur which prevents all possible lines from lasing simultaneously. In the case of the argon laser used in this example, only six of the 10 possible lines will lase simultaneously when broadband optics are employed. Krypton lasers are generally not used in multiline mode but rather, with optics, to select the red (647.1 nm) line alone, both the red and yellow (568.2 nm) lines, or white-light mode, in which three or four lines are allowed to oscillate (see Section 6.5). By selecting only required lines, the output power of the already weak krypton laser is preserved.

TABLE 9.9.1 Comparison of Argon and Krypton Laser Output

Argon Ion (Ar^+)		Krypton Ion (Kr^+)	
Wavelength (nm)	Line Power	Wavelength (nm)	Line Power
454.5	140 mW	406.7 ⎫	
457.9	420 mW	413.1 ⎬	150 mW
465.8	180 mW	415.4 ⎭	(combined)
472.7	240 mW	476.2	50 mW
476.5	720 mW	482.5	30 mW
488.0	1.8 W	520.8	70 mW
496.5	720 mW	530.9	200 mW
501.7	480 mW	568.2	150 mW
514.5	2.4 W	647.1	500 mW
528.7	420 mW	676.4	120 mW

It is also possible to doubly or triply ionize a species. Doubly ionized argon (Ar^{2+}), produced by a higher current density than required for Ar^+, produces several powerful lines in the UV region between 275 and 364 nm, so this laser is an important source of continuous UV output. For doubly ionized argon, however, the ULLs are located about 27 eV above the ion (Ar^{2+}) ground state, which is far above the Ar^+ or neutral ground state. The ULLs for the species are hence 71 eV above ground, so enormous amounts of pump power are required for this laser, with tube currents of 70 A common. Also, efficiency is much lower than for the singly ionized species with power output on the UV lines 20-fold less than the expected multiline visible output. Krypton can be doubly ionized as well but is of even lower efficiency than doubly ionized argon and not commonly available.

Aside from argon and krypton, other noble (inert) gases can be utilized in an ion laser. Xenon-ion lasers, emitting on several lines in the blue, green, and yellow regions of the spectrum, have been demonstrated to operate in pulsed mode. This laser was once popular for precision laser trimming (using the powerful output on several green lines) but has been superseded by more efficient frequency-doubled YAG lasers. Xenon-ion lasers are uncommon today and are usually found only in research labs. Neon, another noble gas, is more efficient than doubly ionized argon at producing UV output. With ULLs centered around 53 eV above ground state, the energy-level situation in a neon-ion laser is more favorable than in a doubly ionized argon (Ar^{2+}) laser with comparable UV output wavelengths, so a neon-ion laser operates at lower currents than does a doubly ionized argon laser. Commercially, there are few manufacturers offering lasers of this type.

Many other gases will operate as ion lasers, such as oxygen and nitrogen. Most were discovered by accident when they were present as impurities in laser tubes intended to lase argon or krypton (in many cases they were left over from an incomplete cleaning of the tubes and electrodes)! It is interesting to note, however, that the fact that the argon ion lases at all was discovered by accident in 1964 when argon was tried as a buffer gas for a mercury-vapor ion laser (the first ion laser discovered). An excellent account of the development of ion lasers can be found in "Ion lasers: the early years" by W. Bridges in *IEEE Journal of Selected Topics in Quantum Electronics*, Vol. 6, No. 6 (Nov./Dec. 2000).

Many materials in addition to gases will operate as an ion laser when vaporized. The aforementioned mercury vapor was the first ion to lase. Buffered with helium at low pressures (and with a high vapor pressure at room temperature), a mercury ion produces pulsed laser output at 615 nm (in the red) and 568 nm (in the green). A number of other metals lase in a similar manner, including selenium, cadmium, silver, gold, nickel, copper, and many others, although many metals require elevated temperatures to increase the vapor pressure and hence the number of atoms available in the plasma. Of the 41 elements that will lase as an ion, only one other than the noble gases (e.g., argon, krypton) became a commercially available laser: cadmium. The cadmium-ion laser, called the helium–cadmium (HeCd) laser, operates much more like a HeNe laser with much lower currents (around 100 mA) and temperatures (around 250°C) than argon or krypton ion lasers. It is thus generally not classed as an ion laser (since it is so different from other ion lasers) despite the fact that the active lasing species is indeed ionic (although one could argue that the ruby or YAG laser is an ion laser, too). Like the HeNe laser, the primary excitation mechanism is transfer from excited helium to the active lasing species. HeCd lasers produce output in the UV at 325 nm, blue at 441.6 nm (the strongest output), and on four other lines in the green and red. These are commercially available from a number of manufacturers and are an important source of continuous UV light.

In any ion laser a high-current density (measured in amperes per square centimeter) in the gas discharge is required to excite the ions to the level required. High currents (up to 40 A for an Ar^+ laser) and a small-diameter (between 0.5 and 2 mm) bore are used in plasma tubes to increase current density since the output power increases as the square of current density. In the case of a 1-mm-diameter bore with a 40-A discharge current, the current density is over $1200 \, A/cm^2$, an extremely large value compared to a typical current density of $0.16 \, A/cm^2$ for a small HeNe tube. Doubly ionized species (Ar^{2+}) require even current densities. These high current densities create a host of problems for designers of ion laser tubes since the plasma runs at an extremely hot temperature and can easily erode tube structures such as electrodes and tube materials.

Finally, revisiting Figure 9.9.1, it is obvious that the efficiency of the argon laser is poor since almost 36 eV of energy must be injected into an argon atom to produce a photon with an energy of about 2.5 eV. The dynamics of a doubly ionized laser (as well as the krypton laser) are even worse: doubly ionized argon (Ar^{2+}) has a ground state 43 eV above ground and upper lasing levels about 28 eV above that! Overall, about 71 eV of energy is required to produce a single photon with 4 eV of energy.

9.10 OPTICS AND CAVITIES

Some small ion lasers have internal optics [notably, Cyonics/Uniphase argon tubes (Uniphase holds the patent)] and resemble overgrown HeNe tubes (with extensive cooling fins), but most lasers, even many small air-cooled tubes, feature external optics. In almost all cases the laser tube has two Brewster windows protruding from the ends of the tube (on quartz stems sealed to the laser tube), so most ion lasers have a polarized output. Like a HeNe, ion lasers have very low gain, so low-loss windows are necessary for operation. Cavity mirrors are mounted on a frame which keeps these aligned. For a longer laser the design of the frame becomes very important, since thermal expansion and mechanical movements can easily misalign the cavity.

Cavities are frequently of plane-spherical type, the OC being spherical and having a radius of curvature slightly longer than the cavity length. This arrangement allows the use of interchangeable flat optics at the HR end. For multiline use a broadband reflector can be installed in the HR position. Selective reflectors may also be used to allow only certain wavelengths to oscillate, as is frequently done with krypton lasers to select only the red line and argon to select the powerful green or blue line (see Section 6.5). Wavelength selectors using a prism and an HR are also an option for single-line operation, and most tunable lasers also allow the addition of an intracavity etalon, allowing single-frequency, narrow-spectral-width operation. To reduce losses at the mirrors, mirrors are made from multiple layers of thin dielectric films.

9.11 LASER STRUCTURE

Because of the high current densities required in an ion laser tube, the temperature of the plasma is incredibly hot, exceeding 5000 K! Glass melts well below this point, so there are a limited number of materials available from which a plasma tube can be constructed: beryllium oxide (a ceramic) and a few high-melting-point (refractory) metals, including tungsten and graphite. In small lasers the bore is sometimes simply made from beryllium oxide, while larger lasers often use a beryllium oxide tube with graphite or tungsten disks inserted into the tube. Holes drilled in the disks form the bore of the laser where the actual discharge takes place. Additional holes around the outside of the disks are used as a gas-return path, to equalize gas pressure in the tube, since cataphoresis will serve to pump gas toward the anode (in older designs, a gas-return path, often external to the tube, was provided). This should give the reader an idea of the engineering required for the relatively complex ion laser tube compared to, say, a HeNe tube.

Even with such exotic materials and construction techniques (such as inserted disks of refractory metal), the energetic plasma of a large ion laser (one with a discharge current of perhaps 30 to 40 A) can easily erode and destroy the tube material on contact. For this reason, magnetic confinement is invariably employed with large plasma tubes. The magnet is coaxial with the laser tube and is water cooled along with the plasma tube itself. Magnetic fields of about 1200 G are employed with visible lasers, which serves to confine the discharge to the center of the plasma tube, away from the tube walls (UV lasers, with higher currents, require even higher magnetic fields). In many lasers, failure of the magnet results in rapid degradation of the plasma tube. By squeezing the plasma to the center of the tube, current density is also increased, increasing the probability of pumping an ion to a high-energy state and hence increasing gain (and ultimately, output power). Generally, small air-cooled lasers do not use confining magnetic fields since their discharge currents are much lower than those of large-frame ion tubes (10 A for a small tube as opposed to 40 A for a large one) as well as the physical difficulties in constructing a magnet around a ceramic tube which must have an extremely large airflow around it to avoid failure (a magnet would serve to insulate the plasma tube trapping heat). With water cooling, of course, this is not a concern.

Whereas the use of a magnetic field enhances output power, too high a magnetic field can actually impair laser output. As the magnetic field is increased, the plasma becomes more confined to the center of the bore, increasing current density and hence output power. However, there comes a point where the Zeeman effect (Section 3.9) of the magnetic field on the ion energy levels becomes significant. As the magnetic field becomes stronger, it causes the energy levels to split into many closely spaced levels. At low magnetic fields this effect serves to broaden laser output spectrally, but as the magnetic field becomes too intense, the gain curve becomes too spread out (the gain curve flattens) with many points at both edges below lasing threshold. Gain is hence wasted and the output power drops. Ion lasers have separate constant-current power supplies for these magnets, and the optimal magnetic field is dependent on the line in operation.

Unlike HeNe lasers, which utilize simple aluminum cold cathodes, heated cathodes are required in an ion laser to prolong the life of the tube. By heating the cathode, electrons are emitted from the surface, which serves to reduce the voltage drop associated with the energy required to pull electrons off the surface of the cathode. With a lower voltage drop, the power dissipated at the cathode drops; the tube runs at lower voltage and there is less damage to the cathode during operation. In most cases the heater is a small coil of tungsten wire coated with oxide to prolong life. A low voltage, around 2.5 to 3 V, at a high current of between 10 and 30 A, heats the coil. The coil itself often acts as the cathode, so the laser tube may have only three electrical connections: two for the cathode/heater and one for the anode (the positive terminal). The cathode must be preheated before initiating the discharge. Most lasers have a built-in delay, which activates the heater for 45 seconds before initiating the discharge.

Some ion laser tubes (especially older ones using graphite disks) feature tertiary electrodes in the quartz stems, extending from the tube holding the Brewster windows, such as those shown in Figure 9.11.1, in which the electrode is seen along with two leads connecting to the cathode. These *stinger electrodes* are used to generate a small charge that repels dust generated from the graphite disks, keeping it off

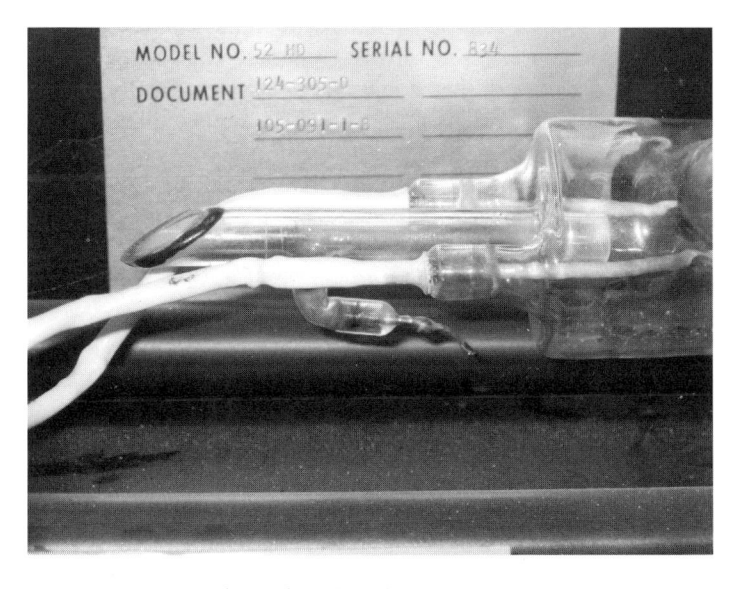

Figure 9.11.1 Stinger electrodes.

the inside surface of the window. Power is provided via a small high-voltage generator, usually inside the laser head itself. These electrodes are also used on some metal-vapor lasers, such as HeCd tubes, to help keep stray metal vapor from migrating toward the windows, where it would be deposited as an opaque film.

As an ion laser operates, it consumes gas through a process called sputtering in which energetic (and heavy) ions strike the internal tube material (both the tube and the electrodes) and are buried into the surface, effectively reducing gas pressure in the tube. As tube pressure is reduced, tube voltage drops, damage to the cathode increases, and laser output drops until it is finally extinguished. Ion lasers hence require a gas replenishment system to keep the tube pressure within a usable range. For a simple ion laser tube such as a small air-cooled type, this gas is usually replenished by means of a large ballast volume connected to the tube. This serves as a buffer to supply extra gas to the tube as required. An external ballast volume (in the form of a tank) is almost always found on krypton laser tubes since the output of these lasers is considerably more sensitive to pressure than is that of argon lasers. On many small argon laser tubes the volume of the cathode and anode is sufficient to permit them to serve as ballast volumes, but with larger ion lasers, an active-replenishment system is used which injects gas into the tube as required. The system monitors the tube pressure either by means of a vacuum gauge directly attached to the tube or by monitoring the voltage across the tube (which drops as the pressure drops). When gas is required, a system of two electromagnetic gas valves releases a small, controlled amount of gas from a high-pressure reservoir into an intermediate volume (the ready condition) and then into the tube (the fill condition). Two valves are opened in sequence (as illustrated in Figure 9.11.2) to increase gas pressure in controlled steps (the small intermediate volume between the valves). This arrangement ensures that the tube is not accidentally overfilled with gas from the high-pressure reservoir. While an underfilled tube can simply be filled with more gas from the reservoir, an overfilled tube is usually impossible to start, and hence an expensive ion laser tube can easily be ruined by overfilling. Large krypton lasers usually employ both a large ballast volume and an active-replenishment system to help stabilize gas pressure as the laser operates.

As an example of a practical ion laser, consider the small air-cooled research laser pictured in Figure 9.11.3. This particular laser is housed in an acrylic box, allowing inspection of the laser components. The laser tube, seen here glowing, is terminated on either end by Brewster windows. The heated cathode is housed in the large bell-shaped piece on the left and the anode on the right. The plasma tube between the electrodes, made of beryllium oxide, is shrouded in a mass of cooling fins. Some light is, however, visible between the fins and the mount; this is emanating from the intense

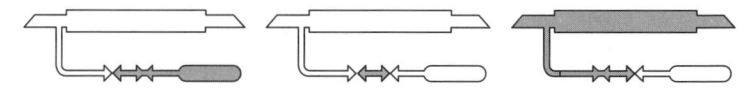

Figure 9.11.2 Ion laser gas replenishment.

Figure 9.11.3 Small krypton-ion laser.

light inside the translucent ceramic bore. Cooling air from a 400-ft^3/min blower enters the laser from the rear, brought in through the white tube behind the laser, and blows directly through the fins to remove the over 1000 W of heat produced. On the top of the entire assembly is a large metal can connected to the anode via a flexible metal line. This serves as the gas ballast, which is quite large in this case because it is a krypton laser, in which the output of the yellow line is extremely sensitive to pressure. Cavity mirrors, as well as the tube assembly itself, are mounted on an optical rail in this case. Many commercially available air-cooled lasers are packaged in a small, compact housing with an umbilical cable connecting the laser head (containing the plasma tube, optics, and some electronics) to the main power supply. Some lasers have huge fans mounted directly atop the laser assembly, while others, like the unit pictured, use a flexible hose to connect the blower. All feature safety interlocks to ensure that airflow is adequate and/or laser temperature is not excessive, since failure of a cooling fan can cause a laser to fail in under a minute!

Figure 9.11.4 Typical large-frame argon-ion laser.

A large-frame ion laser (with the cover removed) is pictured in Figure 9.11.4 together with the remote control operator's panel. This is a standard laboratory-type laser with a water-cooled ceramic plasma tube and magnet. The only parts of the tube that are visible are the heated cathode, in the metal can at the front of the tube, and the Brewster windows mounted on quartz stems protruding from the tube (seen here glowing brightly, with laser light scattered from the windows). In this particular laser an aperture (for mode limiting) is mounted in front of the OC on the left. The HR is mounted on a quick-change bayonet mount, allowing the user to select a multiline optic or a wavelength selector for single-line operation (see Figure 6.5.4). The laser is powered from a separate power supply that consumes 45 A of 208 V three-phase power. The power supply, along with the laser tube/magnet combination, is water cooled.

9.12 POWER SUPPLIES

Whereas HeNe lasers operate at relatively high voltages (1000 to 1500 V) and low currents (3 to 8 mA), ion lasers operate at much high currents (10 to 40 A) and relatively low voltages (90 to 300 V). Even the smallest air-cooled argon-ion laser with 10 A of discharge current and 90 V across the plasma tube represents a power dissipation of 900 W (as opposed to a HeNe tube, which may dissipate only 5 or 6 W). The design of an ion laser supply, then, is radically different from that of a HeNe laser.

The power supplies of most ion lasers, large or small, simply rectify the incoming AC line directly into DC current, sparing the expense of a large transformer. Power supplies for small air-cooled lasers operating in North America, for example, frequently take ordinary 110 V AC wall current and rectify it to about 155 V DC. This DC current is passed through a small ballast resistance (for the same reasons as in the HeNe laser) and through the tube. Large-frame lasers frequently run on three-phase current of higher voltages, which is advantageous since longer tubes present a greater voltage loss (whereas a small air-cooled ion tube might have 100 V across it when operating, a large-frame 7-W argon might require 260 V). So most large ions ($>\frac{1}{2}$ W for an argon) use three-phase power that is rectified into DC. The advantage of three-phase power (aside from lower average currents per phase and higher voltages because of the service—208 V is standard) is that the capacitor bank which serves to smooth the DC current may be much smaller. As illustrated in Figure 9.12.1, when three-phase power is used, the rectified line voltage never crosses zero and varies only 14% or so during the 60-Hz power cycle, whereas single-phase rectifiers cross zero volts 120 times per second. A single-phase system, then, requires larger capacitors to supply current during these zero-cross intervals, whereas a three-phase system can be made with much smaller capacitors.

Because of the low DC resistance of the plasma itself (a tube consuming 10 A at 100 V represents an effective resistance of only 1 Ω), the plasma tube essentially acts as a dead short to the power supply, so DC current regulation is almost always required—this, in addition to a series ballast resistance, which is frequently 1 Ω or

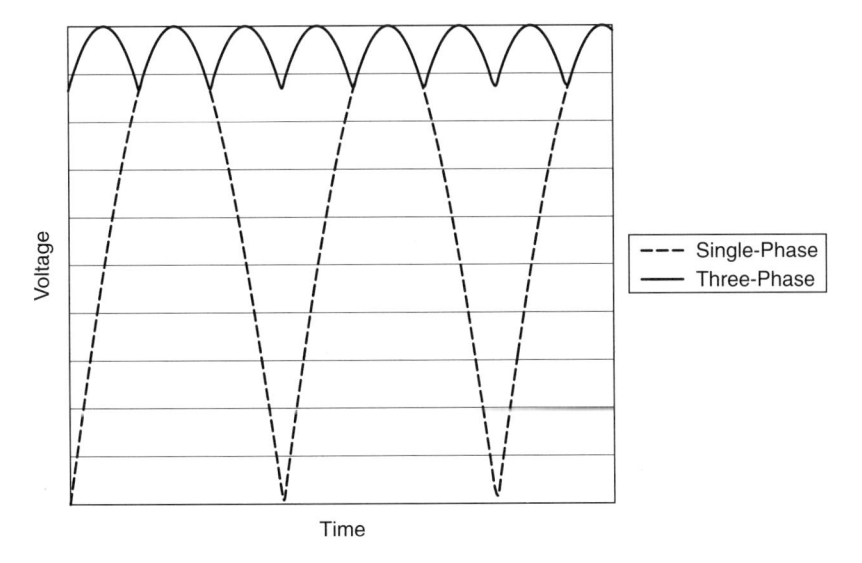

Figure 9.12.1 Single- and three-phase rectifiers.

less. In many lasers current regulation is accomplished by a bank of transistors in what is called the *series pass* or *linear passbank regulator*. Many transistors are used in parallel since it is impossible to find a single transistor that will handle the current required. On a large-frame ion laser, these transistors are mounted on a heatsink cooled by the same water as that used to cool the plasma tube. Some smaller lasers (e.g., air-cooled types) use efficient switchmode regulators.

Having a heated cathode, a low-voltage (2.5 to 3 V), high-current (10 to 25 A) source is required as well as the main supply. This is usually supplied by a center-tapped AC transformer not only because it is inexpensive but also because an AC source (with the main DC from the power supply feeding the center tap of the transformer) ensures that the plasma is distributed over both ends of the filament, prolonging its lifetime. As in the HeNe laser, the main supply voltage (between 155 and 400 V DC) is insufficient to start the plasma; a starting pulse of around 10 kV is required. Because of the high currents in the supply, though, a simple multiplier with blocking diodes will not work here, so a pulse transformer in series with the anode is used to generate the high voltage required. These starting transformers are mounted close to the plasma tube. On a small air-cooled laser they are often mounted on a circuit board beside the tube, while on a large-frame ion laser they are mounted on the side of the magnet or immediately under the tube.

A block diagram showing essential components of an ion laser power supply is depicted in Figure 9.12.2. The rectifier converts AC to DC; in this case a single-phase bridge is shown, but on a large-frame laser using three-phase power, six diodes would be used. The DC current is then smoothed by capacitor C1, providing a constant DC voltage. Current flows from the DC supply through ballast resistor R1 and the secondary winding of the starting transformer. To start the laser tube a pulse

258 VISIBLE GAS LASERS

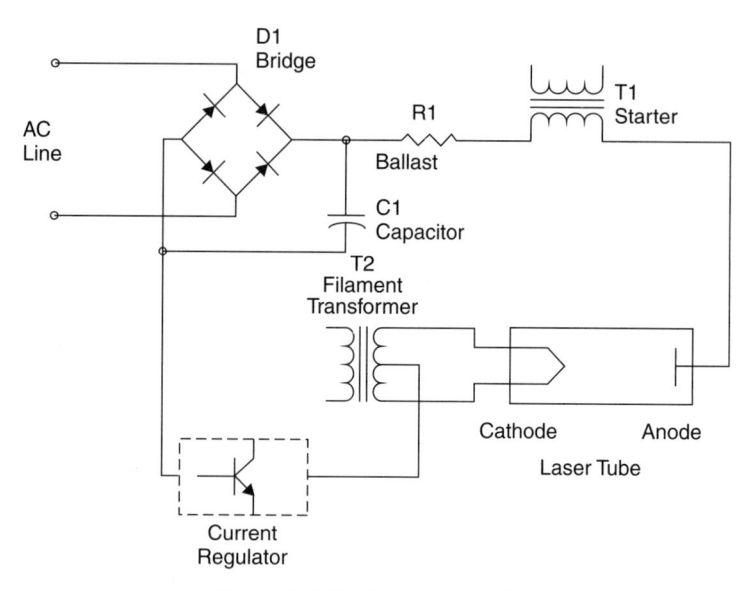

Figure 9.12.2 Ion laser supply.

of a few hundred volts would be applied to the primary of the transformer, which is stepped up to about 10 kV on the secondary, which then appears across the plasma tube, ionizing the gas inside and initiating the discharge. Current flows through the laser tube to the heated cathode, which is powered by a center-tapped transformer. Current passes from the transformer through the current regulator (the series passbank) and back to the main DC supply.

Many laser supplies allow the user to select constant-current or constant-light mode. In constant-light mode, current through the tube is adjusted automatically to obtain a constant output power (which would otherwise drift slowly as the plasma tube heats). A small portion of the output beam from the laser is sampled by a photocell or photodiode and a control loop adjusts current accordingly via the regulator.

9.13 OUTPUT CHARACTERISTICS

In an argon laser (the most popular ion laser), the powerful blue line at 488 nm appears first as current through the discharge is gradually increased, followed by other visible lines. Ultimately, the output power available on the 514-nm green line exceeds that of the blue line, despite the fact that the blue line has the highest gain in this system (almost an order of magnitude greater than any other transition). The quantum processes governing a multiline laser are anything but straightforward.

When operated at high power levels, argon lasers tend to operate in multiple transverse modes. In many cases the TEM_{01} donut mode is favored (probably because the bore is relatively small and hence cannot support particularly high orders). With the aperture installed, the laser can easily be made to oscillate strictly in TEM$_{00}$ mode, giving excellent beam quality. Also, most ion lasers (those with Brewster windows) have highly polarized outputs. Given the high operating temperatures of an ion laser plasma, the spectral width of these lasers is relatively large compared with that of other types. As we calculated earlier, argon lasers frequently exhibit a width of 5 GHz. This, of course, can be reduced through the use of intracavity etalons, allowing single-mode operation with an extremely narrow linewidth of only 1 to 2 MHz.

9.14 APPLICATIONS AND OPERATION

Ion lasers, especially the argon laser, have numerous applications where higher power than that obtainable from HeNe lasers (which are limited to about 100 mW) is required or green-blue light is required. It has become a workhorse lab laser used for all sorts of measurement and excitation purposes, including chemical fluorescence studies and pumping CW dye and solid-state lasers. In the forensics lab, fingerprints are found to fluoresce when illuminated by argon laser light. In medical applications, the wavelength of the argon laser is absorbed readily by red blood cells but passes easily through the liquid filling the eyeball, so it is ideal for certain types of opthalmological surgery, including welding detached retinas. Other medical-lab applications include flow cytometry and DNA sequencing, where the blue and green wavelengths are important to the application.

The powerful beams are also useful for high-speed copiers and printers, where they can be used in the same manner as an IR semiconductor laser is for small laser printers. The more energetic photons and more powerful beam allow faster printing than a semiconductor laser does (many small air-cooled argon lasers are designed precisely for this purpose). In the UV region, ion lasers can be used to expose tiny features on high-resolution printing plates and films (useful since beam quality is excellent and UV wavelengths can be focused to a much smaller spot than visible wavelengths). Other industrial purposes include the mastering of CDs and DVDs.

The krypton laser, which can be designed for white-light output, is a favorite for laser light shows. Multiple wavelengths of the laser can be split using a prism and directed independently, or a single wavelength can be selected from the output by a high-speed AO modulator called a PCAOM. Beams are then scanned using mirrors attached to precision galvanometer movements to create patterns. As in a television set, fast scanning ensures that the eye renders the image continuous.

For years the argon laser remained the only powerful CW source of green and blue laser light available, but at present the market is turning toward simpler (and smaller) frequency-doubled YAG lasers, which feature an output at 532 nm in the green, or frequency-tripled YAGs, which have a UV output at 355 nm. For specific applications requiring the 488-nm blue wavelength, newly developed solid-state lasers with precisely this wavelength threaten the dominance of the argon laser.

Regardless of the appeal of solid-state lasers, the argon ion will remain an important source of blue-green light for years to come, especially since it features a multi-wavelength output which cannot be achieved (easily, at least) with a solid-state laser.

Operation of an older ion laser has been described as more art than science. What with alignment of optics, the necessity to monitor gas pressure manually (in many cases via a thermocouple pressure gauge attached to the tube), perform manual gas fills via "ready" and "fill" pushbuttons as required, and monitor voltage across the passbank and tube to keep each in a safe operating region, these lasers could hardly be called "plug and play" and usually required a skilled operator. Today, electronic controls have greatly simplified the process since most users simply need to press the start button, wait 45 seconds for the laser to preheat and ignite, and then adjust the tube current or light output for the application at hand. Standard features such as interlocks ensure water flow and temperature are adequate to protect the tube from damage due to careless operation.[4]

Ion lasers should not be left in storage for long periods of time, since some of the gas trapped in the tube through sputtering processes (which serve to reduce gas pressure as the tube operates) can be released back into the tube as the laser sits idle, raising the tube pressure. A high-pressure tube can be difficult, if not impossible, to start. Many manufacturers recommend operating an ion laser at least once every two weeks to keep gas pressure in the tube within a nominal range.

Owing to the high cost of replacement (in 2002 a 6-W argon replacement tube listed for $20,000), dead ion tubes are often used for rebuilding, at about half the cost of a new tube. Rebuilding is, however, much more than simply opening the tube, pumping to high vacuum, and refilling with fresh gas (called a "chop and pump" in the industry). Although a simple refill will work to revive a dead tube temporarily, the lifetime of a repumped tube is quite short compared to that of a properly rebuilt tube. Proper rebuilding of an ion tube entails replacement of the entire cathode (which is usually badly eroded by the time the gas is used), and disassembly and cleaning of the entire tube, not an easy task. Laser output is also sensitive to contamination in the tube, so extreme vacuum purity is required. Tubes must be pumped down and cycled to clean them thoroughly before fresh gas is added.

---

[4] I think most people forget to turn on the cooling water at least once in their lifetime!

CHAPTER 10

UV Gas Lasers

Among the most important ultraviolet lasers are the excimer and the nitrogen. These lasers employ a similar technology (it was on the wings of lasers such as the nitrogen that the excimer was made possible), so are covered here, with the somewhat simpler nitrogen laser introduced first and the excimer, which employs similar technology, covered second. Both lasers are molecular lasers in which the lasing species is a diatomic molecule. In the case of the nitrogen laser, the active lasing species is a nitrogen (N_2) molecule; in an excimer a transient molecule consisting of a halide (such as chlorine or fluorine) and an inert gas (such as argon or krypton) is formed within the laser itself. The structure of both lasers is physically similar and features two long parallel electrodes with a low-inductance, fast-discharge path. Excimer lasers are generally much larger than nitrogen lasers, though, and have higher power outputs, producing enormous power outputs in the ultraviolet region of the spectrum.

10.1 NITROGEN LASERS

The nitrogen laser was developed in 1963 by H. G. Heard, who succeeded in producing 10-W pulses of UV light (which sounds impressive enough but with pulse widths in the nanosecond range, the average power of such a laser was incredibly low). Within four years a radical shift in the design methodology of these lasers had been developed, yielding peak powers in the megawatt range! Development continued and TEA (transverse electrical discharge at atmospheric pressure) nitrogen lasers capable of producing megawatt powers using nitrogen at atmospheric pressures appeared. This laser was an important milestone in UV laser development that led directly to the more powerful excimer laser. Among other things, development of the nitrogen laser led to advances in the production of high-speed, high-current discharges required for a high-efficiency nitrogen laser and later for the excimer laser.

Today, the nitrogen laser is found primarily in the lab and is still a useful source of coherent UV light, producing pulses with millijoule energies and pulse widths around 10 to 20 ns. Owing to the ease of construction, it is also a favorite home-built

Fundamentals of Light Sources and Lasers, by Mark Csele
ISBN 0-471-47660-9 Copyright © 2004 John Wiley & Sons, Inc.

laser for both amateur laser constructors and small labs on a budget. These inexpensive lasers are commonly used as workhorses in biology and chemistry labs.

10.2 LASING MEDIUM

As introduced in Chapter 3 and expanded in Chapter 5, diatomic molecules such as nitrogen have vibronic energy levels that combine electronic and vibrational states. The major separation of energy states is caused by a change in the electron energy of the atoms themselves and minor states caused by vibrations of the N_2 molecule itself, with $v = 0$ corresponding to the lowest energy for a given major energy state. The combined effect is that essentially all energy levels consist of a series of closely spaced levels essentially forming a band. As expected, the lasing transition is actually a series of transitions closely spaced to render a primary output centered at 337.1 nm and having a bandwidth of 0.1 nm. These transitions are the result of a change in both electronic and vibrational states of the diatomic molecule.

Excitation of the nitrogen molecule is accomplished by collision of high-energy electrons in a gas discharge. A massive electrical pulse, with currents on the order of thousands of amperes, initiates the process by creating a large flux of electrons with high energies in the laser tube. Electrons strike the nitrogen molecules and excite them to a high vibronic energy state. In the excitation process for this laser, nitrogen molecules do not dissociate into an atomic species but rather, remain as an excited molecular species.

The energy levels of the nitrogen molecule as they apply to this laser are outlined in Figure 10.2.1. Each level shown is actually a series of vibrational levels dependent on internuclear separation as outlined in Chapter 3. The laser begins when a nitrogen molecule is excited by direct collision with electrons in the discharge to enter the ULL, termed the $C^3\Pi$ *energy band* of the molecule. From the ULL the molecule falls to the LLL (termed the $B^3\Pi$ *band*), emitting a photon of UV light in the process. Transitions in a normal nitrogen laser operating at 337.1 nm are in the *0–0 band*, in which the only levels involved are those with the lowest vibrational state ($v = 0$). Other transitions are possible in the *0–1 band*, where the upper lasing level is the same lowest state ($v = 0$), but the lower lasing level involved is the $v = 1$ vibrational state. The transition is a smaller jump than the 337.1-nm transition and hence has lower energy with a wavelength of 357.7 nm, the difference between the $v = 1$ and $v = 0$ levels being about 0.2 eV.

After emitting a photon in the lasing transition, the molecule then falls to a metastable state and finally, to the ground state. Although not a four-level laser in the proper sense (the system lacks a pump level since the ULL is pumped by direct electron collision), it is certainly not a three-level system since it does feature a distinct LLL. In this respect the system resembles that of a copper-vapor laser with the inclusion of a metastable state. Each band then consists of multiple closely spaced energy levels, resulting in many possible transitions. In all, molecular nitrogen lases at 61 known wavelengths in the 0–0 band between 336.4903 and 337.9898 nm, with the vast majority of lines clustered around 337.1 nm (with spacing between lines

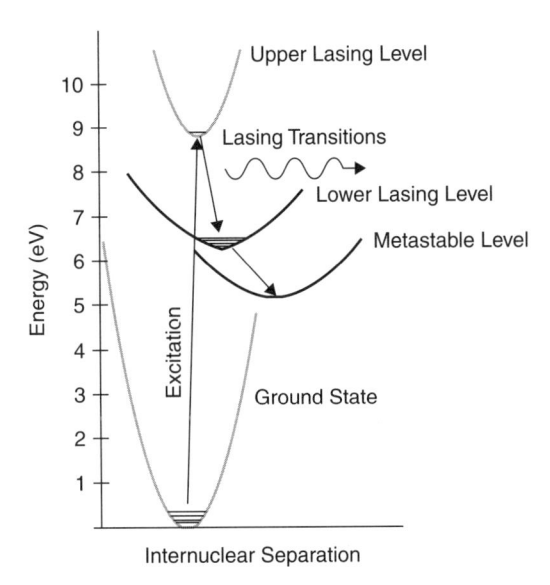

Figure 10.2.1 Molecular nitrogen laser energy levels.

typically below 0.005 nm) and a FWHM of the combined output from these lines of about 0.1 nm. Nitrogen lasers can also operate on transitions involving the ionized molecule (N_2^+) with an output at 427 nm. Such transitions are only possible, though, in a laser operating at high pressures and with large quantities of helium in the gas mixture. This is possible in an excimer laser (also covered in this chapter) by using a helium–nitrogen gas mixture.

The energy levels for molecular nitrogen are far from favorable for lasing, with the upper lasing level having a much shorter lifetime than the lower lasing level. The upper lasing level has a lifetime that is pressure dependent according to[1]

$$t = \frac{36}{1 + p/58}$$

where t is the ULL lifetime in nanoseconds and p is the pressure in torr. So for a nitrogen laser operating at 60 torr, the ULL lifetime is about 18 ns. This is six orders of magnitude shorter than the LLL lifetime of about 10 µs. From the LLL the molecule falls to a metastable state with a longer lifetime yet.

To obtain lasing action from molecular nitrogen there must be a higher population of molecules at the ULL than at the lower level. The level lifetime situation makes CW laser action impossible with this species, but pulsed laser action is still possible provided that the laser mechanism can pump energy preferentially and

---
[1]This relationship and ramifications for TEA laser design are outlined in "Compact high-power TEA N_2 laser" by B. S. Patel in *Review of Scientific Instruments*, Vol. 49, No. 9 (Sept. 1978).

quickly into the upper lasing level to generate an inversion. In a nitrogen laser this is done by a massive electrical pulse where electron collisions cause the preferential population of the upper energy band first. (Were it not for this effect of being able to pump the upper energy level first, this laser would not work at all.) After about 20 ns, though, this energy will decay to the lower level, population inversion is lost, and lasing action will quickly cease. To make matters worse, nitrogen molecules at the lower state absorb UV light strongly, so gas in the laser channel quickly absorbs laser energy.

On average, then, the pulse length of the nitrogen laser is limited to less than 20 ns (the point where population inversion is no longer possible since half of the molecules in the upper energy state have decayed to the lower state). As we have seen, the lifetime of the upper lasing level depends on the gas pressure in the laser tube. The value of 20 ns quoted assumes a low-pressure laser design where lasing gas is at a pressure between 25 and 60 torr (where 760 torr is 1 atmosphere), but in a TEA laser the pressure is much higher, 760 torr or more, so the lifetime is about 2 ns.

10.3 GAIN AND OPTICS

Nitrogen lasers operate without mirrors, a type of laser called *superradiant*, discussed in Chapter 6. The gain medium exhibits such high gain that spontaneous emission from nitrogen molecules at the end of the laser are amplified by the same group of nitrogen molecules in the tube, producing a usable output pulse in a relatively short path length. Gain for a nitrogen laser is on the order of 40 to 50 dB/m or more,[2] depending on the specific laser. Even at the lower figure of 40 dB/m, light is amplified by a factor of 10,000 for every meter of travel through the laser tube! This is an enormously high gain for a laser which also serves to prohibit tuning of the laser through any of the 61 lines in the gain band, since they will lase without any cavity mirrors, and hence a wavelength selector is ineffective.

Although not required for laser action, a single high-gain reflector is frequently installed in a nitrogen laser tube as well as a 4% reflective (one uncoated surface of a window) output coupler. Logically, one expects to double output power in this manner (by adding the rear and forward beams), but the actual increase is over threefold, since light reflected into the laser channel does not simply pass through the tube but is further amplified along the way. Use of a cavity such as this also decreases beam divergence by at least 50%. Optics must be designed for UV, so the coating of the high reflector must reflect UV (aluminum is frequently

---

[2] Gain figures of about 50 dB/m are common for a Blumlein nitrogen laser. The highest gain reported for a nitrogen laser was a whopping 75 dB/m from a multielectrode crossed-field device with short discharge lengths, which serve to give an unsaturated gain (i.e., the gain medium is not saturated, so power output does not increase linearly with length but rather with length squared).

used), and windows on the laser tube must be made of quartz or some other material that is transparent to UV radiation.

10.4 NITROGEN LASER STRUCTURE

The basic requirement for a practical nitrogen laser is to supply a massive electrical current (i.e., a huge quantity of electrons) with a fast rise time and short pulse length to excite the gas. To achieve this, most nitrogen lasers use an electrical configuration called a *Blumlein configuration*, which generates a massive overvoltage of the laser channel (and subsequent large current through the lasing gas) with a rise time of nanoseconds. A Blumlein configuration is shown schematically in Figure 10.4.1, where we find two capacitors essentially in parallel separated by the laser channel itself. Both capacitors charge simultaneously through the charging inductor (which offers little electrical resistance to the charging current) until the spark gap fires when the breakdown voltage is reached (typically, about 15 kV for a small laser). In simplest form, a nitrogen laser may operate as a relaxation oscillator, repeatedly charging and firing; the use of triggered spark gaps or thyratrons allows the laser to be triggered as required. The gap now conducts essentially short-circuiting C1 and draining charge from it, making the top terminal of C1 negative. A massive voltage difference appears quite suddenly across the laser gap since the left side of the tube is now negative and the right side still positive. Charge from C2 flows across the laser channel as a pulse of very high electrical current, in many cases thousands of amperes.

The electrical dynamics of the laser are not as simple as a discharge, though, since the laser tube has long transverse electrodes over which the high current must be distributed. With a short discharge time necessitated by the short upper-lasing-level lifetime, the laser is best excited by a traveling electrical wave which starts at the rear of the laser and moves forward at the speed of light exciting nitrogen molecules as it progresses. To accomplish this, capacitors are fabricated as long, distributed capacitances parallel to the laser tube. The initiating spark gap is placed at

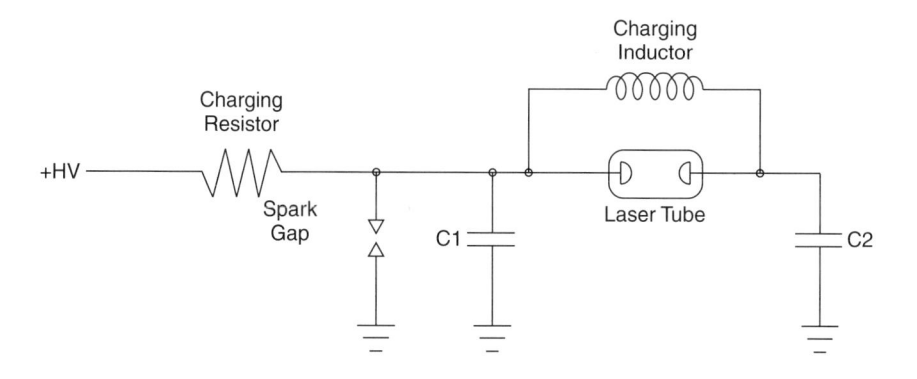

Figure 10.4.1 Electrical schematic of a Blumlein laser.

the rear of one capacitor so that the electrical pulse begins at the rear of the laser first and travels toward the front of the laser. In many cases it is possible to design the laser as a transmission line for efficient transfer of electrical energy into the lasing volume. The sequence of events during firing of the laser is outlined in Figure 10.4.2. In this particular design, common for small laboratory-type nitrogen lasers, the capacitors are fabricated on an epoxy–glass substrate. Two capacitors are formed with copper foil on the top of the board (there is a separation between the two capacitors underneath the laser tube, so they are not visible in the figure), and the bottom of the board is the common terminal for both. The spark gap that initiates electrical discharge is mounted on the left capacitor near the rear of the laser.

In the simplified figure (in which details such as the charging inductor are omitted for clarity) both capacitors are charged to a high voltage equally, so no voltage difference appears across the laser channel. As the capacitors are charged, the voltage across the capacitors as well as the spark gap rises until breakdown occurs (in Figure 10.4.2b) and the spark gap conducts. Charge can be visualized in the figure as moving toward the spark gap in an arc centered around the spark gap. No voltage appears across the laser channel until the charge at the rear of the left capacitor for the laser has been drained and a voltage differential appears at the rear of the laser channel. The discharge in the laser thus begins here. As the traveling wave in the capacitor spreads, the voltage differential travels toward the front of the laser channel at the speed of light, generating a discharge that also travels toward the front of the laser. Light emission follows the discharge and a beam emerges from the laser.

A practical nitrogen laser based on this design is seen operating in the author's laboratory in Figure 10.4.3. The initiating spark gap is visible in the upper-left cor-

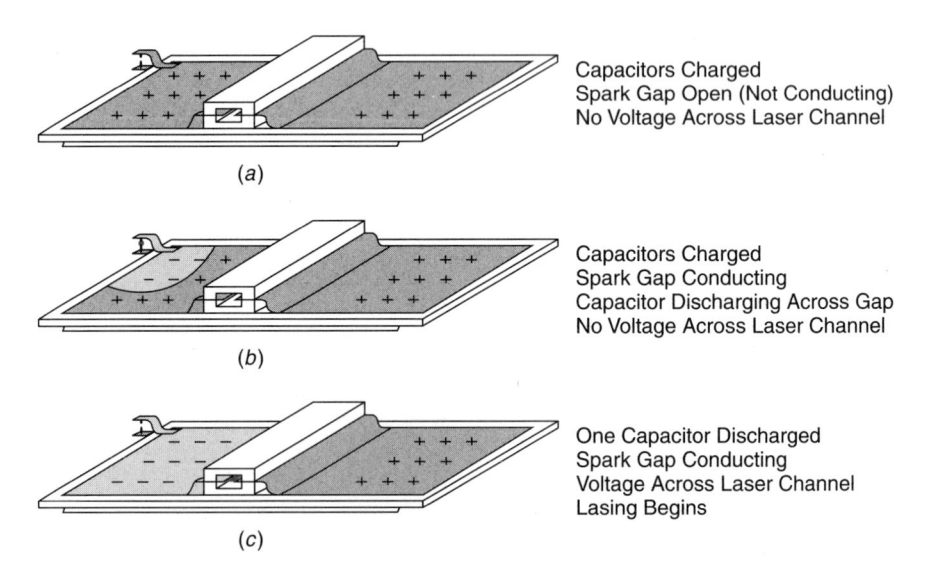

Figure 10.4.2 Nitrogen laser discharge sequence.

Figure 10.4.3 Practical nitrogen laser.

ner. It emits an intense flash of light since high currents pass through it. As well as the gap, other visible features include the charging inductor, which bridges the laser tube and the charging resistor near the spark gap. This is a flowing-gas laser in which nitrogen gas under low pressure flows slowly through the tube. A needle valve used to regulate flow is also seen in the lower left of this laser. After initiation of laser action by the spark gap and discharge through the laser channel, the laser pulse is generated and will continue until population inversion ceases. Laser action ceases regardless of how long the electrical pulse lasts, so it is pointless to design a laser with a longer discharge time than this. This parameter determines the size of the capacitors employed in the laser.

As well as operating at low pressures (generally, 20 to 60 torr), nitrogen lasers may operate at atmospheric pressure in a TEA configuration. These lasers are physically similar to the low-pressure types described in this chapter, and most use a Blumlein configuration. In the case of a TEA nitrogen laser, though, the lifetime of the ULL decreases to about 2.5 ns. The requirements for a fast discharge are even more pronounced in a TEA laser, which must be constructed to keep inductances in the discharge path to an absolute minimum. With this in mind, dielectrics for capacitors are kept very thin (since the intrinsic inductance of a transmission-line capacitor is proportional to the thickness of the dielectric) and the laser channel is mounted directly on top of the capacitors. A practical TEA laser is pictured in Figure 10.4.4, in which the spark gap is located in the upper-right corner of the photo and the long transverse electrodes are visible down the center of the laser.

The laser pictured operates using open air as the lasing gas. This is possible since air is 78% nitrogen, although output power decreased by 80% over the use of pure nitrogen gas. A possible source of inefficiency is evident in that the discharge reveals hot spots or arcs. Unlike the low-pressure laser, which features a consistent and even discharge between the electrodes, discharges in TEA lasers tend to concentrate and resemble individual sparks. For efficiency, measures must be taken to even out the discharge, including dilution of the nitrogen gas with helium, use of an electrode structure consisting of multiple points, and preionization of the discharge channel

268 UV GAS LASERS

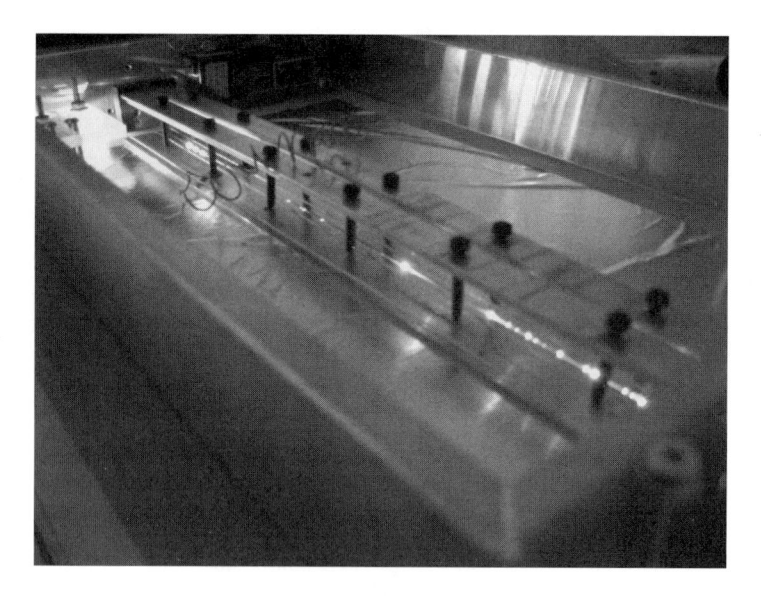

Figure 10.4.4 TEA nitrogen laser.

with a high-voltage corona or ultraviolet radiation before the main laser discharge ensues.

Most small commercially available nitrogen lasers use spark gaps for their simplicity. These spark gaps are filled with nitrogen gas for reliable, predictable firing. In the case of a TEA laser, the gap may be placed inside the same pressure vessel as the rest of the laser. Filling with nitrogen also eliminates the objectionable production of ozone when the gap fires. In some larger lasers, thyratrons are used instead of spark gaps. Thyratrons are switching devices that use mercury vapor or hydrogen gas and feature incredibly fast rise times, many times faster than spark gaps. As well as faster switching times, thyratrons also allow triggering on command instead of simply triggering when the spark gap is overvoltaged, an important feature when this laser is used in a laboratory experiment requiring synchronization and precise timing.

Because of the low inductance discharge path required, essential elements of the nitrogen laser, such as storage capacitors and switches (thyratrons or spark gaps), are an integral part of the laser itself. To charge the capacitors, a basic high voltage supply is required, the supply current of this supply limiting the maximum firing rate for the laser. Designs for high-voltage power supplies vary from simple neon-sign transformers to efficient and compact switching power supplies. These supplies are usually housed in the main laser housing for safety.

In many low-pressure nitrogen lasers, gas flows continually through the lasing channel. This helps to eliminate impurities generated during the discharge as well as cool the laser. Gas flow is quite slow, so consumption is minimal. Many small commercial nitrogen lasers are of the TEA variety and use a sealed laser channel, so a gas supply and vacuum pump are not required, making a much simpler laser for laboratory use. With a fast enough discharge time, some low-pressure nitrogen lasers can also operate using hydrogen gas (with a transition in the extreme UV at 160 nm) or neon gas (with a visible transition at 540.1 nm). Both of these alternatives have shorter upper-level lifetimes than nitrogen does, so an extremely fast laser discharge is required (much faster than normally required in a nitrogen laser). Another alternative is the use of a carbon dioxide gas mixture consisting of helium, nitrogen, and carbon dioxide. With suitable optics the laser operates as a TEA laser, producing fast, short pulses of light at 10.6 μm in the infrared.

10.5 OUTPUT CHARACTERISTICS

Beam quality is poor in a nitrogen laser. Being a superradiant laser with enormous gain, the output really consists of highly amplified spontaneous emission. In a lower-gain laser, photons traverse the cavity many times, stimulating the emission of many more photons of exactly the same wavelength. In a nitrogen laser, photons often make only one pass through the amplifier before exiting. Collimation is hence poor and divergence is quite large compared to other types of lasers. Coherence length is also poor, since the spectral width of the laser output is quite wide. Molecular nitrogen lases at 61 known wavelengths between 336.4903 and 337.9898 nm, an extraordinarily wide range for a laser, but the vast majority of lines are clustered around 337.1 nm, so the FWHM of the combined output from these lines is about 0.1 nm. Since gain is high, wavelength selectors are ineffective in allowing single-line operation.

10.6 APPLICATIONS AND PRACTICAL UNITS

Given their intense UV output and short pulse length, nitrogen lasers make excellent pump sources for pumping dye lasers (covered in detail in Chapter 14). The 337.1-nm UV emission is absorbed readily by many laser dyes, and in this application the wide spectral bandwidth and poor coherence of the laser beam are unimportant. In addition, the short pulse length does not allow triplet production in the dye, so the conversion efficiency of the laser-pumped dye laser is very high. Nitrogen lasers are also useful for exciting fluorescence in substances other than laser dyes, allowing studies of these molecules. As an excitation source, the molecules are commonly used to produce ions for time-of-flight spectroscopy. In the MALDI (matrix-assisted laser desorption/ionization) technique, the laser vaporizes and ionizes nonvolatile biological samples, which are then analyzed by a time-of-flight mass spectrometer and detected based on their mass-to-charge ratio. Nitrogen lasers may also be used for small microcutting procedures on individual biological cells or for trimming thin films for the semiconductor industry. With a typical pulse energy of less than 1 mJ and a repetition rate under 100 Hz, nitrogen lasers yield a low-cost source of intense UV light.

A few commercial manufacturers of nitrogen lasers build both low-pressure and TEA variants. Many manufacturers of nitrogen lasers also offer matching dye lasers designed to be mated with the same brand of nitrogen pump laser to form a compact package. As well as a laser designed exclusively to operate as a nitrogen laser, many excimer lasers (see Section 10.7) can also use nitrogen gas, although this is an expensive option for obtaining UV at 337.1 nm since excimer lasers are considerably more expensive. Since an excimer laser operating as a nitrogen laser operates in TEA mode with pressures averaging over 2 atm, measures must be taken to ensure that the discharge does not occur in localized hot spots along the long transverse electrodes. Aside from preionization employed in all excimers, the gas mixture also consists of a small amount of nitrogen diluted in helium. One popular manufacturer of excimer lasers specifies a gas mixture for use as a nitrogen laser (N_2 at 337.1 nm), consisting of 2% nitrogen and 98% helium. To lase the N_2^+ species at 428 nm, the same manufacturer specifies a mixture consisting of only 0.2% nitrogen, with the balance helium.

Owing to the simple mechanism of this laser, many are constructed in-house by both amateur laser constructors and commercial laboratories. Low-pressure lasers can be constructed easily using thin printed-circuit board for capacitors and a laser channel manufactured from acrylic plastic or glass. TEA types are constructed using lower-inductance materials (and hence faster discharge times), using capacitors fabricated from thin polyethylene or Mylar sheets and foil electrodes.[3]

10.7 EXCIMER LASERS

Excimer lasers[4] produce intense pulsed output in the ultraviolet. The excimer is unique because the lasing molecule is one consisting of a halogen and an inert gas. This molecule is transient and exists for only a moment in time. Like the nitrogen laser, a fast, high-current discharge is required to produce the excimer molecule, but excimers lasers are considerably more complex since they operate at high pressures and one of the active gases is highly toxic and corrosive. The first discharge excimer lasers were simply TEA CO_2 lasers (which, like nitrogen lasers, feature long transverse electrodes) with an excimer gas mix and UV optics. The problem with this scheme was the halogen gas in the excimer mix, which quickly attacks laser electrodes. Manufacturers quickly adapted their lasers to handle these halogens

---

[3]This author has personally constructed several nitrogen lasers in the lab from components available in a hardware store. My favorite "junkbox" laser was built from aluminum foil and polyethylene sheeting (from leftover vapor barrier when an addition was built on my house), with a piece of metal used as a transition between hardwood and carpet floors used for electrodes. The entire apparatus was held down using scuba weights (which compressed the transmission lines) and was charged to 5 kV using a power supply scavenged from an old laser printer destined for a dumpster. Details of this "under $10" laser may be found on the web site for this book.

[4]Two essential articles for those wishing to delve into the development of UV laser technology are "From N_2 to high-order harmonic generation: 40 years of coherent source development in the UV and VUV" by J. G. Eden in *IEEE Journal on Selected Topics in Quantum Electronics*, Vol. 6, No. 6 (Nov./Dec. 2000), and "Excimer laser technology development" by J. J. Ewing in the same volume.

and excimer technology progressed especially in the electronics used to produce high-current pulses in the laser.

Modern excimer lasers produce pulses with energy ranging from 0.1 to 1 J and can (for a large industrial laser) produce these pulses at a rate of over 300 per second. Pulses are fast, with a FWHM of 10 to 30 ns. Fast pulses combined with high peak powers (a 0.5-J pulse with a 20-ns width corresponds to a power of 25 MW) serve to ablate target material without heating the surrounding material, a desirable effect for many processing purposes.

10.8 LASING MEDIUM

Energy levels in an excimer laser are defined by the state of the atomic components. When unbound, the energy of the system depends purely on the separation between the individual atoms; as the atoms move closer together, energy rises. This is illustrated by the lower curve in Figure 10.8.1. The lower energy level in an excimer system is defined as a separation of the halogen and inert gas atoms. This is the normal state for an inert species such as argon, krypton, or xenon, which will not, under ordinary circumstances, form compounds with any other atom. The upper

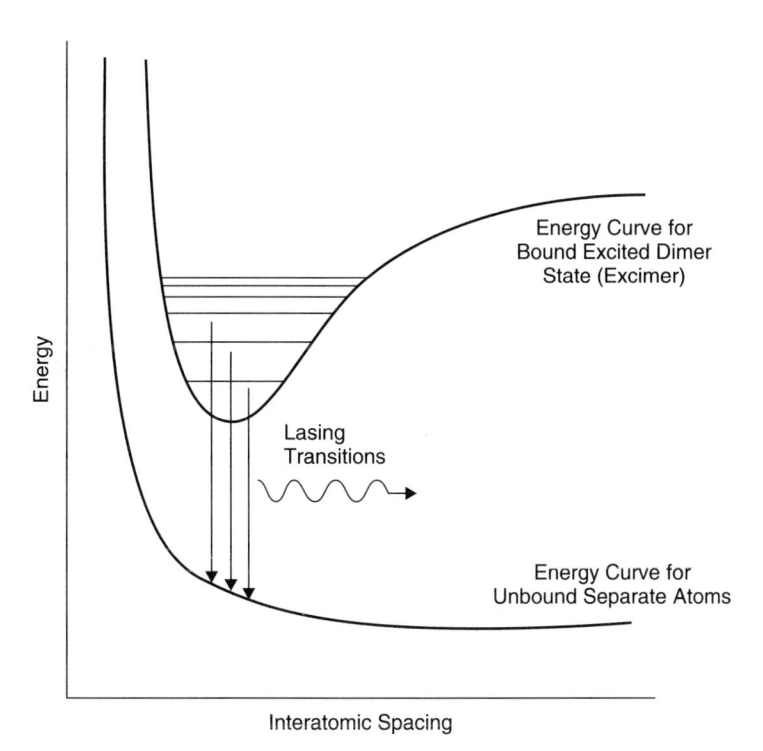

Figure 10.8.1 Excimer energy levels.

energy state is formed when the inert atom and the halogen form an excited dimer (or *excimer*) molecule. This is a temporary state that forms only for an instant in time when the excited atoms combine to form a molecule. The energy of the excimer molecule is much higher than that of the unbound individual atoms and also depends on the interatomic spacing in a similar manner to that of any other molecule, such as hydrogen (described in Chapter 3). As with the hydrogen molecule example, numerous vibrational levels are possible. Collectively, these closely spaced levels form a band serving as the upper lasing level.

The laser tube is filled with a mixture of an inert gas and a halogen such as fluorine or chlorine. When an enormous, fast-rising electric current (generated by a mechanism similar to that of a nitrogen laser) passes through the mixture, gas atoms become extremely excited and excimer molecules form. This forms the ULL for the species, which has a short lifetime, on the order of tens of nanoseconds. In this respect the excimer laser resembles that of a nitrogen laser and a fast discharge is required to generate the excimer molecules effectively and ensure an inversion before the lifetime of the species has elapsed.

Various excimer species are outlined in Table 10.8.1, which also lists the relative power output relative to KrF, the lost powerful excimer laser. Although KrF produces the most powerful output, other gas mixtures, such as XeCl, are popular for use in excimer lasers. Shortcomings of the KrF laser include the output wavelength of 249 nm, which is absorbed readily by air, and the extremely corrosive nature of fluorine, which shortens the useful life of the gas mix in this laser. Worse yet from a beam-management perspective is ArF, which produces a wavelength so short that it produces ozone gas from atmospheric oxygen as it passes through air. When using ArF, beam paths must be enclosed and flushed with dry nitrogen, helium, or argon. XeCl, on the other hand, has a longer wavelength, allowing better transmission in air and the use of considerably cheaper optics. The gas mixture also has a much longer useful lifetime (up to 10 times longer, by some estimates). The useful lifetime of lasing gases may also be extended by using a cryogenic gas processor in which the lasing gas mixture is passed through coils immersed in liquid nitrogen to trap impurities (some of which adsorb from internal components and others produced by the hostile environment inside the laser) in the gas mixture. A cryogenic gas processor such as that shown in Figure 10.8.2 is a standard option on most commercial excimer lasers. In the figure two copper lines, evident on the top of the processor, connect to the laser vessel itself, and laser gas is recirculated continually by a pump in the pro-

TABLE 10.8.1   Excimer Species

Laser Species	Wavelength (nm)	Relative Power Output
ArF	193	0.5
KrF	249	1.0
XeCl	308	0.7
XeF	350	0.6

Figure 10.8.2 Cryogenic gas processor.

cessor through these lines and into the processor's cold trap. Many large excimer lasers (i.e., intended for prolonged use) have gas ports allowing connection of a gas processor. The cost of a gas fill is quite pricey (for many lasers, well over $100 per fill), so gas processors pay back quickly for many industrial users.

The actual gas mixture employed consists of a small quantity of fluorine combined with a moderate amount of inert gas, with the balance helium or neon. Pure fluorine is such a corrosive and toxic gas to handle that it is not available in pure form but only in diluted form, as 5% fluorine and 95% neon gas (or possibly, helium). Typical tube pressures are around 2 to 3 atm. Table 10.8.2 lists typical

TABLE 10.8.2 Excimer Gas Mixtures

Excimer Species	Halogen	Inert Gas	Balance[a]
KrF	0.2% fluorine	5% krypton	Helium
	0.1% fluorine	2% krypton	Neon
ArF	0.23% fluorine	14% argon	Helium
	0.1% fluorine	4% argon	Neon
XeF	0.39% fluorine	0.75% xenon	Helium
	0.15% fluorine	0.35% xenon	Neon
XeCl	0.06% hydrogen chloride[b]	1.5% xenon	Helium
	0.06% hydrogen chloride[b]	1.5% xenon	Neon

[a]The use of neon as a buffer gas does not altogether eliminate the need for helium, since helium is used, along with a small percentage of fluorine, to passivate (condition the surface of) system components in the laser.

[b]In both cases, hydrogen gas is present in the supply cylinder at about one-half the concentration of hydrogen chloride to prolong gas life. Even so, halogen gas shelf life is only about six to nine months.

excimer gas mixtures for two different excimer lasers, as specified by a manufacturer of such lasers. In all cases, only a tiny amount of halogen (fluorine or HCl) is used in the gas mixture, along with a moderate amount of inert gas, with the vast majority of the mixture an inert buffer gas such as helium or neon. Aside from assisting in excitation of the molecule and helping to distribute the discharge, helium also helps to conduct heat away from the discharge so that the gas mixture can be cooled by a heat exchanger inside the laser housing.

10.9 GAIN AND OPTICS

Like nitrogen lasers, the gain of excimer lasers is extremely high, so the output is superradiant. A single rear mirror is employed and an output coupler of 4 to 8% which transmits in the region of interest (which is in the UV or vacuum–UV region, depending on the gas mixture used). Divergence of the beam is reduced when a full optical cavity is used, and alignment is easy since the laser operates even when cavity mirrors are completely misaligned. The beam is rectangular in profile. A stable resonator consisting of a totally reflecting rear mirror and an output window yields the highest output pulse energies and uniform energy distribution throughout the beam, although divergence is somewhat high (usually, a few mrad). Unstable resonators are often used with excimer lasers to improve the high divergence. This type of arrangement also increases the brightness at the center of the beam, making the beam more focusable for cutting applications. For use as an illumination source for photolithography (a primary use for excimer lasers), the effects of laser speckle must be eliminated, so it is desirable that the output beam be multimode, a simple matter since the beam profile is quite large compared to most lasers. Unlike nitrogen lasers, quartz cannot be used for most excimers since fluorine attacks the material. Output couplers are hence made primarily of magnesium fluoride (which absorbs less UV at most wavelengths than quartz does, making it a better choice regardless).

10.10 EXCIMER LASER STRUCTURE

Excimer lasers resemble TEA lasers in that they feature two long, transverse electrodes and the gas pressures are at atmospheric pressures or greater (most excimer lasers operate at many times atmospheric pressure). Unlike the nitrogen TEA laser, the lifetime of the ULL is on the order of tens of nanoseconds, so the requirements for a low-inductance (and hence fast) electrical discharge path are somewhat relaxed (although by no means trivial). Most excimer lasers use a discharge circuit consisting of a single large capacitor and a low-inductance thyratron switch, as outlined in Figure 10.10.1.

The main capacitor (C1) charges with high voltage of about 40 kV through the charging resistor and the charging inductor. To fire the laser, the thyratron is triggered, shorting the left side of C1 to ground. Current then flows from capacitor C1 through the laser tube, where the discharge occurs. This discharge occurs very

EXCIMER LASER STRUCTURE

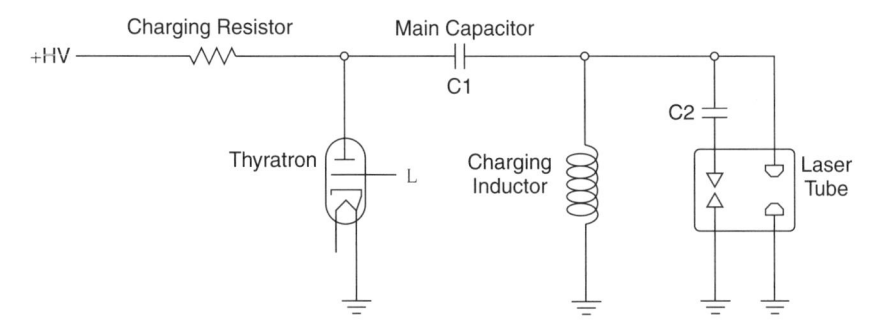

Figure 10.10.1 Excimer laser discharge circuitry.

rapidly, on the order of around 40 ns, so the charging inductor is essentially an open circuit during the rapid discharge. Essential components of the excimer laser high-voltage section are annotated in Figure 10.10.2. The capacitor is the large white block and is connected to the thyratron via a thick copper tube.

Current flowing from the capacitor flows into the laser tube but discharge does not occur immediately since the pressure of the laser tube is high (many atmospheres) and gas inside the laser channel is not yet ionized. Ionization is performed by current flowing through small capacitors C2 and jumping small preionization spark gaps inside the laser tube immediately adjacent to the main laser discharge channel. UV radiation produced from these sparks ionizes gas in the laser channel, which then conducts the main discharge current, producing a laser pulse. A photograph of the laser channel is shown in Figure 10.10.3. The white cylinders on either

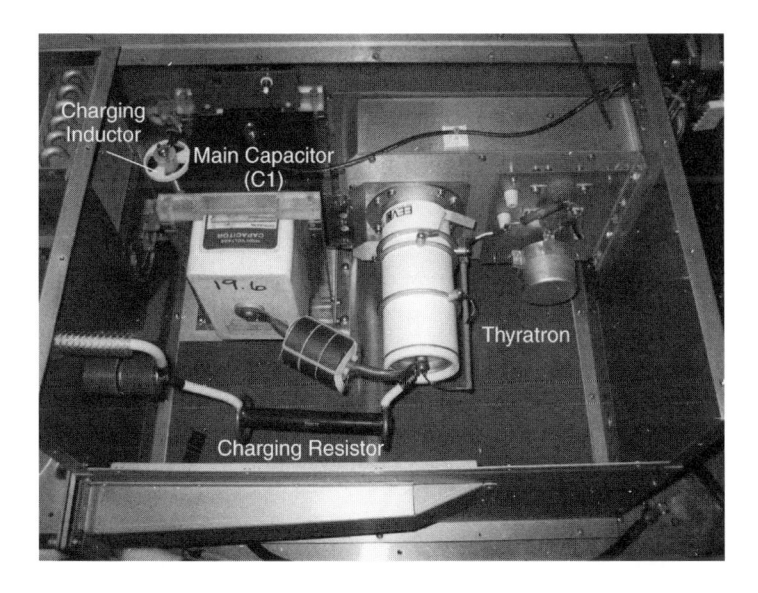

Figure 10.10.2 Excimer laser high-voltage section.

276 UV GAS LASERS

Figure 10.10.3 Excimer laser discharge channel with preionizer.

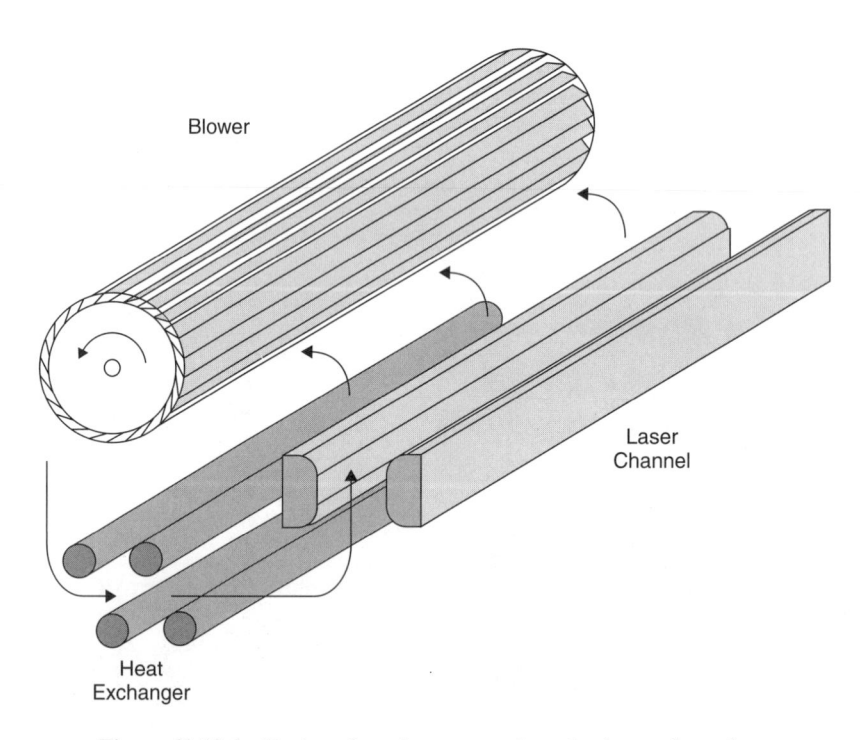

Figure 10.10.4 Excimer laser heat removal mechanism and gas flow.

Figure 10.10.5 Excimer laser heat exchanger and fan.

side of the laser discharge channel are the preionization capacitors (C2) which are inside the actual laser housing. A detail in the right corner of the figure shows a single preionization spark gap. There are two rows of 13 in this laser, one row on either side of the laser channel. It is evident that the preionization gaps are much smaller than the gap for the laser channel, ensuring that these gaps fire despite the high pressure in the tube.

Heat removal from the discharge is a major issue with an excimer laser. Given that the input energy to the tube is several kilowatts, there is a large amount of heat that must be extracted from the high-pressure lasing gas. This is accomplished by using a large squirrel-cage blower and water-cooled heat exchanger tubes within the laser vessel itself. Laser gases are forced around the vessel at high speeds: through the lasing channel and passing through water-cooled coils where the gas is cooled and recirculated through the laser. The heat-removal mechanism is diagrammed schematically in Figure 10.10.4 and shown in Figure 10.10.5, where the laser channel and capacitors visible in Figure 10.10.3 have been removed to reveal the blower and heat-exchanging tubes (four of them in this case).

10.11 APPLICATIONS

With high average powers (commonly over 100 W for many commercial lasers) and an output in the ultraviolet region of the spectrum, excimer lasers are useful for many applications, ranging from dye laser pumping to cutting and materials processing applications. The largest commercial application for excimers are use in eye surgery to correct the shape of the cornea to reduce the need for corrective lenses (commonly known as *lasik surgery*). For these applications ArF with an output wavelength of 193 nm and a pulse energy of under 0.5 J is used. The 193-nm wavelength is used since it is readily absorbed by tissue on the surface of the cornea, which is ablated without producing significant heat to damage surrounding tissue. Since the laser light is absorbed readily by tissue on the front of the eye, no laser energy is transmitted to the retina, sparing it damage. By 2003, over 2 million such corrective surgeries have been performed using the excimer laser.

The largest current industrial use for excimer lasers is photolithography. When manufacturing integrated circuits, multiple layers of semiconductor material are "grown" onto a wafer of silicon by diffusion at high temperatures. Features (desired areas) are then produced on the layer by coating the wafer with photoresist, exposing the photosensitized wafer to UV light passed through a mask outlining features desired on that layer, developing the resist, and etching away unwanted silicon material with hydrofluoric acid. To expose the photoresist, UV light from a mercury lamp, the standard light source for such uses until the mid-1990s, is passed through a photographic mask. Whereas mercury lamps feature short-wavelength UV light, which allows for finer features to be resolved, the spectral linewidth is broad, leading to chromatic aberration in the lens system of cameras used to focus the image onto the photosensitized chip. The net effect is to defocus the image at the edge features on the mask. In contrast to this, excimer lasers have a relatively narrow linewidth, producing a sharper image (despite the fact that excimers have a much broader linewidth than do many lasers). The KrF laser has become the workhorse of the semiconductor industry for photolithography.

Because of the extremely short wavelength, ArF excimers are also used for glass-marking applications, since the beam is readily absorbed by glass (which is transparent to longer wavelengths). Along these lines, ArF (and sometimes KrF) is used to manufacture fiber Bragg gratings for optical fiber communications. During manufacture, a pattern is cut into a single-mode fiber using a phase grating. Other applications for excimer lasers include wire stripping (especially for ultra-fine wires used in the microelectronics industry), surface-mount component marking, drilling inkjet printer nozzle holes, and marking wires.

10.12 PRACTICAL AND COMMERCIAL UNITS

Most commercial excimer lasers start in the low $10,000 range and are priced up to $200,000. A major additional cost is that of installation and operation. These costs vary depending on the ultimate use of the laser; a laser designed for industrial processing must often operate at pulse rates of 300 Hz for extended periods of time, generating an average power of over 100 W, whereas a small excimer designed for eye surgery may be required to fire at most once per second. Clearly, there will be huge differences in the way such a variety of lasers will be built. The industrial laser is often designed for high throughput and reliability, with many components (such as optics and the main capacitor) designed to be changed in the field in much less than an hour, while the low firing rate of the surgery laser may well mean that a laser capable of only 1 million shots in its entire lifetime is more than adequate for the application! An excimer of this type does not have to be designed for fast serviceability, but it may be required to generate extremely consistent pulse energies to yield predictable tissue removal.

Installation of most large excimer lasers requires expensive gas-handling apparatus, including gas cabinets and leak detectors. The alternative employed by one manufacturer (Oxford Lasers, which holds a patent on the method) is to generate halogens in the laser itself by heating a fluorinated salt:

$$K_3NiF_7 \longrightarrow K_3NiF_6 + F$$

This material is uncommon, though, and must be specially prepared. In the laser the salt is heated to over 300°C to release fluorine. Although a fluorine generator does eliminate the need for a gas cabinet as well as a bottle of fluorine, these generators are not cheap either, and it is claimed (by competing manufacturers) that exposure of the generator to water vapor in air essentially ruins the generator, which must then be reprocessed. When dealing with a gas such as fluorine, there is no such thing as a cheap or convenient source; most major excimer manufacturers use bottled supply gases.

In addition to the halogen gas (the most difficult gas to handle in the system), buffer and inert gases are also required in high-purity form. A typical excimer laser installation is depicted in Figure 10.12.1. Most excimer lasers can use a variety of gas mixtures with only a change of optics to suit the wavelength emitted (if required—many materials transmit over a large enough range that they work with many gases emitting in the UV). For ultimate flexibility (e.g., in a research environment) an excimer laser would be supplied with two halogens (fluorine and hydrogen chloride), three

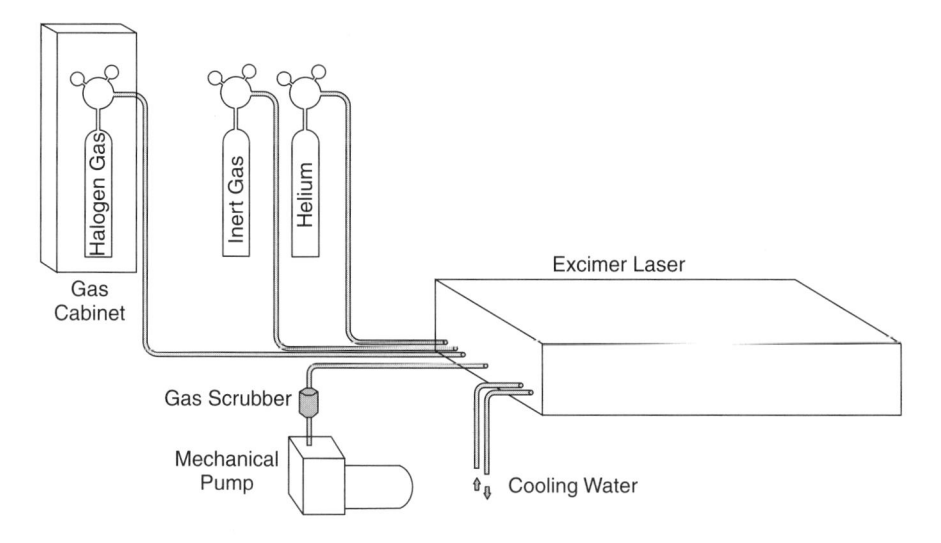

Figure 10.12.1 Typical excimer installation.

rare/inert gases (argon, krypton, and xenon) and two buffer gases (helium and neon). Many commercial lasers have large manifolds attached, allowing connection of many gases to the system; the smallest manifolds usually have four ports whereas many research-class lasers feature up to eight. In industry many excimers are dedicated to lasing a single excimer species, so the gas requirements are simplified as shown in Figure 10.12.1, which depicts a minimal excimer system. The gas scrubber, a cylinder filled with soda lime and a molecular sieve material, is essential for removing active fluorine (or chlorine) from exhaust gas (given its toxicity) before pumping into the atmosphere. Of course, where the excimer is used to lase a carbon dioxide or nitrogen laser line, the gas requirements are much simpler since none of the gases used are toxic or corrosive.

Some excimer systems, used for a fixed purpose such as eye surgery, use a convenient premixed gas supply containing argon and fluorine to simplify the system. One popular manufacturer of lasik surgery systems employs only two gases—premixed ArF and a buffer gas—simplifying the gas system for the laser. The laser is operated at only 40% of its maximum pulse energy, which is stabilized by adjusting capacitor voltage. When voltage reaches a preset limit, fresh premixed gas is injected into the laser (called a *boost*) to raise the output power to acceptable limits at minimum capacitor voltage. The laser then operates with this fill, and voltage is raised again to compensate for gas loss. The goal of this methodology, of course, is to produce consistent energy in repetitive pulses so that the surgeon will know precisely how much tissue will be ablated on each pulse.

Aside from gas supply (the biggest concern with this type of laser), most excimers also require a supply of cooling water and three-phase power. Whereas a small excimer with a 40-W output may require 20 A at 110 V, a larger excimer producing 150 W of output requires 30-A, 208-V, three-phase service. Typical efficiency for a commercial KrF laser (the most efficient excimer) is about 1.5%, so a KrF laser with a 100-W optical output would consume almost 7 kW of electrical energy. Operation of most excimer lasers is simplified greatly by computerized controls which mix gases automatically and refill the laser with fresh gas as required. A single gas fill can last millions of shots but must eventually be discarded (first by scrubbing the halogen from the mixture) and then refilled with a fresh fill. To extend the gas-fill lifetime, extra halogen gas can be injected into the mixture during operation, which also serves to increase output power, which decreases as the laser is run and halogen concentration decreases. One manufacturer claims gas-fill lifetimes of 50 million shots using KrF with automatic replenishment. Many large excimer installations also feature a cryogenic gas processor: The same KrF laser boasts 120 million shot gas lifetimes on a single KrF fill using a gas processor. Operating costs of a modern excimer laser, with features designed to extend the gas-fill lifetime, are claimed to be about $15 per hour for a laser operating continually at 150 Hz. With advances made to extend gas-fill lifetime, the largest operating expense is no longer gas supply, but inevitable replacement of electrical components.

Before operating an excimer that has been opened to the atmosphere, left unused for a period of time (ranging from a week to over a month, depending on the manufacturer's recommendations), or when changing gas mixtures from a fluoride to a chloride gas mix (or vice versa), passivation of the laser is required. In passivation a layer of metallic halogen is deposited on internal metal components to prevent further absorption and reaction with halogens in the laser gas mix. In this operation a mixture consisting of a small percentage of the halogen used (fluorine or hydrogen chloride) in a buffer of helium is either left in the (not operating) laser overnight, or the laser is operated for about an hour with a passivation gas mixture. This process is somewhat involved and time consuming, so most excimer users prefer to keep the laser operating with one type of excimer (fluoride or chloride) and not switch between the two.

Aside from gas supply and the occasional passivation, other maintenance required for an excimer laser includes cleaning of optics. Unfortunately, cleaning the internal surface of cavity optics (which must be done at intervals of every 50 to 100 million pulses, which may be as short as two days for a system under continual use) may well force the user to bring the laser vessel to atmosphere, which would require passivation again before use (unless the vessel is constantly purged with a supply of inert gas during the process). Some manufacturers (e.g., GSI Lumonics) have solved this problem by offering gate valves between the tube and the optics so that optical elements may be removed while the laser channel stays pressurized with the excimer gas mix. In such a design passivation is not required when optics are changed since air never enters the laser pressure vessel and the gas fill in the laser is preserved. Finally, electrical components such as the main capacitor and thyratron have finite lives and must be replaced periodically. Although late-model lasers employ optimized discharge paths extending thyratron lifetime to over 1 billion pulses, so reliability of electrical components has improved drastically over earlier designs.

CHAPTER 11

Infrared Gas Lasers

Of all infrared gas lasers, the carbon dioxide laser is by far the most commonly used, with other gases, such as nitrous oxide (N_2O) and carbon monoxide (CO), used less frequently. Most mid-IR molecular lasers operating in the wavelength range 2 to 20 μm involve vibrational energy levels that result when bonds between atoms in these molecules bend or stretch (see Chapter 3). Longer wavelengths are possible in a molecular laser as well, but these involve purely rotational transitions with correspondingly lower energy levels.

11.1 CARBON DIOXIDE LASERS

The carbon dioxide laser is about as close to the "death ray" of classic science fiction as it gets. At an infrared wavelength around 10.6 μm, the beam is readily absorbed by most materials and readily converted into heat. Power outputs range from big (compared to most common lasers, such as the HeNe) to enormous, with power outputs in excess of 50 kW possible! Even the smallest lasers output several watts, enough to burn many materials on contact with the beam. Aside from high power outputs, the other distinguishing feature is high efficiency, with typical efficiencies of industrial lasers being over 10%. These features combine to make the CO_2 laser the de facto standard materials processing laser, with applications including cutting and welding of such materials as hardened metals and ceramics. There are a variety of forms of CO_2 lasers, based on tube design. Many small lasers resemble any other gas laser, being a glass tube with two cavity optics, while larger lasers may resemble an excimer in structure, with long, transverse electrodes.

11.2 LASING MEDIUM

CO_2 lasers use a mixture of carbon dioxide, nitrogen, and helium in the approximate ratio 1 : 2 : 8, with each gas in the mixture assuming a specific role in this laser. The quantum system of CO_2 laser uses a scheme similar to that of the HeNe laser, in

Fundamentals of Light Sources and Lasers, by Mark Csele
ISBN 0-471-47660-9 Copyright © 2004 John Wiley & Sons, Inc.

284 INFRARED GAS LASERS

which the pump level for this four-level system is in a separate species from the lasing atom. Nitrogen (N_2) becomes excited with energy from the discharge and the first vibrational energy level of that molecule provides a pump energy level that matches very closely the ULL in the CO_2 molecule (the first asymmetric stretch mode, 001). This is identical to the role of helium in the HeNe laser. A large quantity of nitrogen (i.e., a higher percentage than CO_2) ensures that CO_2 molecules in the ground state are pumped rapidly to the ULL.

Lasing occurs as a result of a transition between two vibrational energy levels in the carbon dioxide molecule, the levels resulting from the various modes of vibrations possible (see Section 3.16). Transitions can terminate at possible lower levels, as shown in Figure 11.2.1, with the most common (and powerful) transition resulting in the production of radiation at 10.6 μm. From that level, depopulation takes place in a two-step process by which either LLL decays to a lower energy state, corresponding to the bending motion of the molecule (010) and finally, to ground state. The addition of helium to the gas mixture ensures that CO_2 molecules at the LLL are depopulated quickly—required for a sizable population inversion. Helium also serves to conduct heat from the discharge to the walls of the tube since helium conducts heat much better than most gases do. This provides a means of decreasing the thermal population of energy levels of the CO_2 molecule (which lie quite close to ground state), again helping to ensure that an inversion occurs.

As well as "purely" vibrational levels, rotations of the CO_2 molecule are responsible for the output spectrum of this laser, since rotational levels serve to split each major vibrational level into a cluster of multiple closely spaced levels. As a result, the actual laser output is a series of closely spaced wavelengths covering the range 9.2 μm. to almost 11 μm, centered around 9.6 and 10.6 μm. The 10.6-μm transition in a normal CW laser, for example, consists of over 20 transitions in the wavelength range 10.44 to 11.02 μm. With a diffraction grating added to the cavity for tunability, the large range of outputs makes the laser useful as a source for IR spectroscopy.

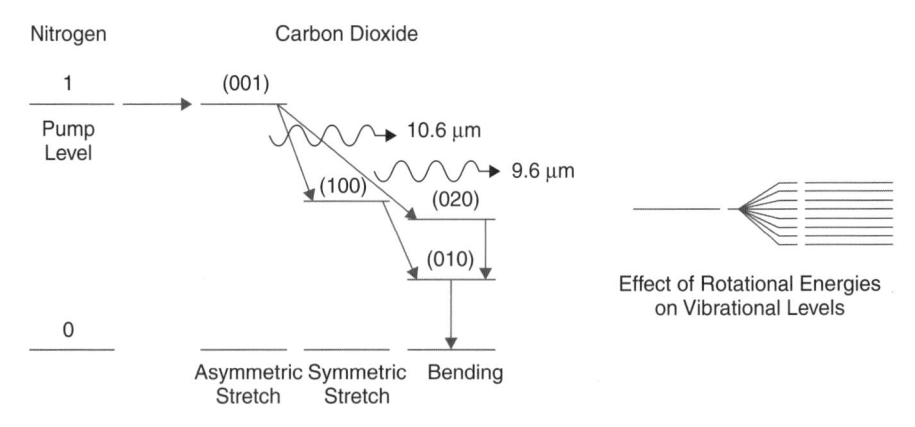

Figure 11.2.1 Energy levels in the carbon dioxide laser.

Water cooling is required for most CO_2 lasers not just to remove discharge heat but also to reduce the thermal population of the lower energy levels, which are very close to ground level. The output power of most CO_2 lasers is quite sensitive to plasma temperature, and a blocked or restricted cooling water line can easily result in a decrease in output power. To this end, many lasers have thermal sensors on the water cooling jacket of the tube, designed to shut down the laser should the temperature reach 40 to 50°C. While the plasma tube would probably tolerate much higher temperatures, laser output would drop drastically at these temperatures.

11.3 OPTICS AND CAVITIES

In the IR band where the CO_2 laser operates, there are few materials that can be used for cavity optics. For reflectors, a glass substrate overcoated with a thin film of copper or gold may be used. This is the most inexpensive option, but these films do absorb a tiny fraction of cavity radiation, heating the glass substrate underneath, so their use is limited to low-powered (usually under 20 W) lasers. Higher-powered lasers generally use solid metal mirrors (in most cases with a film of copper or gold on base metal) or a mirror of solid silicon or molybdenum (both of which reflect well in the IR). The advantage of solid metal mirrors or mirror substrates is that they have excellent thermal dissipation; in many cases, they may be water cooled, allowing even higher-power use.

For an OC one cannot simply use dielectric-coated glass since the glass substrate itself would absorb 10.6-μm radiation, resulting in its destruction. A material transparent to IR is required. Three common materials are germanium, gallium–arsenide, and zinc–selenide. Germanium is the least expensive but has the highest absorption of the three materials, so its use is limited primarily to lasers of under 100 W. Even then, it is often water cooled since it has a low damage threshold temperature. Gallium–arsenide has lower absorption (about two-thirds that of germanium) and can better handle thermal stresses. Zinc–selenide has a very low absorption (only 15% that of germanium) and is yellow in color, allowing visible light to pass through it. This feature allows the use of a coaxial HeNe alignment beam, so for higher-powered lasers it is the OC material of choice. Note that all three materials have high indexes of refraction, so the interface between these materials and air results in a large reflection. To suppress this, an antireflective coating is required on all optics. The same is true for focusing lenses used for the output beam—antireflection coatings are required.

Where a laser uses external optics and a window is required, sodium chloride (salt) may be used. However, this material is hygroscopic; that is, it absorbs water from the atmosphere. Better choices for windows include zinc–selenide. Where a window is employed, it is frequently mounted on metal bellows so that it may purposely be misaligned from the laser tube axis, to prevent feedback from the window surfaces and not the cavity optics. A bellows mount suitable for either windows or mirrors is shown in Figure 11.3.1. The plasma tube is attached to the rear circular plate, which is bridged to the front plate, with the metal bellows forming gastight

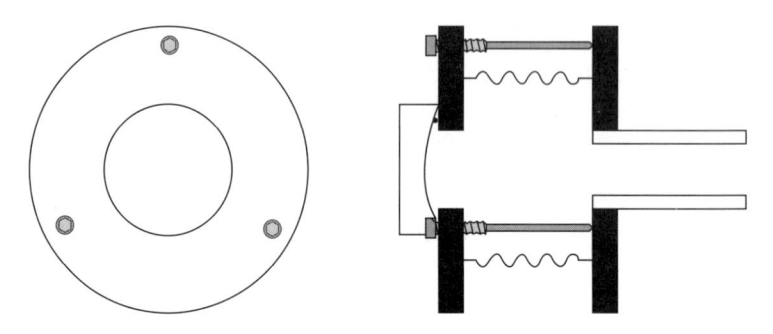

Figure 11.3.1 Bellows mount for optics.

seals. Adjustment screws between the two plates allow angular adjustment of the optic with respect to the axis of the laser tube.

A variety of cavity configurations may be used in this type of laser, but the most popular (a trade-off between utilization of lasing volume and ease of alignment) is the spherical-plane configuration, often utilizing a spherical mirror of much longer radius than the cavity length. Extremely high powered lasers often use an unstable resonator consisting of two solid metal water-cooled mirrors, with the resulting beam shaped like a donut. At extremely high powers, solid metal mirrors are the only reflectors capable of tolerating the large powers involved. Aside from mirrors, the other problem that arises with extremely high powered lasers is absorption by windows, which will shatter given the huge energies involved: If a window absorbs 1% of an exiting 100-W beam, only 1 W of heat needs to be dissipated, but if 1% of a 10-kW beam is absorbed, that represents 100 W of heat! In these cases aerodynamic windows may be employed, which are simply an open hole in the laser cavity. By passing a fast flow of gas across the face of the hole, contaminants can be prevented from entering the tube.

11.4 STRUCTURE OF A LONGITUDINAL CO_2 LASER

Used primarily for lower-power lasers (*low power* being a relative term, since it is used commonly with lasers up to 1 kW), a longitudinal CO_2 laser resembles any other type of gas laser: a glass plasma tube of reasonably large diameter (often 10 to 15 mm) with an integral water cooling jacket and either internal or external optics. In some lasers the cavity reflectors are mounted directly to the laser tube, like that of a HeNe laser, only mounted on flexible metal bellows, allowing adjustment. A laser with external optics will often feature windows (IR transparent, of course) on the ends of the tube, mounted in a similar bellows arrangement.

In longitudinal lasers gas pressures are low, usually between 15 and 60 torr. There are several variations on this type of laser based on the gas flow used. Some lasers of this type are *sealed*, regenerating lasing gas dissociated in the process (below) within the operating plasma tube itself. The gas mixture usually has a finite lifetime, with one gas fill lasting from one to eight hours of operation. The tube is then evacuated and refilled with a fresh gas mixture. Often, an hour-meter is installed on such a laser, indicating the amount of time that a fill has been used.

As the CO_2 laser operates, carbon dioxide molecules dissociate, that is, are literally ripped apart, leaving oxygen and carbon monoxide. Soon the partial pressure of CO_2 drops in the laser tube, as does gain. To overcome this problem one can (1) use flowing gas to remove CO and replace it with a fresh supply of CO_2 (as done in flowing gas lasers, described next); (2) use a hot nickel catalyst within the tube, which recombines CO and oxygen into CO_2; or (3) add water vapor to the tube.

In the case of water vapor a reaction occurs in which CO combines with H_2O to produce CO_2. Water vapor is also formed when hydrogen and oxygen recombine in the tube. This approach is useful in sealed lasers, but care must be taken in designing the tube so that windows and other optical components are not hygroscopic. Hygroscopic materials absorb water vapor and in doing so, become cloudy and absorb IR radiation (with disastrous consequences). Salt crystals (sodium chloride) transmit IR radiation well at 10.6 μm but are quite hygroscopic and so would be a poor material for a tube employing water vapor. Other materials, such as zinc selenide or germanium, would be a better choice. Although an option, addition of water vapor is not a popular approach for commercial lasers, but the alternative, incorporating a nickel catalyst into the tube, is. This approach is used with many sealed tubes simply by fabricating the tube electrodes from nickel. Combined with the heat from the discharge, this catalyzes CO back into CO_2. Sealed lasers are typically restricted to power levels of about 100 W.

Flowing gas lasers admit fresh gas at one end of the tube and evacuate it with a vacuum pump at the other. Flow rates may vary from slow (1 L/min) to moderate (20 L/min) with control of gas pressure usually accomplished by restricting the flow rate via a needle valve. The required rate of flow of an axial flow laser depends on the tube diameter and the power levels involved. Flowing gas both replaces dissociated gas molecules (CO) and removes impurities generated through the discharge in the tube (especially in a very high power discharge, where the electrodes can liberate large quantities of gases as they heat). This approach is most often used with lasers under 1 kW. Higher-powered longitudinal lasers are often *circulating gas* lasers in which gas continually flows through the tube and is recycled, passing it through a heat exchanger in the process. Fresh gas is often leaked into such a system as well to replenish. Figure 11.4.1 depicts the variations of longitudinal laser structures.

Water cooling of any CO_2 laser is the rule—not only to prevent damage to the discharge tube but also to remove the thermal population from the lower lasing levels. In a sealed or flowing-gas laser, cooling is accomplished by conduction of heat through the tube walls, a coaxial water jacket around the plasma tube serving to remove the heat from molecules of gas colliding with the tube walls (conduction of heat being one of the primary roles of helium in the tube). In a circulating gas laser an integral heat exchanger (often resembling a miniature automotive radiator) performs this task, with convection as the heat removal mechanism. Circulating gas

Figure 11.4.1 Longitudinal laser structures.

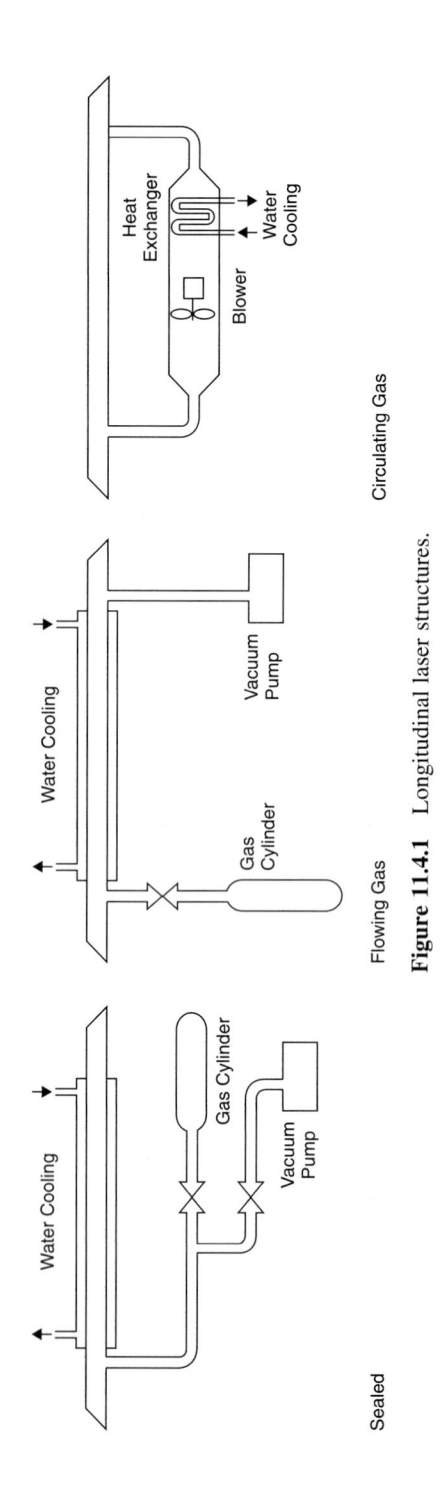

Figure 11.4.2 DC excited laser.

lasers can utilize larger tube diameters than flowing types since the tube walls are not used as a heat exchanger.

Laser gain is also a function of tube diameter since the entire system relies on helium to remove heat from the system and depopulate the lower lasing levels. Use of a larger-diameter tube requires that helium atoms travel farther from the center of the tube to the tube walls in order to dump heat there. Higher gains thus occur with small-diameter tubes, although even the smallest tubes are still much larger than other gas lasers, such as ion lasers.

Figure 11.4.2 shows a twin-tube research laser capable of power outputs of 100 W. Each tube is about 1.5 m long and the arrangement is a folded one in which the tubes are optically in series with the HR and OC at the same end of the laser. Two steering mirrors at the far end of the laser bend the beam between the two tubes so that they are optically connected (this also means that four mirrors must be aligned on this laser, not just the OC and HR). This folded approach is often done to decrease the overall length of the laser, with some industrial lasers featuring four or more tubes in series. The other reason is to allow the use of shorter electrical discharge paths in these plasma tubes to decrease the operating voltage (covered in Section 11.7). Another approach to lower operating voltage is to use multiple electrodes in a single tube.

11.5 STRUCTURE OF A TRANSVERSE CO_2 LASER

The goal of a circulating gas laser is to remove a hot molecule from the discharge path as soon as possible after lasing, but there are, of course, limits to how fast gas can be pushed through such a structure. For extremely high flow rates a transverse-flow arrangement is used (sometimes called a *gas transport laser*), in which gas flows across the discharge electrodes in a manner similar to that of an excimer (see Chapter 10). Gas speeds are very high here in order to quickly replace warm gas (which does not lase well, due to a thermal population of lower energy levels) with cooled gas. Transverse CO_2 lasers operating at low pressures can produce enormous CW power outputs, but with long, transverse electrodes it is difficult to ensure that the discharge spreads out. Preionization may be employed (as it is with many excimer lasers) to sustain the discharge, or alternately segmented anodes may be used.

Another use of a transverse arrangement, not specifically for cooling purposes but for the sake of electrode arrangement, is the TEA configuration. TEA lasers operate at very high pressures, usually several atmospheres, so allow the extraction of a large amount of energy from the lasing volume. These are strictly pulsed systems, and in such lasers, which are often built in a manner similar to excimer lasers, the need for transverse electrodes is apparent when considering that at such high tube pressures 20 kV can discharge only a few centimeters. In most cases, commercial excimer lasers can use a CO_2 gas mix with only a simple change in optics. Excimer lasers, too, employ transverse gas flow for cooling purposes (see Chapter 10 for details of a similar system), lending themselves well to CO_2 TEA laser use.

11.6 ALTERNATIVE STRUCTURES

Alternative structures for a CO_2 laser include the *waveguide structure*, in which discharge is confined in a small-bore ceramic (usually, beryllium oxide) tube. The tube acts much as a fiber optic does, with total internal reflections confining radiation inside. The advantage of this arrangement is the high power extraction possible. Small waveguide lasers, the length of a typical HeNe tube, can be built which provide multiwatt power outputs. Frequently, this configuration is used for a compact laser design, so most lasers of this type incorporate internal mirrors. Waveguide lasers can be sealed or flowing-gas types and usually incorporate a water-cooling jacket around the ceramic tube. These are usually small lasers with power outputs in the under-50-W range.

Finally, no discussion of CO_2 laser structures would be complete without mentioning the largest, the *gas dynamic structure*. These lasers produce power outputs of well over 10 kW and are used primarily for military research purposes. Gas dynamic lasers do not use electrical excitation but rather, rely on the thermodynamic properties of gases as they are compressed and expanded. Resembling a rocket engine with an optical cavity perpendicular to the output, the laser works by burning fuel at high temperatures in a combustion chamber. One of the products of this combustion is CO_2 gas, and the hot CO_2-rich exhaust is mixed with nitrogen gas, which becomes excited and transfers energy to the CO_2 molecules, raising them to a high vibrational energy state. The gas mixture is then allowed to expand rapidly, cooling the entire mixture. Excited CO_2 molecules remain at the high-energy state, but the low temperatures depopulate the lower energy states, so massive population inversions are achieved. Lasing occurs in this region of the laser around which an optical cavity is fabricated. Exhaust gas is then released.

11.7 POWER SUPPLIES

Most CO_2 lasers (the exceptions being the gas dynamic, which does not need electrical excitation, and the TEA, which is pulsed) are driven from either a DC or an RF pumping source which generates a plasma. A DC excited laser requires only a simple high voltage supply providing enough voltage to initiate and sustain the discharge at moderate currents. Typical figures might be 10 to 30 kV at about 15 to

100 mA for a unit operating with a power output below 100 W. Most lasers can use low-frequency AC as well as DC.

Like the HeNe laser, CO_2 lasers exhibit maximum output at a finite discharge current, and any increase of current beyond that point results in a decrease in output power. Although output naturally depends on pumping rate, too much pumping results in an increase in heat in the tube, thermal population of the lower lasing levels, and a decrease in output for the laser. As tube length increases, so does the voltage required to excite the discharge. Whereas a 1.5-m tube requires about 30 kV to operate, a longer tube requires even higher voltages. A 3-m tube would probably require 60 kV! Power supplies operating at this voltage become quite costly (more than double that of a 30-kV supply). Voltages this high are also difficult to manage, due to the requirement for better insulation and problems such as corona and arcing. By shortening the active discharge length, voltage is reduced and power supply design is greatly simplified, which is the electrical reason for preferring shorter tubes.

Another approach to lower discharge length is to install multiple electrodes in a tube: for example, a single anode in the center of the tube and two cathodes. By splitting the effective discharge length in half, voltages across these portions of the tube are reduced to a manageable level (this approach is also used in some long HeNe lasers as well). The complexity and cost of the power supplies are much lower since it is easier and cheaper to obtain electronic components that operate at higher currents and lower voltages than the other way around.

Regardless of discharge length, the CO_2 laser plasma, like other gas discharges, has a significant negative resistance, which may be overcome using either a ballast resistance (as in an HeNe) or some form of current regulation in the power supply. For a laser operating at a current of 60 mA, the ballast resistance would be required to dissipate an enormous amount of power, making the supply inefficient at best (i.e., a large space heater!). For this reason, most power supplies use some method of active current regulation, but regulating current at the high voltages required by this laser is not an easy task.

The simplest approach is to use a saturable core reactor to limit current through the tube. These are incorporated into most neon-sign transformers (a popular power source for small lasers) for precisely this reason, since neon signs exhibit similar discharge characteristics and require current limiting. Unfortunately, saturable reactors (called *magnetics* by some in the industry) work only with AC currents, and the resulting laser would be pulsed, with the discharge extinguishing 120 times a second (for 60-Hz power) at the zero-cross point of the AC line voltage. Because of this, most industrial lasers use continuous DC discharges. Many older industrial lasers (and quite a few of these are still in use today) use vacuum tubes as current regulators since these can handle the high voltages involved. These regulators are often installed between cathodes (often, multiple cathodes, each with a separate regulator, with a single anode in the tube) and the negative output of the power supply.

The problem of high voltages, and specifically, increasing voltage with tube length, is solved in many small lasers (the majority under 100 W) by using RF excitation of the plasma. These laser tubes often feature long, transverse electrodes driven with RF power; the configuration resembles that of a nitrogen or excimer laser but the cavity optics are quite different. RF excitation solves some problems of DC excitation, such as high voltages (for longer tubes) and sputtering of electrodes, but the power supply electronics are more complex and expensive. For a small, sealed laser, RF is preferred (since sputtering consumes gas, which is a big concern in a sealed system).

TEA lasers require a unique discharge circuit to generate fast pulses. The circuitry is identical to that described in Chapter 10 for the excimer laser, in which a capacitor is charged to a high voltage and suddenly discharged through the laser channel using a thyratron switch. Preionization electrodes ensure that gases in the laser channel are ionized to form a pathway for current from the main capacitor.

11.8 OUTPUT CHARACTERISTICS

Most CO_2 lasers are used for their high output powers in materials-processing applications where beam quality is not important. In applications such as ceramics cutting, longitudinal modes or the presence of multiple output wavelengths in the output beam is hardly a concern. A carbon dioxide laser operating with standard, broadband optics, for example, has output on many lines between 9 and 11 μm. Where single-frequency operation is required (e.g., in spectroscopy applications), a wavelength selector such as a diffraction grating may be included to tune a single line.

In most lasers, single-line operation usually implies that multiple longitudinal modes will oscillate, but in most CO_2 lasers, only a single longitudinal mode will usually oscillate. Consider that a 1-m laser cavity has a free spectral range (FSR) of $c/2L$ or 150 MHz, so that longitudinal modes are spaced that far apart. Now the natural linewidth of this laser, like that of most gas lasers, is broadened primarily by Doppler shifting, and equation (4.7.3) allows us to calculate the linewidth of this laser at around 60 MHz (the atomic mass of the CO_2 molecule is 44 atomic mass units or 7.3×10^{-26} kg). This means that the FSR of the cavity is larger than the linewidth of a single emission line, so only a single longitudinal mode will oscillate—an etalon is not required. Exacting wavelength control and stabilization can be done by adjusting the cavity length in minute amounts using piezoelectric positioners. This method of active cavity-length control is especially important in longer-wavelength lasers operating in far-IR regions.

11.9 APPLICATIONS

Commercially available CO_2 lasers vary from small sealed units with integral RF power supplies under 40 cm in length, featuring output powers in the several-watt range, to behemoth industrial monsters with multiple plasma tubes several meters in length utilizing three-phase 600-V AC power with output powers in the 10-kW range. Since CO_2 lasers are always high powered (compared to other gas lasers), most are used for materials-processing applications. The main competition is from

YAG lasers, which can also yield outputs of several hundred watts. While YAG lasers compete with CO_2 types in the lower-power end of the scale (<100 W), the CO_2 laser can easily be scaled to power levels in the tens-of-kilowatts range, much larger than a YAG, so it is the dominant laser in the high-power area.

The relatively long IR wavelength of the CO_2 laser is readily absorbed by most organic materials (such as plastics), as well as glass, water molecules, and many common materials. For cutting materials such as cotton (used in making jeans), the CO_2 laser is ideal. It is also used in surgical applications since the wavelength is readily absorbed by flesh vaporizing it; the heat also serves to cauterize the cut, thus reducing bleeding.

Most metals are somewhat reflective at this long wavelength, so higher powers are required to ensure that enough energy is absorbed to cause vaporization of the material (easily provided, though, by these types of lasers). Cuts made into metals are hence somewhat rougher than those made with other lasers (e.g., the YAG), but the high average power available from the CO_2 laser (which no other commercially available laser can generate) usually means that higher-speed operations are possible, so for many large-scale cutting applications, this is the dominant laser. This laser is used extensively to cut stainless steel and titanium, which are difficult to cut by any other means. For drilling applications, the YAG is usually preferred for controllability.

TEA types generate short pulses ideal for marking such items as plastic pop bottles (expiry dates and batch codes). The direct competition for a TEA laser is a Q-switched YAG laser, which produces similar pulses (although at a much shorter wavelength). Directing the beam is often accomplished by movable mirrors, and most large CO_2 lasers incorporate a coaxial HeNe alignment laser which passes through the OC (usually zinc selenide, which passes red light) allowing targeting of the laser.

11.10 FAR-IR LASERS

Gases other than carbon dioxide may be made to lase in the infrared, many in the far-IR region of the spectrum (generally defined as wavelengths greater than 20 μm). While a few are pumped by an electrical discharge, most are excited optically by a carbon dioxide or, less frequently, a nitrous oxide laser, so far-IR lasers represent an application for IR gas lasers. Most far-IR lasers use molecules such as alcohols (e.g., CH_3OH) or other organic compounds. It is rumored that Scotch whiskey will lase when used in such a configuration. Scotch, like other liquor, contains a number of organic compounds that can lase.

Far-IR laser tubes are simple structures: a glass tube into which the lasing species (most of them liquids) are injected at low pressures of around 0.1 torr. Liquids vaporize readily at low pressures, forming the gaseous lasing medium. Because of the low pressures involved, far-IR lasers require a high-vacuum system using a diffusion or turbomolecular pump to evacuate the tube of all contaminants. Optics for a typical far-IR laser consist of a pair of mirrors that are movable, to produce a variable cavity length. The laser is pumped in an end-on configuration in which the pump beam (usually from a CO_2 laser) passes through the rear optic to excite molecules in the tube. The front optic (the OC) must reflect 10.6-μm pump light back into the laser tube but pass far-IR radiation.

In the example of the CO_2 laser, we found that the FSR of the cavity exceeded the linewidth of a single emission line, so only a single longitudinal mode will oscillate when a wavelength selector is used with the laser; an etalon is not required for single-mode operation. In a far-IR laser, the FSR of the cavity (the distance between resonant peaks) can be many times greater than the gain curve of the lasing species, so it is likely that the cavity will not be resonant for emission at the required wavelength. For this reason, the cavity length must be adjusted until a resonant peak corresponds to a wavelength at which the lasing species exhibits gain.

CHAPTER 12

Solid-State Lasers

The oldest technology, but one reborn recently and becoming increasingly important, is that of the optically pumped solid-state laser. Solid-state lasers (not to be confused with semiconductor lasers) consist of a crystal of glasslike material doped with a small concentration of a lasing ion such as chromium (in the case of ruby) or neodymium (in the case of YAG). The ruby laser, the first ever, was a reasonably simple structure with integral mirrors on a rod of synthetic ruby (chromium ions embedded into a host crystal of Al_2O_3) pumped with a helical flashlamp. For many solid-state lasers the technology has not changed much, but in recent years more efficient materials with lower pumping thresholds have been used, and compact solid-state lasers have been developed that are pumped by semiconductor laser diodes instead of lamps. Many solid-state lasers have integral harmonic generator crystals to produce visible, even UV light. This technology promises to replace gas lasers for many applications. Although ruby lasers continue to be used in a few niche markets, most modern solid-state lasers use more efficient neodymium-doped crystals such as Nd:YAG or Nd:YVO_4. There are other solid-state materials of importance as well, such as erbium (used in fiber amplifiers for communications systems), holmium, and titanium (useful in a tunable solid-state laser).

12.1 RUBY LASERS

The very first laser discovered (by Theodore Maiman in 1960), the ruby laser is a three-level system having a high pumping threshold and requiring high pump energies. With the development of more efficient lasers, many of which are capable of CW operation, the demand for ruby lasers has diminished steadily, but it is still a useful source of intense pulsed red light for applications such as holography. The ability to Q-switch this laser results in fast, intense pulses of red light that cannot be generated by other means. Being confined somewhat to a narrow application market, many modern ruby lasers utilize EO Q-switches and often use a MOPA configuration in which pulses generated via a Q-switched oscillator are amplified by

---

Fundamentals of Light Sources and Lasers, by Mark Csele
ISBN 0-471-47660-9   Copyright © 2004 John Wiley & Sons, Inc.

a separate rod outside the cavity of the oscillator laser. Energies of 1 J per pulse with a pulse length of 10 ns are common from a Q-switched ruby laser.

12.2 LASING MEDIUM

Ruby is synthetically grown aluminum oxide (Al_2O_3) doped with chromium ions (Cr^{3+}) at a concentration of around 0.05%. It appears light pink in color, with the chromium ions giving the characteristic color of the material. High purity is required, so crystal growth is a critical matter accomplished by only a few companies skilled in this area. Unlike the musings of several old science fiction movies in which a laser is built with natural gemstones, natural ruby (which has the same chemical composition) is unsuitable for use as a lasing material since it is not homogeneous enough and probably contains impurities that would inhibit lasing action.

The ruby laser is a three-level system and as such, exhibits high pumping thresholds. The dynamics of ruby are poor for lasing, but its broad absorption bands and relatively long upper lasing level lifetime allow ruby to operate in pulsed mode, in which inversion is achieved only temporarily. The energy levels of ruby are outlined in Figure 12.2.1. Broad absorption bands in the violet and green portions of the spectrum absorb light, usually from a xenon flashlamp, pumping chromium ions (Cr^{3+}) to the pump levels. Pump levels have very short lifetimes (about 1 µs), and a fast decay occurs from those levels to the upper lasing level, which has a much longer lifetime, of 3 ms. From there, ions decay to ground state, emitting a photon of 694.3-nm light in the process. The long lifetime of the upper lasing level allows ruby to store energy in that level, making lasing possible (as a pulse) and allowing Q-switching of the laser to produce massive pulses.

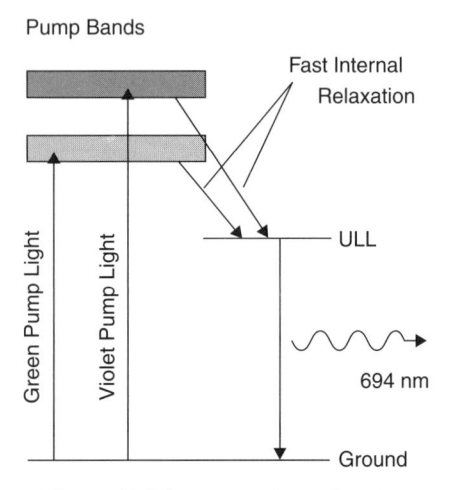

Figure 12.2.1 Energy levels in ruby.

The lower lasing level is the ground state itself, but this isn't one discrete level but a collection of closely spaced energy levels, normally all thermally populated. If these levels are depopulated by cooling the rod to, say, liquid-nitrogen temperatures (77 K) ruby can be made to operate as a four-level system, and hence CW laser action is possible, although a CW ruby laser is purely a laboratory curiosity, since other, more convenient and much more efficient CW laser sources (e.g., the YAG laser), exist.

12.3 OPTICS AND CAVITIES

Some compact ruby lasers use integral mirrors fabricated directly onto the ends of the rod. This is often a coating of silver (as used in the original laser developed by Maiman) or, more recently, a dielectric coating. The rear of the rod is coated to reflect 100% (or as close as possible), and the front of the rod is coated as an output coupler with a partial transmission, the optimal transmission being a function of the length of the rod and the pumping rate. Larger ruby lasers as well as those for special-purpose applications (the type of ruby laser primarily found nowadays) use external optics.

All solid-state laser crystals exhibit, to a varying degree, an effect called *thermal lensing*. Imagining a ruby laser pumped with an intense light and cooled with flowing water, a thermal gradient will develop since the outside of the rod will be cool and the inside of the rod, relatively warm. Changes in rod temperature result in minor variances in the indexes of refraction of the material. These changes result in a spherical lensing effect, which affects the intracavity beam exiting the rod. Localized heating may also cause thermal lensing effects in a square crystal, such as those used for many diode-pumped solid-state lasers (discussed later). As such, a laser rod exhibits a focal length that can be measured. Cavity reflectors used in a ruby laser are often slightly concave to compensate for this effect.

Most ruby lasers manufactured at present are purpose-built, so many have special optics designed for a specific purpose. Double-pulse lasers used for holography, for example, often take extreme measures to ensure beam quality. Such lasers may feature a standard HR with an etalon for an OC, which reflects only certain wavelengths, those wavelengths separated by the FSR of the etalon. This scheme allows single-frequency operation of the otherwise spectrally wide ruby output and hence a longer coherence length required for holography.

Often, a MOPA configuration is used with two optically pumped rods in which one rod is used as an oscillator and a second as an amplifier. The oscillator is a complete laser with HR, OC, Q-switch, and intracavity optics as required. The oscillator usually produces as clean a beam as possible, which then passes through an amplifier to increase the output power by up to 10 times that of the oscillator output. Often, the amplifier has a longer rod and more pump power than the oscillator. Having a long upper lasing level lifetime, Q-switching is easily done with a ruby laser. AO or EO Q-switches may be used with a ruby laser, with faster (and more controllable) EO modulators a popular choice for this laser.

12.4 LASER STRUCTURE

The structure of a ruby laser can vary from a simple design with integral mirrors deposited directly onto the faces of the rod, to a complex design featuring a plethora of optical elements, such as Q-switches, mode-limiting apertures, and single-frequency selectors within the cavity. The optical train of a relatively complex ruby laser, a double-pulse ruby laser used for holography, is depicted in Figure 12.4.1. The cavity resonator itself consists of a dielectric high reflector and an etalon for an output coupler, the etalon being used to ensure single-frequency operation of the oscillator. Whereas the etalons covered in Chapter 6 function as transmission filters with transmission peaks corresponding to the FSR of the etalon, this particular etalon is a reflector, reflecting wavelengths separated by the FSR. In that respect, it simply replaces the broadband OC normally used in a laser. The single-frequency operation of this laser increases the coherence length to about 10 m, desirable when using this laser as a source for holography. Aside from frequency stability, spatial quality is also ensured through the use of a variable aperture in the optical train, which serves to limit the transverse mode of the laser to TEM_{00} mode.

The laser is Q-switched and uses a Pockels cell EO modulator to generate fast pulses. For this laser, two pulses are produced in rapid succession by opening the Q-switch twice. An EO switch is used since it is faster than an AO modulator and allows true modulation: It can be opened partially, allowing the first pulse to be produced without draining the entire energy of the rod. Energy left in the rod is then used to generate the second pulse. Pulses in a laser of this type must usually be balanced, so that they have the same energy. This balancing procedure also illustrates the utilization of energy stored in the rod. The procedure begins by setting the EO modulator to dump all energy from the rod during the second pulse (i.e., the switch is opened fully during the second pulse and is closed completely during the first pulse). At this stage only the second pulse appears in the output. The switch is opened gradually for the first pulse, and the laser is test fired, with the energy of each pulse monitored. The process is repeated with the switch opened slightly more for the first pulse, until eventually the first pulse extracts one-half of the energy stored in the rod and the resulting two pulses are balanced.

Although an EO Q-switch usually employs two polarizers (Section 7.5), this particular laser uses only one polarizer, consisting of a stack of quartz plates at

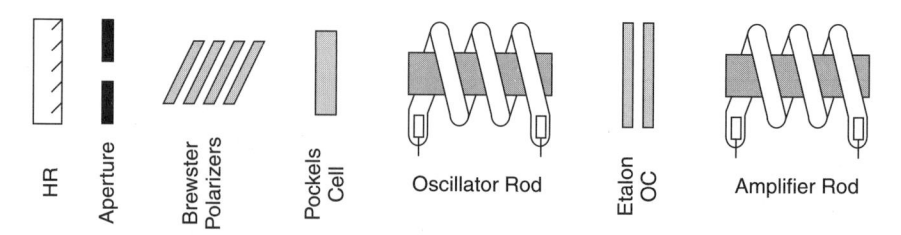

Figure 12.4.1 Optical train of a ruby laser.

Brewster's angle to enhance the degree of polarization. The second polarizer is the ruby rod itself, in which the gain is highly dependent on the orientation of light passing through it, so gain is highest when amplifying light of only one polarization. By careful alignment of the rod (which has a definite crystal axis easily determined by rotating it while viewing through a polarizer), so that the optical axis is rotated 90 degrees from that of the polarizing stack and placing the Pockels cell between these polarizing elements, the cell can be used to switch intracavity light. This particular laser also incorporates a separate amplifier rod, which boosts the output power by a factor 5- to 10-fold. The amplifier rod is twice the length of the oscillator rod and is pumped with correspondingly higher energy than the oscillator.

12.5 POWER SUPPLIES

Ruby lasers are almost always pulsed, with only a few research lasers operated in CW mode (and these require extreme cooling); flashlamp pumping is hence the rule for ruby lasers. Being a three-level system, pumping thresholds are quite high, with pump energies of 1000 J or more common in ruby lasers. Often, the flashlamps used with ruby lasers are helical in shape, with the ruby rod at the center of the helix. Helical flashlamps have a comparatively larger volume than linear flashlamps, so can handle the higher energies required for this laser (as opposed to the YAG, which has much lower thresholds and hence usually uses smaller lamps).

A flashlamp is designed to produce an intense pulse of light, usually in a short time frame ranging from microseconds to 1 ms. The lamp itself consists of a glass tube filled with low-pressure (about 450 torr) xenon gas. Electrodes at either end of the glass tube deliver current to the lamp, and triggering is accomplished either by applying an external high-voltage pulse to the surface of the glass tube or superimposing a high-voltage pulse across the main terminals in the same manner that HeNe tubes are ignited. When the lamp fires, it exhibits a very low resistance consuming all energy from the storage capacitor, producing an intense pulse of light in the process, the spectra of the light emitted being characteristic of the gas used. In the case of ruby, xenon is used since the output is rich in blue light, which is readily absorbed by ruby.

Figure 12.5.1 shows the circuit for a typical flashlamp discharge circuit; all flashlamps, including photographic types, are quite similar in design. Capacitor C1 charges with energy from the power supply until reaching the terminal voltage, usually between 500 and 1000 V. This voltage is present across the flashlamp, but the lamp does not ignite, since the voltage is not sufficient to cause the gas inside to ionize and conduct current. In this particular circuit a trigger pulse of between 4 and 10 kV is applied externally to the glass envelope (via a wire wrapped around the lamp) to ionize gas in the tube and initiate the discharge. To generate the high-voltage trigger pulse capacitor C2, a relatively small capacitance, charges from the main power supply through R2 and through the primary of the trigger transformer T1 itself. When the pushbutton is pressed, the left side of C2 is grounded and current flows through the primary of T1. Being a step-up transformer, a high potential

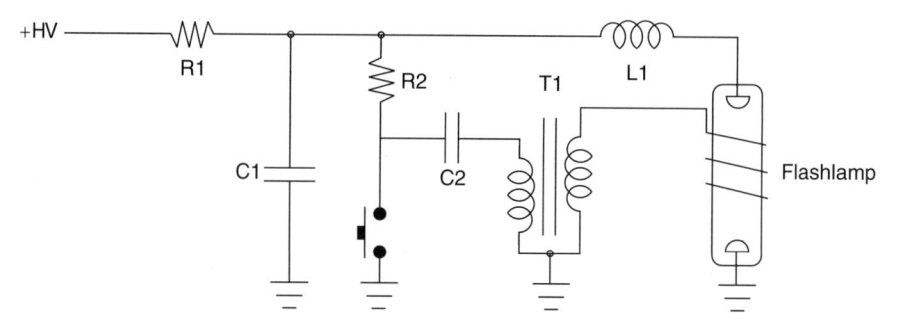

Figure 12.5.1 Flashlamp circuit.

appears across the secondary of T1 sufficient to ionize gas in the lamp. Once the lamp is ionized, current flows from the main storage capacitor, C1, through inductor L1, which helps form the pulse into one suitable for the lamp in what is called a *critically damped LC circuit*, and through the lamp, producing an intense pulse of pump light. When the capacitor has dumped all energy into the lamp, the voltage across the lamp falls to a level insufficient to sustain discharge and the discharge current through the lamp ceases. The capacitor then recharges again for the next pulse.

The external triggering method shown is used with most air-cooled flashlamps, including photographic types as well as lasers, with a very limited firing rate (e.g., one pulse each 10 s). Larger ruby lasers often use water-cooled flashlamps, in which the lamp and rod are both bathed in deionized water for cooling, with deionized water used since it is an insulator and the lamp electrodes are often immersed in cooling water. This somewhat precludes the use of external triggering, so series triggering in which the secondary winding of a trigger transformer inserted into the anode lead is used for water-cooled lamps. This triggering scheme is described in Section 12.12.

12.6 OUTPUT CHARACTERISTICS

The output from an unrestricted ruby laser employing broadband optics is somewhat wide, spectrally speaking, spanning a range of 0.5 nm. Use of an etalon in the cavity (like that employed in the previous example) can reduce the spectral width drastically into the region 20 to 40 MHz, increasing the coherence length to up to 10 m in length. Like many other lasers, ruby lasers tend to operate in high-order transverse modes which can extract the highest powers from the rod. Restricting the mode by using an intracavity aperture can yield a TEM_{00} mode with minimal divergence for optimal beam quality, but energy extraction is not as efficient, so power output is decreased. Q-switching decreases the energy available in a single pulse, in some cases by a factor of 100, but the peak power available is increased drastically. Where a non-Q-switched laser produces pulses of about 1 ms, pulses as short as 10 ns are possible when Q-switching is employed. Peak powers of 100 MW to over 1 GW are possible, especially when a separate amplifier is used.

12.7 APPLICATIONS

Once the "only game in town," the dominance of the ruby laser has diminished in years following the discovery of more efficient lasers such as the YAG. The primary applications for ruby lasers nowadays are research lasers and as sources for holography. The double-pulse ruby laser, for example, is used to record deformation of a test material by using each of the two closely spaced pulses to record a holographic image. Any deformation or movement is recorded as an interference pattern between the two images.

One military application (which has now been replaced by newer technology) is the U.S. M-60 tank rangefinder, in which a compact ruby laser was used to produce a fast, narrow pulse of light which could be reflected from a distant target to determine the range by measuring the time of flight of the laser pulse to the target and back. This laser, perhaps one of the smallest commercial ruby lasers built, consisted of a small ruby rod (8 mm in diameter by 75 mm in length) pumped by a linear flashlamp driven with 125 J of energy supplied by a small capacitor and pulse-forming network. A unique aspect of this laser was the rotating-mirror Q-switch, in which one cavity mirror was spun at 30,000 rpm by an internal motor. A magnetic pickup coil sensed the mirror position and triggered the flashlamp just before cavity alignment occurs in a scheme similar to the manner in which most automotive engines sense camshaft position in order to fire spark plugs at the appropriate time. A delay could be imposed by an electronic circuit to optimize the pulse energy before the cavity became resonant and oscillation occurred. This small laser had an output energy of 50 mJ, and since the entire assembly was air-cooled, repetition rates were low.

12.8 YAG (NEODYMIUM) LASERS

Although it should be called a *neodymium laser* since the active lasing ion is the rare-earth metal Nd^{3+}, this laser is usually named after the host material, so it is often called a YAG or glass laser, depending on the host (I suppose this makes as much sense as calling the ruby laser an aluminum oxide laser, but it is, indeed, standard). While many host materials may be used for the ion (with new ones appearing all the time), all of these lasers operate in a similar manner, so the term *YAG* will be used to describe a generic neodymium laser in this chapter; keep in mind that other hosts exist.

Used originally as a replacement for the ruby laser in many applications, YAG lasers feature much higher efficiencies than ruby does, have lower pumping thresholds, and can oscillate in CW mode. Traditionally pumped by either a flashlamp or, more commonly, a CW arc lamp, these lasers are pumped most efficiently by semiconductor diode lasers. Having separate cavity optics, SHG crystals can be inserted into the cavity to produce powerful output in the green, for many applications replacing the argon laser as a powerful source of green light. The laser can be Q-switched, making it useful for many materials-processing applications.

TABLE 12.9.1 Common Nd^{3+} Hosts and Wavelengths

Common Name	Chemical Formula and Name	Wavelength (nm)
YAG	$Y_3Al_5O_{12}$ (yttrium aluminum garnet)	1064
Vanadate	YVO_4 (yttrium *o*-vanadate)	1064
Glass	Various phosphate and silicate glasses	1060/1054
YLF	YLF (yttrium lithium fluoride)	1053
LSB	$LaSc_3(BO_3)_4$	1062

12.9 LASING MEDIUM

The active lasing ion is Nd^{3+} embedded in a host crystal in a manner almost identical to the way in which chromium ions are embedded in an aluminum oxide host in a ruby laser. The most common host crystal is YAG (yttrium–aluminum–garnet), but other host materials, such as vanadate (YVO_4) or glass, may also be used. The wavelength of the resulting laser beam depends on the host material itself, which modifies the energy levels of the neodymium ion embedded in it. Common host materials and resulting lasing wavelengths are listed in Table 12.9.1. Although not an extensive list (many other materials exist or are under development), it should give the reader an idea of the types of hosts that can be used and the variations in wavelength (which are minimal and all emit in the near-IR). Of all the materials listed, YAG is the most common material, especially for medium- to high-power units, with vanadate being the favored material for low-power (<1 W), compact solid-state lasers.

Nd:YAG (and related materials, such as Nd:YVO_4 and Nd:glass) is a four-level system featuring distinct upper and lower lasing levels. Multiple pump levels allow the material to absorb pump light at a variety of wavelengths in the red and near-infrared region of the spectrum. The absorption spectrum of YAG in Figure 5.1.1 reveals the multitude of wavelengths where pumping is possible.

Materials other than neodymium will also lase in an almost identical configuration, including other rare-earth metals, such as holmium and erbium. Ho:YAG lases at 2060 nm and Er:YAG at 2840 nm. None of these lasers is particularly common, although Er:glass is used extensively in fiberoptic communications systems as an amplifier for weak signals at 1549 nm.

12.10 OPTICS AND CAVITIES

Most YAG lasers, especially those used industrially or in the lab, feature separate optics with all components mounted on a rail for stability. The optics for YAG lasers are usually straightforward, consisting of two mirrors of which one or both are slightly spherical. Spherical mirrors are usually employed to compensate for the thermal lensing effect of the rod, which is quite pronounced in a CW arc lamp–pumped YAG laser since pump energies are quite large (2 to 4 kW even for a small CW laser). The dielectric reflective coatings employed on cavity mirrors are frequently transparent to visible light (so much so that they can be mistaken for uncoated optical flats at first glance), allowing the use of a coaxial HeNe targeting laser to locate the infrared beam. The HeNe targeting laser can be mounted on the rail behind the laser, but frequently is mounted parallel to the rail (with mirrors used to steer the beam toward the laser axis) to make the entire assembly more compact. The red HeNe beam passes through both the HR, laser rod, any other components in the system, and the OC. Transparency of cavity mirrors in the visible region also helps facilitate alignment of cavity optics.

YAG lasers frequently include a Q-switch, allowing the production of fast, intense pulses (many applications depend on these type of pulses). Q-switches are simply attached to the mounting rail between the rod and optics, and most have adjustment screws that allow alignment with the intracavity beam. Q-switches are usually of the acoustooptic (AO) type, using inexpensive quartz or similar glass (which is quite transparent at this wavelength). Another popular option is a second (or third) harmonic generator, generally a simple crystal in a holder which also attaches to the rail.

Beam expanders are another common component found in YAG lasers. Placed between the rear optic and the rod, these components help fill the entire cross-sectional area of the rod with the intracavity beam for higher power extraction from the lasing volume. Since the wavelength of the beam is in the near-IR region, optics may be fabricated using commonly available materials and coatings. Antireflective coatings are usually deposited on the faces of rods as well as on the surfaces of optical components in the system, such as Q-switches and intracavity beam-expanding optics.

12.11 LASER STRUCTURE

Lamp pumping (occasionally, flashlamps, but in many cases CW arc lamps) is used by the largest YAG lasers, most of which use YAG instead of a different host since YAG has better thermal properties. With few absorption peaks in the visible regions of the spectrum, YAG is not well suited for pumping with xenon lamps, which emit primarily in the violet to green region, although for many flashlamp-pumped YAG lasers, xenon lamps are used, due primarily to their lower cost. Krypton, with an output rich in the red region, is a better match and is used extensively for CW arc lamps used to pump YAG lasers.

CW arc lamps operate at low voltages and high currents. In most YAG lasers a linear lamp is employed in which pump light from the lamp is coupled to the YAG rod via an elliptical reflector, as shown in Figure 5.10.2. By placing the YAG rod and the lamp each at a focus of the ellipse pump, light is effectively coupled to the rod. Reflectors are frequently machined from a block of stainless steel coated with pure gold since gold reflects red and IR wavelengths (i.e., wavelengths at which YAG absorbs) quite well. Gold is also resistant to corrosion, important since the reflector is usually bathed continuously in cooling water. Larger lasers such as the one shown in Figure 12.11.1 may use two or more lamps. This particular high-power YAG laser (used for welding) uses a ceramic reflector that can take thermal shock well.

CW arc lamps (and even many flashlamps) must be water cooled to remove the kilowatts of heat produced by the lamps. The entire lamp and rod are usually immersed in flowing deionized water since it is an insulator and will not short the electrical terminals of the lamp (nor corrode the metal reflector). Such lasers usually have a closed water-cooling loop in which deionized water is recirculated through the laser housing and heat is exchanged with a supply of city water. The water-cooling system for a typical YAG laser is shown in Figure 12.11.2.

Large lasers such as these feature separate optics, with the laser head and optics all mounted on a rail, allowing adjustment. Tubing for cooling water as well as wires to power the lamp are usually run between the separate power supply and the laser itself. Use of an optical rail also allows the addition of intracavity components such as Q-switches and harmonic generators as required.

While lamp pumping is used for the highest-powered YAG lasers, the highest efficiencies are achieved when these materials are pumped using a semiconductor laser. Output powers are lower than possible with lamp pumping, with most diode-pumped solid-state (DPSS) lasers limited to under 5 W. Whereas both YAG and vanadate exhibit a large absorption peak at 808 nm, precisely where semiconductor lasers can emit, many DPSS lasers use vanadate since it has a much lower pumping threshold than YAG. Mere milliwatts of pump power are sufficient to pump

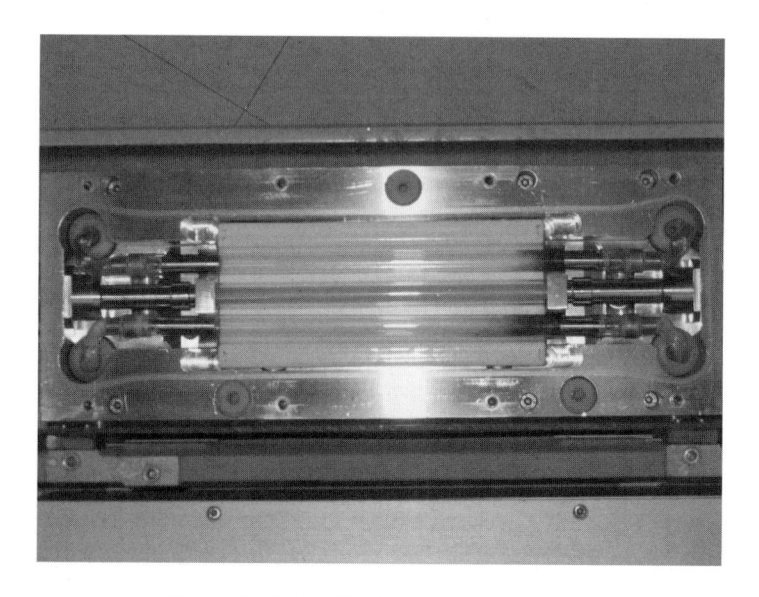

Figure 12.11.1 YAG rod and pump lamps.

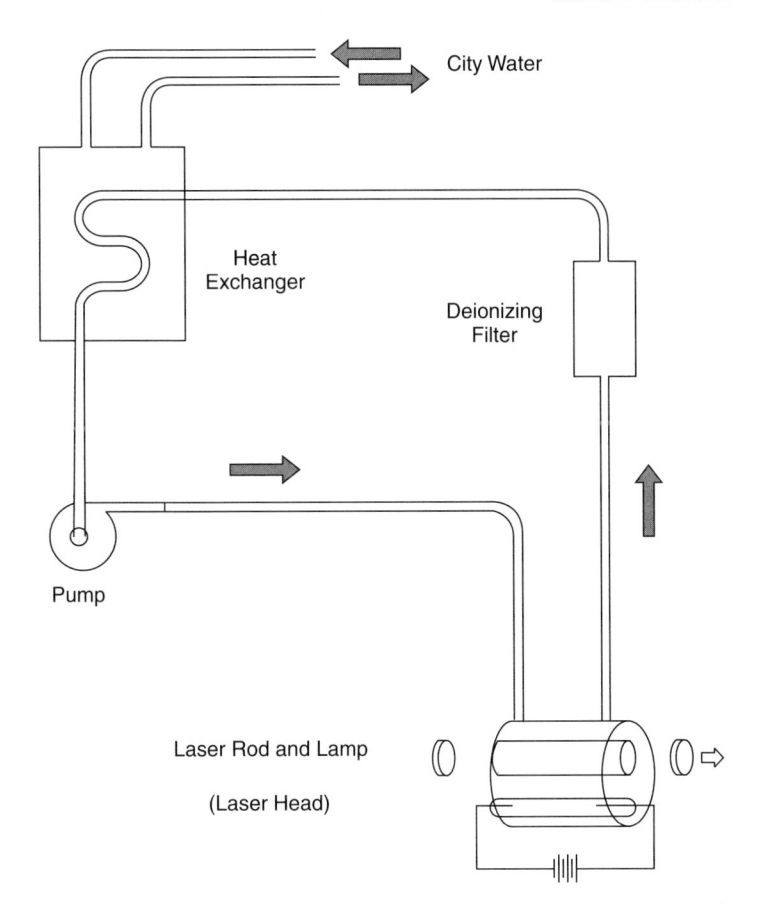

Figure 12.11.2 Water cooling for a YAG laser.

such a laser to threshold (important when the pump laser is limited to milliwatts), so this is the material of choice for compact solid-state lasers. Like ruby, it is a self-polarizing material, making design of compact SHG DPSS lasers much easier than with YAG, which is randomly polarized.

On the small end of the scale are green laser pointers. These compact units usually consist of a diode-pumped vanadate crystal coupled with a KTP SHG crystal. Dielectric mirrors are deposited directly onto the crystal faces, the rear (HR) coating allowing 808 nm pump light from a semiconductor laser to pass through to excite the vanadate. The entire assembly (vanadate, KTP, HR, and OC) is pre-aligned as a single package, the only external component being the pump laser with associated power supply. Figure 12.11.3 details the components of a small DPSS laser used in a green laser pointer.

Upon exit from the laser, the beam passes through a collimation lens as well as a filter, which removes any 808-nm pump light or 1064-nm IR light remaining. Powers of up to 5 mW are common from such laser pointers, with larger versions

306 SOLID-STATE LASERS

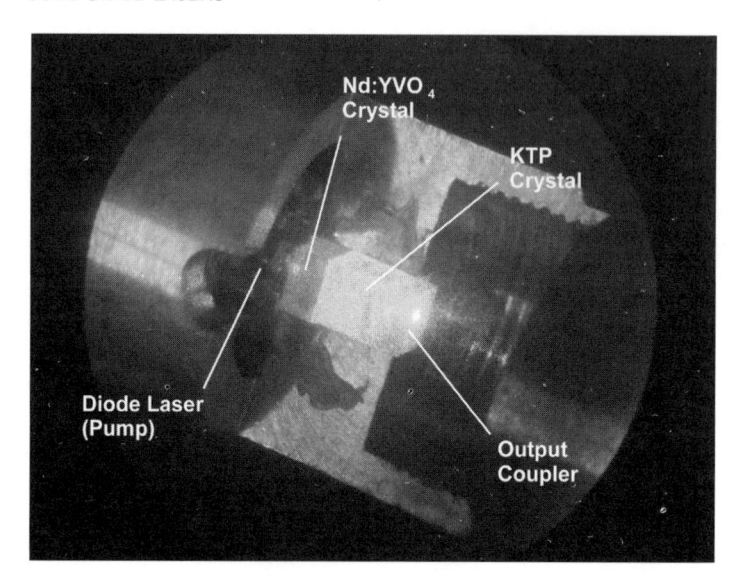

Figure 12.11.3 DPSS laser components.

(used primarily in the laboratory as a general-purpose laser) available which use the same basic configuration (i.e., a single-crystal assembly) boasting powers of up to 100 mW. Larger DPSS lasers tend to use separate components and optics, which also allows the inclusion of intracavity devices such as Q-switches.

12.12 POWER SUPPLIES

The specific power supply involved depends on the pump source. Although often pumped by a CW arc lamp or a semiconductor diode, some small YAG lasers use flashlamp pumping similar to that used in ruby lasers (Section 12.5). The main difference between ruby and YAG flashlamps relates to the pump energies involved. The four-level YAG laser features a much lower pump threshold than the ruby does, so lamps tend to be smaller with 250 J of input common (instead of over 1000 J for the ruby). Since energies are usually lower, linear flashlamps (with lower energy-handling capabilities than helical lamps) are often used, with the pump light focused onto the rod using an elliptical reflector. Both the rod and lamp are placed at the foci of the ellipse.

Most industrial YAG lasers are pumped by a CW arc lamp to produce CW laser output or, alternatively, high repetition rates when used as a Q-switched laser. These lamps have radically different power supplies from flashlamps since they require a large continuous current through the lamp. Like a flashlamp, a high-voltage trigger is required to initiate the discharge.

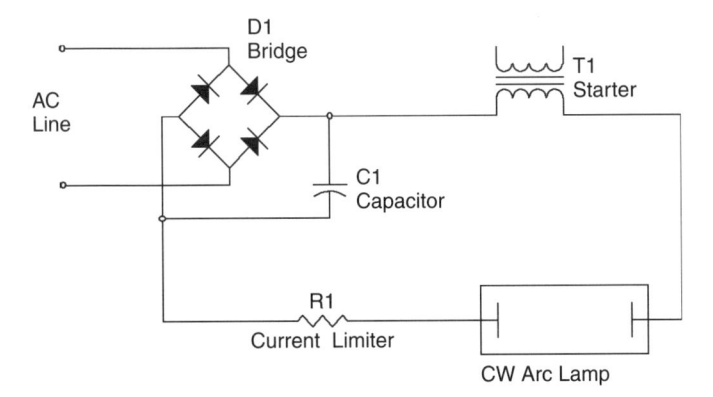

Figure 12.12.1 CW arc lamp circuit.

CW arc lamps have characteristics similar to those of an ion laser discharge (see Chapter 9) with a relatively low operating voltage and high operating currents.[1] These lamps are filled with high-pressure gas (usually, krypton) through which an arc is sustained. Power supplies for these lamps usually involve rectification of the ac line, filtering with a capacitor, and regulation of current via a large ballast resistor or an active electronic regulator circuit, as shown in Figure 12.12.1, which outlines the essential components for a commercially available arc lamp supply. This particular supply uses a rather large variable resistor (R1, which in this example is rated at 5 Ω with a maximum power dissipation of 225 W!) to regulate current through the lamp. Many newer arc lamp supplies use active current regulation.

Like most other discharge tubes (including flashlamps and ion laser tubes) an arc lamp requires a high trigger voltage to initiate discharge. Essentially all lamps of this type used for laser pumping are bathed in deionized water for cooling, with deionized water used since it is an insulator and the lamp electrodes are often immersed in cooling water. This somewhat precludes the use of external triggering, as shown in the circuit of Figure 12.5.1, so series triggering, with a trigger transformer (T1 in the figure) inserted into the anode lead, is the rule for these lamps. This arrangement also saves the need for a third terminal to be brought into the water-cooled laser head. Discharge of a capacitor (not shown) through the primary of T1 results in production of a high voltage across the lamp, which ignites it. In some cases, a high-frequency AC signal is used instead of a DC capacitor discharge to ignite the lamp. This triggering scheme is identical to that used for ion lasers discussed in Chapter 9.

---

[1] Small ion laser tubes can, in fact, be powered from modified CW arc lamp supplies, which often incorporate the same features as required for the ion laser, such as current regulation (and/or series ballast resistance) and a high-voltage trigger generator. Modification of such a supply must, however, incorporate the addition of required safety interlocks.

308 SOLID-STATE LASERS

Aside from lamps (flashlamps or more commonly CW arc lamps), YAG can be driven from a semiconductor diode laser in what is called a diode laser–pumped solid-state (DPSS) laser. Semiconductor lasers utilize their own unique power supplies, which are detailed in Chapter 13.

12.13 APPLICATIONS, SAFETY, AND MAINTENANCE

YAG lasers are a workhorse for many applications involving cutting, drilling, and trimming. In competition with a carbon dioxide laser for many applications, the short pulse length possible with a Q-switched YAG laser makes it ideal for many applications where the CW carbon dioxide laser is not optimal. Most YAG lasers for materials processing have Q-switches already installed at the factory. For drilling applications, especially in metals, the fast pulse ablates materials without creating heat in the substrate being drilled. For semiconductor processing, the fast pulse allows a tiny amount of resistor material to be obliterated without heating the surrounding material and hence affecting the properties of the material. YAG lasers also work well for many marking applications.

In the entertainment industry, frequency-doubled YAG lasers have been used for numerous laser light displays, especially high-power applications such as cloud writing, where they offer an alternative to argon-ion lasers. The lack of an inexpensive solid-state CW red or blue laser, though, as well as a laser that oscillates on several wavelengths simultaneously, precludes the use of solid-state lasers for full-color displays, where the krypton-ion laser is still commonly used (although this situation may change as the development of solid-state lasers progresses and new materials operating at new wavelengths are found).

Maintenance involves the usual cleaning of laser optics (required with essentially all lasers) as well as periodic lamp changes and maintenance of the cooling-water system. Cooling-water systems usually use deionized water in an inner loop that floods the lamp and rod. Heat from the inner loop is then changed with a supply of city water. A deionizing filter keeps the inner water clean since the accumulation of ions results in conduction that will short the lamp and corrode metal parts such as reflectors. Filters must be changed periodically, and many lasers have a safety feature that shuts down the system if the conductivity of the cooling water exceeds a safe value.

Although the danger involved in most lasers (including high voltages in the power supplies and ocular hazards from powerful beams) is quite apparent, there are a few precautions worth noting with regard to the YAG laser. This is perhaps one of the most dangerous lasers from an eye-safety standpoint. Because the wavelength penetrates the eye readily and Q-switched laser pulses can damage tissue rapidly (the blink response being ineffective), this laser is probably responsible for more eye injuries than any other! Combine this with an equally (if not more) dangerous SHG output at 532 nm and you've got a laser worthy of great respect. Safety glasses are mandatory since even a specular reflection can be damaging. Unfortunately, glasses used with these lasers must block both IR and green light with a large attenuation factor [or optical density (marked OD on the glasses)], so it is usually required to obtain glasses specially designed for use with a SHG YAG laser. Broadband glasses designed for use with many lasers are generally not designed to block both wavelengths effectively. Be careful around these lasers!

A further warning regarding the handling of high-pressure arc lamps: Because of the high pressures, the lamps have a tendency to explode during lamp changes. This author has had numerous lamps explode, especially when old lamps are removed, since they tend to be weakened, due to stress generated during operation. Protection from flying glass is mandatory when relamping a lamp-pumped YAG laser.

12.14 FIBER AMPLIFIERS

A solid-state amplifier used extensively in the communications field to boost weak signals in long runs of fiber optic cables is the erbium-doped glass fiber amplifier. As described in Chapter 4, this amplifier consists of a long (10- to 20-m) section of glass fiber doped with erbium ions (Er^{3+}), making an Er:glass medium. A pump laser at 980 nm is coupled to the amplifier fiber via a coupler. This pump radiation is absorbed by the erbium atoms in the fiber (which have a strong absorption peak at this wavelength), exciting them to an upper level which rapidly decays to a level 0.80 eV above ground state. This level, which has a relatively long spontaneous radiative lifetime (τ_{sp}, about 10 ms long) can amplify incoming signals via stimulated emission, producing a net optical gain at 1549 nm. A diagram of this amplifier is shown in Figure 12.14.1.

Where no input signal is present, erbium ions eventually emit spontaneous radiation which is amplified by the fiber and appears as broadband noise in the output [called amplified spontaneous emission (ASE)]. Where an input signal is present, the extraordinarily long lifetime of the upper lasing level gives the ions a good chance of emitting by stimulated instead of spontaneous emission, so the amount of broadband noise in the signal is reduced drastically, as the ions are now coaxed to produce output on the single wavelength of the signal. The output from such

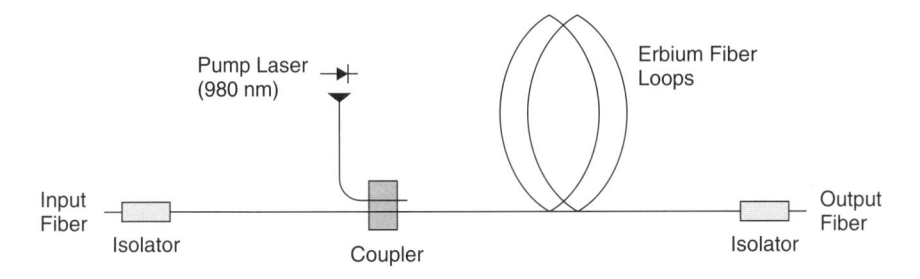

Figure 12.14.1 Erbium fiber amplifier.

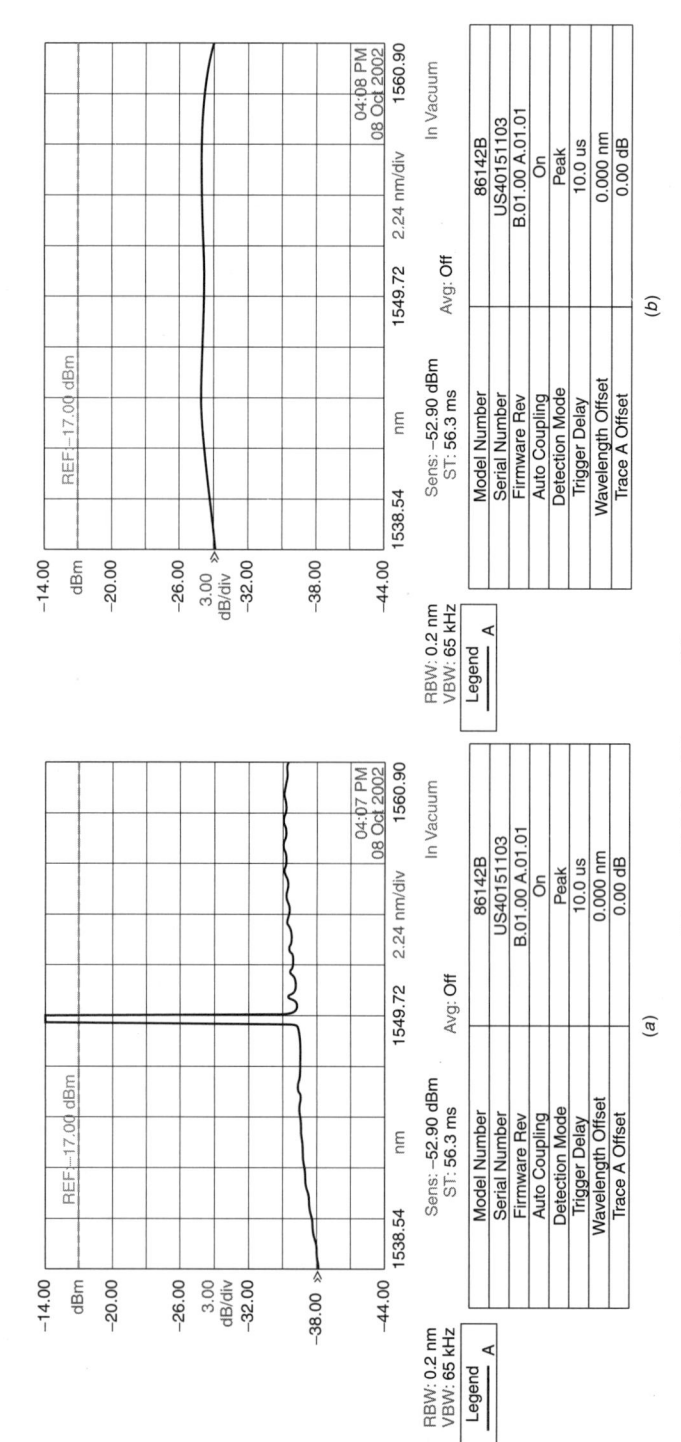

Figure 12.14.2 Fiber amplifier output.

an amplifier, as analyzed on an optical spectrum analyzer (OSA), is depicted in Figure 12.14.2.

Although the fiber amplifier is just that—an amplifier, not a laser—Er:glass will lase if provided with a suitable feedback mechanism. Er:glass may also be fabricated in the form of a rod and used in the same manner as an Nd:YAG laser. Because the wavelength is considered eye-safe (meaning that it will not penetrate the eye to reach the retina), applications include rangefinders, where ocular exposure is likely (the U.S. military M-1 tank reportedly uses an Er:glass laser in the rangefinder, replacing the more dangerous Nd:YAG lasers used previously).

CHAPTER 13

Semiconductor Lasers

No laser has gained such widespread applications as the semiconductor (diode) laser. Found in applications ranging from laser pointers to DVD players, these tiny, efficient lasers have made possible many of the optical devices we take for granted. Consider that the very first CD players (built in the early 1980s) used a HeNe gas laser. Such an arrangement is hardly portable and does not lend itself well to, say, an in-dash player in a car. As well as being tiny, these devices are also inexpensive and require only a simple power supply to operate.

Although most laser diodes operate in the infrared or red regions of the spectrum, new diodes are being developed that can produce output in the blue and violet region of the spectrum, driven primarily by the need for a shorter wavelength for higher-density optical storage than is currently possible with an infrared wavelength. Although the output characteristics of most laser diodes is not impressive compared to those of a gas laser, they do make excellent pump sources for solid-state lasers such as YAG or YVO_4. Combined with harmonic generation techniques, high-quality diode-pumped solid-state (DPSS) lasers are possible which are compact and make the green-to-ultraviolet regions of the spectrum accessible.

13.1 LASING MEDIUM

In Sections 2.9 and 2.10 we examined light-emitting diodes, which produce light as a result of the recombination of electrons and holes in the active region, where a junction is formed between p- and n-type semiconductor materials. Diode lasers work in a similar manner, with the requirement of an optical cavity for feedback and, of course, conformity to the criteria for stimulated emission and laser gain (so that a laser diode is not simply an LED with a cavity).

Consider a simple p-n junction manufactured from a material doped to have excess holes and a material doped to have excess electrons. When a junction is formed with these two materials in equilibrium, a voltage develops that prevents electrons in the conduction band of the n-type material from diffusing across the barrier and combining with holes in the p-type material. When a voltage equal to this

---

Fundamentals of Light Sources and Lasers, by Mark Csele
ISBN 0-471-47660-9   Copyright © 2004 John Wiley & Sons, Inc.

potential is applied across the device, current flows and electrons combine with holes, producing photons in the process.

In most laser diodes, degenerately doped semiconductor materials are used. *Degenerately doped* means that the Fermi levels (the statistical point where 50% of the electrons will be found) are actually within the valence (for p-type material) and conduction (for n-type material) bands themselves. Application of a voltage across the gap causes the Fermi levels for each type of material, aligned at equilibrium, to split into two distinct levels separated by the applied voltage. Electrons in the conduction band of the n-type material now lie just below the Fermi level of that material, and holes in the valence band of the p-type material lie just above the Fermi level of that material. An inversion is hence generated since there are more electrons in the upper energy band than in the lower band.

Of course, electrons and holes involved in the recombination process can lie anywhere in these bands, so a range of wavelengths are possible, with the longest wavelengths corresponding to the bandgap energy. A sharp red cutoff is then expected on the spectral output curve of such a device (as with the LED described in Section 2.10). Emission on the blue side of the curve also has a limiting factor brought about by the nature of the material itself. Electrons are confined to bands, the top band being defined by the Fermi level for the n-type material and the bottom of the conduction band E_c, and the lower band being defined by the Fermi level for the p-type material and the top of the valence band E_v. Photons with energies corresponding to jumps within these bands encounter amplification by stimulated

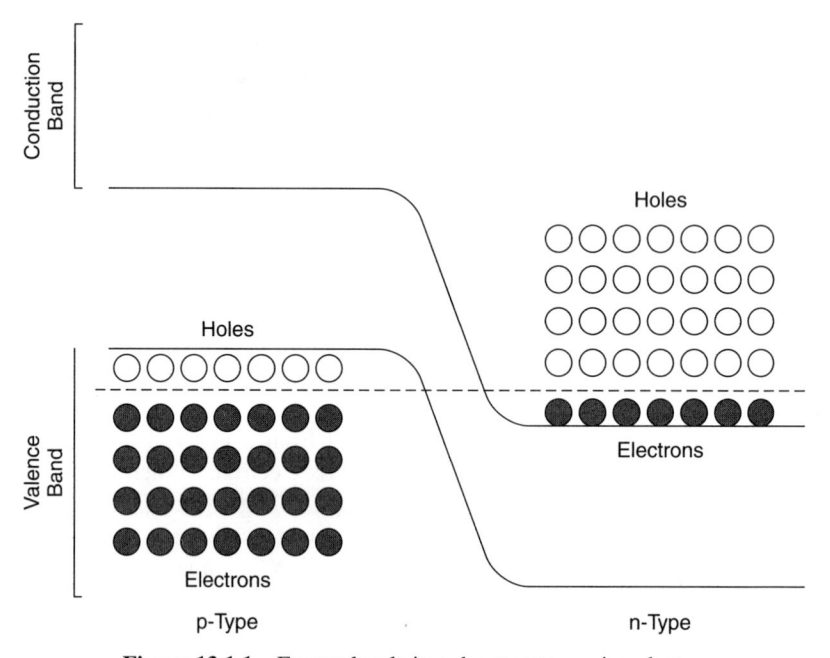

Figure 13.1.1 Energy levels in a degenerate semiconductor.

emission since a population inversion exists, but when the energy of an incident photon exceeds the energy corresponding to the difference between the Fermi levels, it is absorbed rapidly by electrons in the valence band of the p-type material (not the holes at the top of that band) to promote these to the conduction band. The material is thus strongly absorbing at wavelengths shorter than the energy corresponding to the difference between the Fermi levels, as illustrated in Figure 13.1.1. Optical gain by stimulated emission can, therefore, occur only for photons with a specific range of energies. Shorter wavelengths are absorbed and longer wavelengths simply lack the energy to make the transition.

13.2 LASER STRUCTURE

The simplest (and oldest) structure for a laser diode is the *homojunction laser diode*, which uses a single junction. These are fabricated of a single junction between two direct-bandgap materials of the same type, one p-type and one n-type, that is called a *homojunction* since both materials are of the same type. Light is emitted by electron–hole pair recombinations in the thin active region formed by the junction of the two materials (the depletion region). Usually, gallium arsenide (GaAs) is used, with each part of the device doped slightly differently: one part with an electron donor and one part with an electron acceptor. Mirrors for the laser cavity are fabricated simply by cleaving the crystal at right angles to the laser axis. Having an index of refraction of 3.7, the reflectivity of each mirror may be calculated to be 33% by using the Fresnel equations (Section 4.9). This represents a large loss in the cavity; however, most semiconductor laser materials have ample gain, to allow such a simple configuration. Improved performance may be achieved by fabricating a single dielectric mirror, composed of alternating quarter-wavelength-thick layers of high- and low-index-of-refraction materials, at the HR end of the laser diode. Improved mirrors are used on almost all modern laser diodes. A diagram of a simple homojunction structure is shown in Figure 13.2.1. Homojunction lasers are characterized by large threshold currents with a typical device requiring tens of amperes to lase. Such currents prohibit continuous operation at room temperature, so CW homojunction devices require cryogenic cooling, making them impractical for many applications.

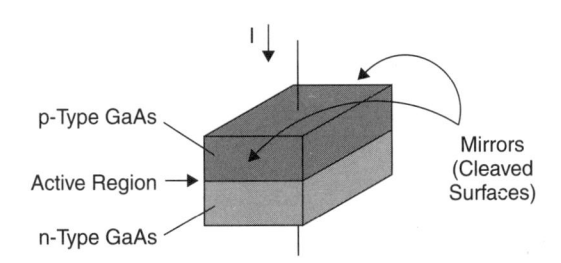

Figure 13.2.1 Homojunction laser structure.

Improvements in the structure of the laser may be made by confining the intracavity beam in a dielectric waveguide structure formed from the semiconductor material itself. Such a structure requires two interfaces of different indexes of refraction, one on top and one below the active region, so two junctions are formed in what is called a *heterostructure laser diode*, or in this case, a *double heterostructure*, since there are two confining interfaces. To obtain different indexes of refraction, two different materials are required. GaAs is generally used as the higher-index material and aluminum–gallium–arsenide (AlGaAs) as the lower-index material. As depicted in Figure 13.2.2, AlGaAs is doped to form p- and n-type materials which essentially have identical indexes of refraction given that dopant concentration is small. Between layers of these materials, GaAs is sandwiched as the active-region material, from which laser light is emitted. Differences between the indexes of refraction occurring at each interface form a reflector that confines light inside the GaAs layer, which drastically improves efficiency, and more important, lowers threshold current for the device by increasing gain. The active region (GaAs) is typically only 0.1 μm in thickness.

Usually, a stripe contact is used on the top of the structure to make an electrical connection to the device. This further limits the area of the active region in the GaAs laser (since current is not spread out over the top surface area of the entire structure), which serves to increase current density and further lower threshold current. In a real laser diode of this type, more than three layers are generally required, and a layer that serves as an electrical interface between metal contacts and each AlGaAs layer is usually employed. Double-heterojunction laser diodes commonly operate at room temperatures with low threshold currents, in the tens-of-milliamperes range. A commercially available diode laser of this type is shown in Figure 13.2.3, in which the diode itself is the small crystal (in reality, the size of a grain of salt) mounted atop a block of metal that serves as a heatsink. Behind the laser crystal is a photodiode that serves to supply the driver circuitry with feedback about the actual light level emitted from the laser, allowing the circuit to stabilize the laser output power.

The double-heterostructure arrangement confines intracavity light in only one direction (top and bottom) of the GaAs layer, but a further improvement in performance can be made by manufacturing the device so that a confining dielectric interface exists on all four sides of the active region in a *buried heterostructure*

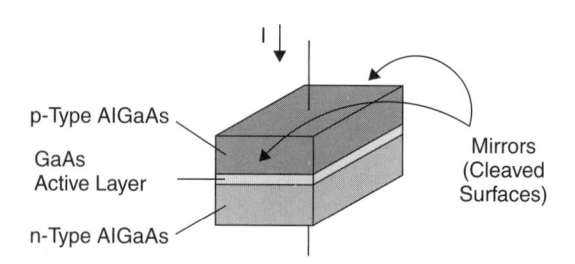

Figure 13.2.2 Heterojunction laser structure.

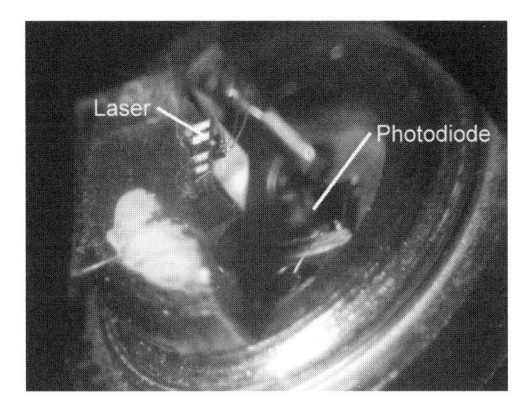

Figure 13.2.3 Typical laser diode.

laser, depicted in Figure 13.2.4. In such an arrangement the entire stack of three layers of a typical heterojunction laser (p-type AlGaAs, GaAs active region, and n-type AlGaAs) is confined on each side by an n-type AlGaAs layer. The interface between the GaAs material in the active region, and this lower-index-of-refraction material on the sides of the active layer, serves to further confine light in the laser cavity. Such lasers are often called *index guided*, since light inside the cavity is guided in a manner similar to that of an index-graded fiber optic.

One of the newest structures for semiconductor lasers is the vertical cavity surface-emitting laser (the VCSEL). In a laser of this type, light is not emitted from the edge of the device but rather through the entire top layer of the semiconductor crystal itself. While edge-emitting laser diodes produce an elliptically shaped output beam that has high divergence (discussed in Section 13.3), requiring an external lens to collimate it into a usable beam, a VCSEL produces a round beam of much higher quality. Rather than emission from the edge of the diode, light is emitted from the surface of a VCSEL. In addition to a better beam shape, VCSELs feature single longitudinal mode operation with a narrow spectral linewidth.

A VCSEL features a thin active layer (100 to 200 nm) like that of a conventional laser diode, but whereas the gain length of a conventional diode is 200 to 500 μm

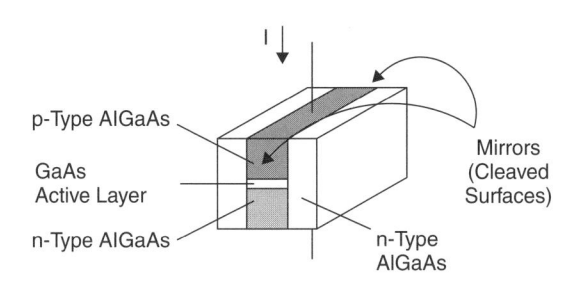

Figure 13.2.4 Buried heterojunction laser structure.

318 SEMICONDUCTOR LASERS

(the length of the structure), the gain length in a VCSEL is the length of the active layer. Resonator optics are fabricated above and below the semiconductor crystal. With a short active layer and low gain, cavity optics must be fabricated from multiple layers of dielectrics—alternating quarter-wavelength-thick layers of high- and low-index-of-refraction materials—for high reflectivity. Current in the device flows along the optical axis through electrodes on the top and bottom of the device instead of perpendicular to it. These electrodes can be fabricated so that current flows through the mirror structure itself or through two contact layers close to the junction, the latter approach offering a lower electrical resistance. The typical structure of a VCSEL is depicted in Figure 13.2.5.

The output from the VCSEL is preferred for coupling to a fiber since emission occurs in the form of a circular beam which is easily focused, as opposed to the elliptical output beam from an edge emitter. Aside from applications in which the output is coupled to optical fibers for communications use, VCSELs are also used in some newer designs, in which a frequency-doubling crystal is placed inside the cavity (or at least an extended cavity with three reflectors) of the semiconductor laser. The structure of the device allows the inclusion of wavelength-selective Bragg reflectors for resonator optics.

VCSEL lasers usually feature low threshold currents, often below 1 mA. Since the device does not require precision-cleaved ends to form cavity mirrors nor the deposition of multiple dielectric layers on the edges of the crystal, they may be fabricated as multiple devices on a single wafer in a manner similar to the method in which microchips are manufactured. Given the quality of the output beam as well as the low threshold of the device this is likely the most important up-and-coming device for optical communications.

Finally, it is possible to *optically-pump* some solid-state materials. One design uses an 808-nm pump diode to optically pump a 946-nm semiconductor laser which is, in turn, frequency doubled to produce 488-nm light. The arrangement, called a VECSEL (for vertical external cavity surface-emitting laser), has external optics allowing inclusion of a nonlinear crystal inside the cavity to accomplish frequency doubling. This particular laser, which has more in common with a diode-

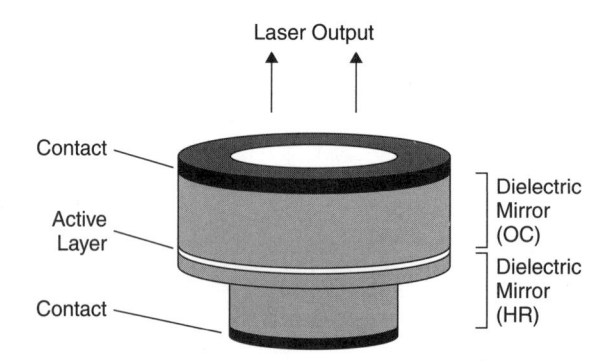

Figure 13.2.5 VCSEL laser structure.

pumped solid-state laser than a semiconductor laser, is designed as a replacement for the blue argon-ion laser.

13.3 OPTICS

The simplest conventional laser diodes can use the cleaved surfaces of the semiconductor crystal as cavity reflectors. The difference of the index of refraction of the semiconductor material to the surrounding air forms a reflector with an approximately 33% reflection. Although this may seem low, semiconductor lasers typically exhibit a high-enough gain to overcome such losses. For higher efficiency, the rear surface of the semiconductor laser can be coated with a multilayer dielectric mirror to reflect almost 100% of the light emitted in that direction. The front optic on such a laser is still usually an uncoated, cleaved surface. With a short active layer, the gain is much lower in VCSELs than in edge-emitting lasers, so that dielectric mirrors fabricated from multiple layers of dielectrics, alternating quarter-wavelength-thick layers of high- and low-index-of-refraction materials, are the rule for both cavity optics for these lasers. Mirrors fabricated in such a way are highly wavelength selective, with the wavelength corresponding to the maximum gain of the semiconductor device.

The inherent spectral width of semiconductor lasers is quite large compared with, say, a gas laser with a typical multimode diode having a spectral width of 2 to 4 nm. To reduce the spectral width, wavelength-selective optics may be employed in a manner similar to that used for other lasers: namely, the use of a grating in place of the HR. Given the tiny dimensions of a semiconductor laser, though, implementation of a discrete diffraction grating is difficult, so the grating may be constructed as an elongated structure called a *distributed Bragg reflector* (DBR). Such a reflector resembles a corrugated surface (like the center of cardboard) manufactured from dielectric materials (Figure 13.3.1). Reflection of waves at the interface between two materials of different indexes of refraction leads to constructive interference at

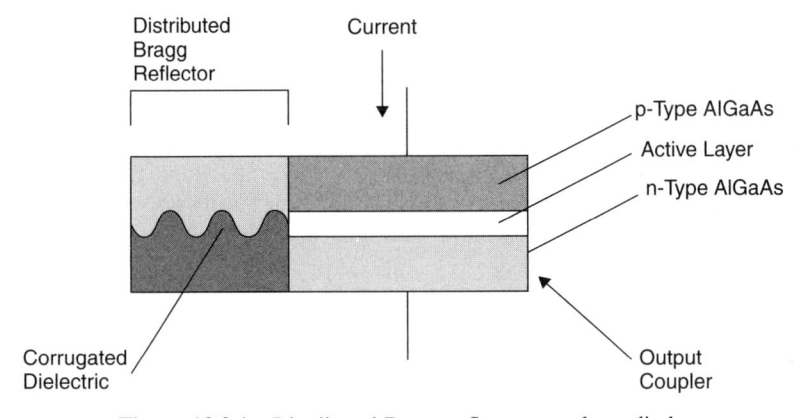

Figure 13.3.1 Distributed Bragg reflector on a laser diode.

320 SEMICONDUCTOR LASERS

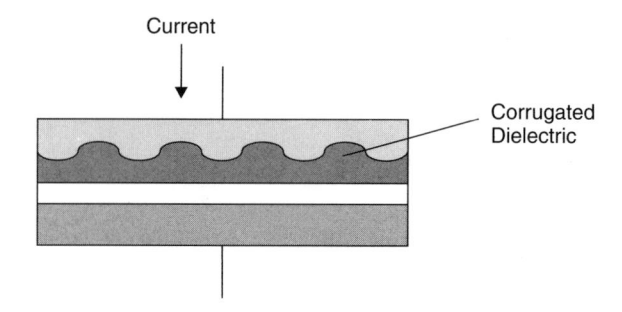

Figure 13.3.2 Distributed feedback laser diode.

a single well-defined wavelength (determined by the distance between peaks in the corrugation). Such a reflector acts much like a high-performance dielectric mirror, with the specific wavelength of maximum reflectivity called the *Bragg wavelength*.

Taking the DBR scheme one step further, the confining layer (top of bottom) on one side of a regular laser diode can be replaced with a Bragg reflector fabricated in a manner similar to that of the DBR. In such a manner, light confined within the cavity is wavelength-selected so that only a single wavelength (where the reflector is resonant) is amplified within the laser. This corrugated structure reflects partially at each interface between the materials of differing refractive index, so optical feedback is distributed throughout the cavity [hence the term *distributed feedback* (DFB)]. The grating, again a corrugated surface composed of dielectric materials, is fabricated into the structure of the diode itself and is distributed over a considerably longer length than a traditional grating stretching over the entire length of the device (Figure 13.3.2). A discrete HR and OC are not required on such a device, and like the DBR laser, the wavelength of the grating determined by the spacing of the corrugations and the resulting spectral width of the output can be as low as 0.1 nm.

13.4 POWER SUPPLIES

Power supplies for laser diodes are relatively simple, providing current regulation, and often, regulation of light output. Although current regulation is sufficient, electrically, laser diode characteristics change as they heat during operation, with light output dropping as the device becomes warmer. For this reason, light feedback is often employed. Many laser diodes (e.g., the one pictured in Figure 13.2.3) are fitted with a photodiode on the side of the device opposite the output, allowing the drive circuit to compensate for temperature by varying the drive current in order to maintain a constant output. Such devices are easily identified since they feature three terminals: one for the laser diode, one for the photodiode, and a third terminal common to both.

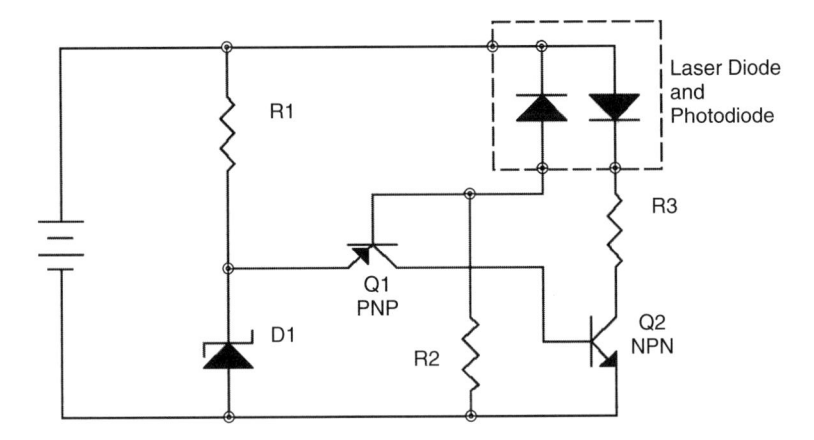

Figure 13.4.1 Laser diode driver.

A simple power supply for a laser diode is diagrammed in Figure 13.4.1. A simplified circuit, current through the laser diode, is regulated to keep constant the amount of light falling on the photodiode. As light falls onto the photodiode (the left diode in the package), current through the base of transistor Q1 will decrease. In turn, current through the collector of Q1 falls, as does current through the base of Q2. Current through the collector of Q2 is hence decreased and laser diode output falls. As light output decreases from the laser, diode current through the base of Q1 increases and the reverse process occurs. The circuit reaches equilibrium and current eventually reaches a constant value.

The simplified controller shown lacks features to protect the diode from damage. Advanced diode controllers feature preset limits on laser diode current since excessive current invariably leads to destruction of the laser diode. As well, current through the laser must be damped (usually, by including a capacitor from the base of Q2 to the emitter) to ensure that it does not oscillate, which might cause current to spuriously reach a damaging value. In addition to control of laser diode current, a temperature controller is frequently desired, which utilizes a thermoelectric cooler to keep laser diodes at a constant temperature for wavelength stability. Temperature feedback is provided, enabling the controller to sense the actual laser diode temperature and control current through the cooling module accordingly. Temperature controllers generally use PID (proportional–integral–derivative) control algorithms for precise control.

13.5 OUTPUT CHARACTERISTICS

Spatially, the output beam from a laser diode consists of an elliptically shaped beam which is diffraction limited. Thinner dimensions, such as the thickness of the active layer, have higher divergence since gain in that direction comes primarily from pho-

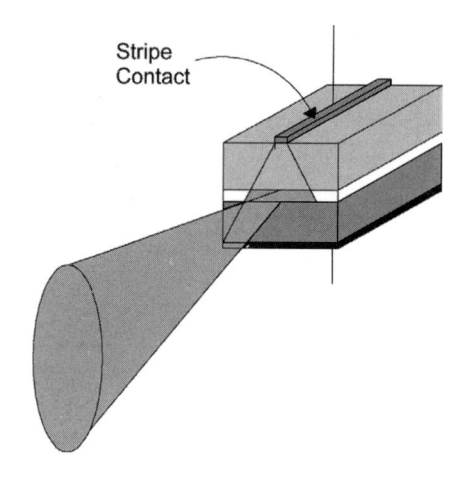

Figure 13.5.1 Typical diode laser output beam.

tons which are confined by the top and bottom layers of the structure and have been reflected many times. The wider active layer of the device means fewer reflections and hence lower divergence in that direction. A typical beam profile is shown in Figure 13.5.1. In this particular diode, a stripe contact on the top layer restricts the flow of current through the device, increasing current density in the center of the active region. Of course, VCSELs feature a circular beam that is easier to focus to a point than is the elliptical beam shown here.

The wavelength stability of a laser diode depends highly on the temperature of the diode. As the temperature of a laser diode increases, the refractive index of the semiconductor material itself changes. Since the resonant wavelength of a cavity depends on the refractive index of the material, the wavelength shifts toward longer wavelengths as the temperature of the device increases. The output wavelength of a conventional laser diode increases more or less linearly as temperature does. Figure 13.5.2 shows the spectral output of a typical laser diode at a current of 83.9 and 111.9 mA, respectively. The wavelength shift is obvious in this case, about 0.36 nm toward a longer wavelength as the current is increased.

Conventional laser diodes have cavity lengths (typically, 200 to 500 μm in length) which are long enough to allow several longitudinal modes to oscillate simultaneously. In the spectrum of the typical diode shown in Figure 13.5.3, modes spaced 0.288 nm apart are evident in the wavelength sweep. In this case, modes are made evident by operating the diode at a current just above threshold, where the output consists of a good deal of broadband (spontaneous) output. The resonant wavelengths of the cavity separate the output spectrum into many modes, as evident in the figure. As the current through the device increases, a dominant mode appears and the output resembles that of Figure 13.5.2.

The output of a conventional laser diode consists of many closely spaced longitudinal modes which lead collectively to a large spectral bandwidth in the output of

OUTPUT CHARACTERISTICS 323

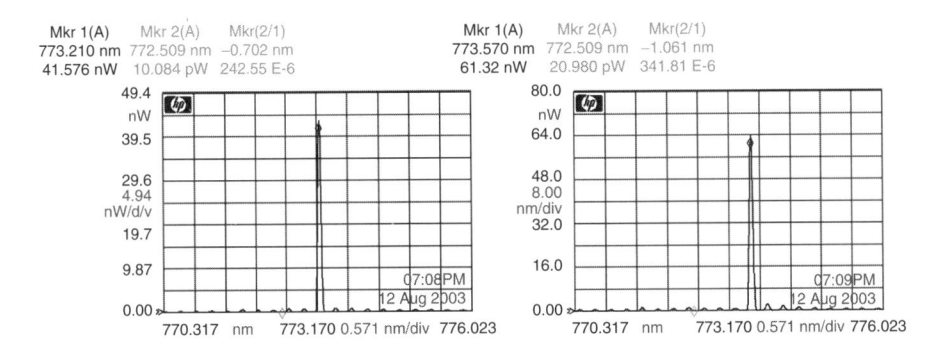

Figure 13.5.2 Laser diode wavelength at increasing current.

the device. Even with a single-mode laser, temperature can affect the output spectra, since as temperature changes, the wavelength at which the cavity is resonant shifts and the laser can hop between allowed longitudinal modes. Again, temperature control is required to ensure wavelength stability.

While the output spectral width can be restricted by using a wavelength selector such as a DFB grating, another approach, used in VCSEL lasers, is to reduce the length of the cavity so that only a single mode is supported. VCSELs feature an extraordinarily thin active layer (less than 100 nm), with resonator optics fabricated above and below the semiconductor crystal. The entire cavity (usually less than 5 μm) is thin enough that only a single longitudinal mode is supported. Some

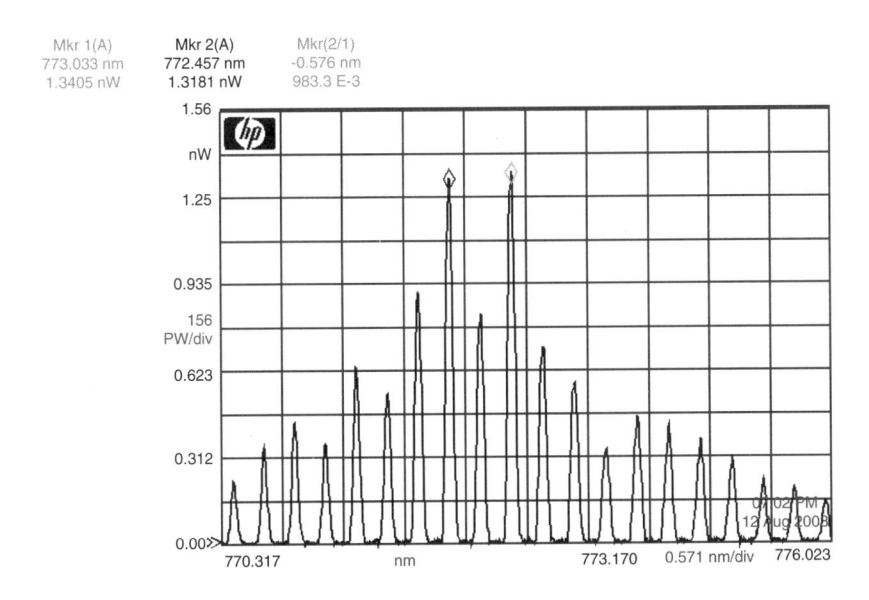

Figure 13.5.3 Laser diode modes.

324 SEMICONDUCTOR LASERS

VCSELs have cavities large enough to support several modes, but spectral output is still narrow compared to an edge-emitting diode since modes are separated by a larger amount than in a conventional diode. The thin dimensions of the cavity result in resonant peaks which are spaced farther apart than in a longer cavity (as outlined in Section 6.4).

As we mentioned, the wavelength stability of a laser diode depends highly on the temperature of the diode. As the temperature of a laser diode increases, the refractive index of the semiconductor material itself changes. Since the resonant wavelength of a cavity depends on the refractive index of the material, the wavelength shifts toward the red as temperature of the device increases. The output wavelength of a normal multimode laser diode increases more or less linearly as temperature does. For a single-longitudinal-mode (i.e., single-frequency) laser diode, though, the output wavelength can shift abruptly as the temperature fluctuates in what is called *mode hopping*. This is caused by the nature of the cavity, which is resonant only at discrete wavelengths (separated by the FSR of the cavity). As temperature increases and resonant wavelength increases, the output can shift from resonance at one mode to the next. For this reason, laser diodes require temperature stabilization, often in the form of a thermoelectric cooler built into the diode to stabilize the output wavelength.

Semiconductors have a wide gain width (output can span over several nanometers), so are relatively broadband sources compared with gas lasers. The output of a typical diode laser is multimode, consisting of a series of peaks (longitudinal modes) spaced by the FSR of the relatively small cavity, although at high drive currents a dominant mode often appears. As discussed in Section 13.3, optical structures can be implemented which serve to reduce the spectral width of the output.

As well as longitudinal modes, transverse modes may also appear in the output, but these may be restricted in the same manner as used with other lasers, by limiting the size of the active region. In an index-guided laser (those utilizing heterostructures), selection of layer size can effectively limit the gain aperture and hence force the laser to operate strictly in TEM_{00} mode. Unfortunately, small apertures lead to large divergences in the output beam, so most lasers with restricted apertures require the use of an external lens to collimate the exiting beam.

13.6 APPLICATIONS

Solid-state lasers are by far the most commonly used today. Low-powered infrared lasers are employed for many optical storage applications, including CD and DVD players. Visible laser diodes are also common and used for applications ranging from laser pointers and levels to scanning applications, often replacing helium–neon lasers. For storage applications a shorter wavelength is preferred since it can be used to yield a smaller spot size and hence increase the density of optical storage media. While short-wavelength diode lasers have recently been developed, their use in optical storage devices is relatively new. In addition to short-wavelength diodes, another alternative is the DPSS laser, in which a diode laser is used to pump a solid-state laser usually employing a YAG or vanadate crystal. High-powered arrays of semiconductor lasers allow the construction of DPSS lasers yielding second-harmonic powers in the tens-of-watts range, usually in the green region of the spectrum.

CHAPTER 14

Tunable Dye Lasers

In a dye laser the active lasing medium is an organic dye dissolved in a solvent such as alcohol. These lasers may be pumped by either flashlamps (like a solid-state laser) or by another laser. Laser-pumped dye lasers normally employ nitrogen or excimer pump lasers and hence are pulsed, but continuous dye lasers are possible using a CW argon-ion laser as a pump source. The major advantage of this laser over other types is continuous tunability over a wide range. A laser employing rhodamine-6G, for example, can be tuned continuously through a range of wavelengths spanning visually from a shade of green–yellow to a shade of red. By changing the dye employed, the range can be selected. Hundreds of dyes are known to lase.

14.1 LASING MEDIUM

In a dye laser the active medium is a fluorescent organic dye dissolved in a solvent (usually, an alcohol). The dye is pumped optically by a laser or by a flashlamp to produce a population inversion followed by stimulated emission to produce a laser gain. The dyes involved are large molecules with molecular weights in the range 400 to 500 amu. Most laser dyes belong to one of several families of dyes such as rhodamines or coumarins. Lasing begins when incident energy is absorbed by the dye, exciting it from the lowest singlet state to a high-energy level within the upper singlet band, as shown in Figure 14.1.1. From the high-energy level the dye falls to a slightly lower state within the same singlet band, which serves as an upper lasing level. A laser transition can then occur between the upper lasing level and the lower singlet state, which serves as a lower lasing level.

Dye lasers feature extremely broad energy bands allowing both absorption and emission over a wide range using wavelengths. Large absorption bands ensure efficient pumping of a dye, especially when a broadband source such as a flashlamp is used. In the case of laser output, tunability across a large range of wavelengths may be accomplished by employing a wavelength selector such as a diffraction grating in the cavity.

Fundamentals of Light Sources and Lasers, by Mark Csele
ISBN 0-471-47660-9 Copyright © 2004 John Wiley & Sons, Inc.

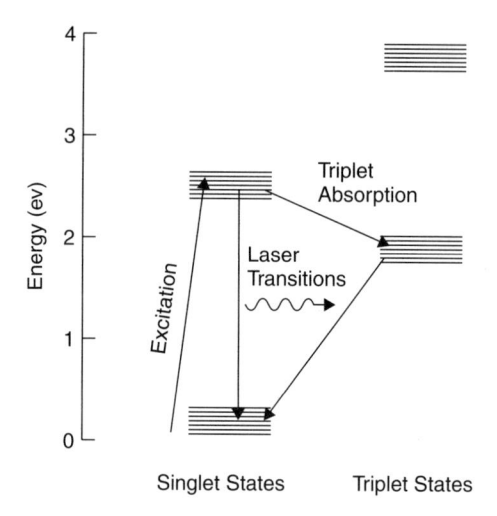

Figure 14.1.1 Laser dye energy levels.

An alternative pathway exists to foil laser action in the triplet states of the dye, also shown on Figure 14.1.1. You will recall from Chapter 3 that triplet states originate when excited electrons in the dye molecule spin in the same direction as that of the remaining electrons in the dye molecule (the singlet states result when the excited electron spins in the direction opposite to the lower-energy-state valence electrons still in the dye molecule). Because triplet states have lower energies than corresponding singlet states, dye molecules can easily migrate to those states and in doing so depopulate the upper lasing level. To make matters worse, triplet states are metastable and have much longer lifetimes than the singlet levels. When a short pump pulse such as that from a nitrogen laser (at 10 ns) is employed, triplet states do not form and do not present a problem for lasing, but when a flashlamp is used (which generally have pulse widths of over 1 μs), triplet states can form. For this reason, flashlamps must be designed to discharge as quickly as possible. Ordinary photographic strobes, for example, often have pulse lengths of 1 ms and will not work for pumping most dye lasers. In addition to a fast pump pulse to prevent triplet formation, triplet quenching additives can be mixed with many dyes, such as cyclooctatetraene (COT). These additives work by providing a deexcitation pathway from the triplet states, allowing the dye molecule involved to reenter the lasing process.

As evident from Figure 14.1.2, an enormous number of dyes are commercially available which span the entire spectrum from UV to IR. Dyes may be pumped by flashlamp or by a pump laser. Pump lasers include excimer, frequency-doubled or frequency-tripled YAG, nitrogen, or CW ion lasers. Regardless of the source, large power densities, typically around 100 kW/cm^2, are required to pump a dye to a level where lasing is possible. It should also be noted that some success has been achieved by dissolving dyes into a host of resin which is solidified into solid acrylic. Such materials have been made to lase when pumped by another laser.

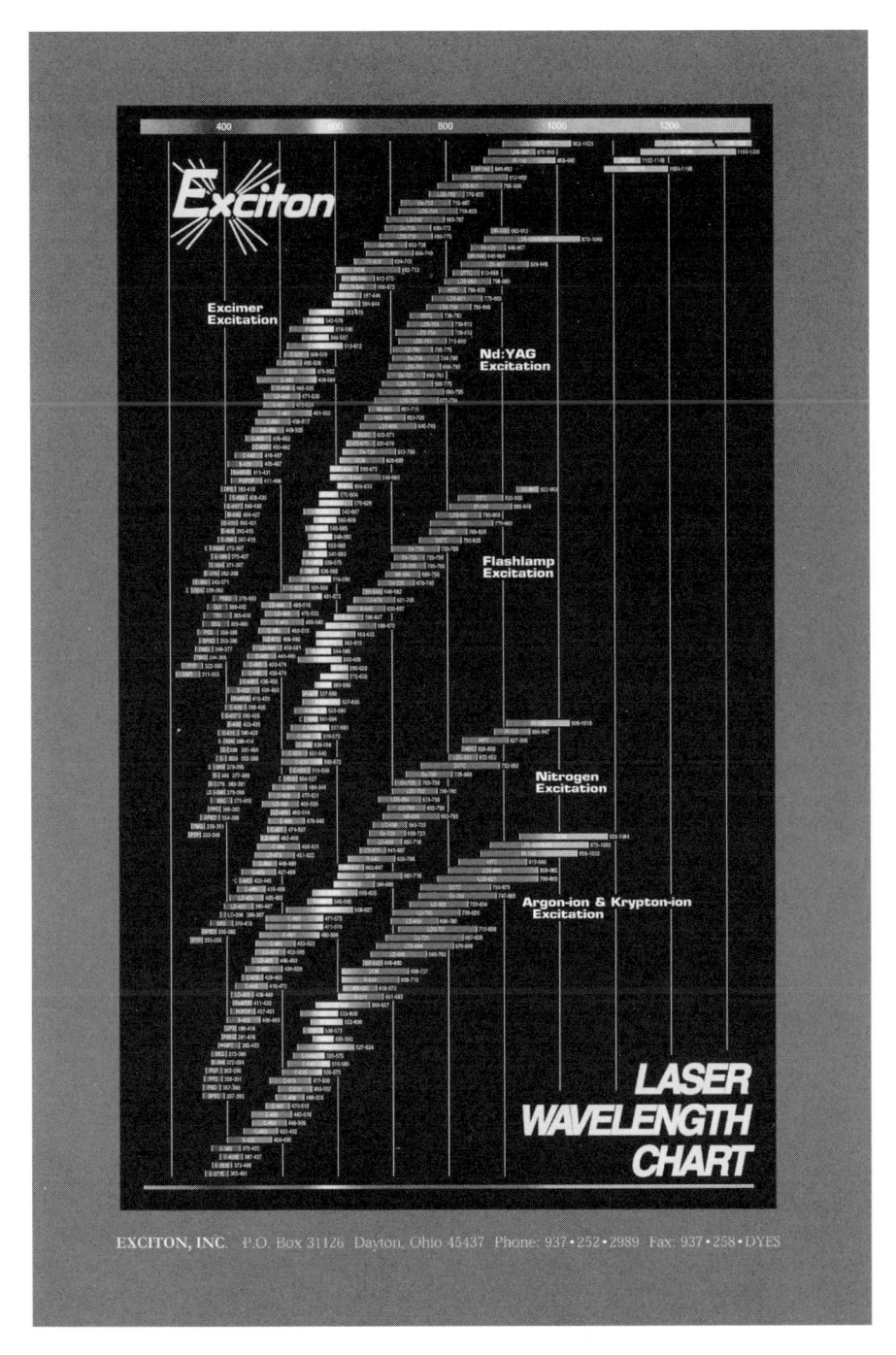

Figure 14.1.2 Laser dyes. (Courtesy of Exciton, Inc.)

14.2 LASER STRUCTURE

Perhaps the simplest form of a dye laser is the flashlamp-pumped laser, which closely resembles a solid-state laser with a liquid-filled cell instead of a solid crystal rod. Flashlamps may be either linear, with light from the lamp focused onto the dye cell using an elliptical reflector (in the same manner as that used for the YAG laser described in Chapter 12) or using a coaxial lamp. The coaxial configuration used in the first lasers of this type[1] is constructed of a space created between the dye cell at the center of the laser and a larger outer jacket, as shown in the cross section of Figure 14.2.1.

This space between the two quartz tubes is evacuated and filled with a low-pressure gas such as xenon. The coaxial approach offers an extremely low inductance path for current if the discharge capacitor is also coaxial (i.e., the entire dye cell and lamp assembly is inserted into the center of a circular capacitor). This approach allows extremely fast discharges, with rise times of well under 1 μs, to be generated. The approach is not limited to the use of a coaxial capacitor, but that configuration certainly produces the fastest discharges.

Other configurations for a flashlamp-pumped dye laser include a slab configuration in which the dye cell is formed between two slabs of glass that have a different index of refraction than the dye solution between the slabs. Total internal reflection confines light within the cavity, producing a long optical path and hence large amplification. This is similar to the confinement provided by various guiding layers in a practical semiconductor diode laser (Chapter 13). In a flashlamp-pumped dye laser, circulation of the dye is required to keep the temperature of dye across the cell consistent. If one region of the dye is warmer than another region, a thermal gradient develops, with the result being a difference of indexes of refraction of liquid in the dye cell which spoil the Q of the cavity, preventing laser oscillation.

Flashlamp-pumped dye lasers feature large pulse energies and are a useful source for opthalmological procedures, where they may be used for retinal photocoagula-

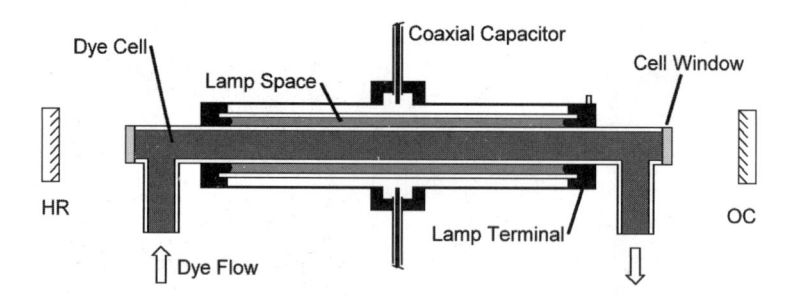

Figure 14.2.1 Flashlamp-pumped dye laser configuration.

---

[1]The original report of a flashlamp-pumped dye laser may be found in "Flashlamp-pumped organic-dye lasers" by P. P. Sorokin et al. in *Journal of Chemical Physics*, Vol. 48, No. 10 (1968). An extremely fast flashlamp was required to excite the dye without the interference of triplet states. The coaxial capacitor provides an intrinsic inductance of under 1 nH.

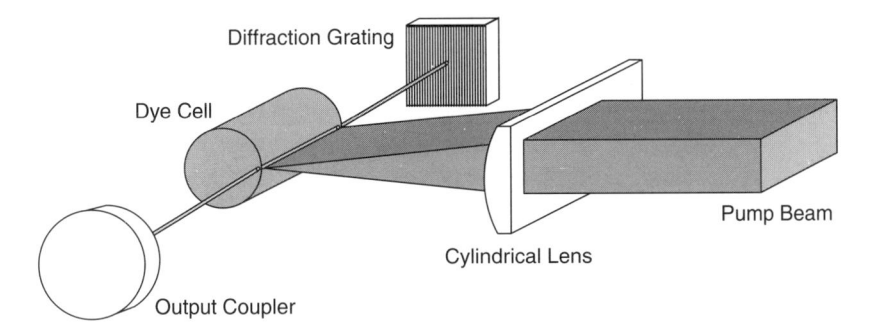

Figure 14.2.2 Laser-pumped dye laser.

tion. However, pulse rate is severely limited to the rate at which dye can flow through the dye cell to remove heat and suppress thermal gradients that may occur. In addition, electrical discharge paths in a flashlamp-pumped dye laser are critical and must be manufactured with extremely low inductance. The other type of dye laser is pumped by another laser, such as an excimer or nitrogen laser as in Figure 14.2.2. These lasers are fairly simplistic and, compared to flashlamp-pumped types, exhibit unusually high gain, allowing the use of a short dye cell.

In this arrangement, the pump laser beam is focused to a line on a dye cell using a cylindrical lens, as shown in Figure 14.2.3, in which the focused beam from a nitrogen or excimer pump laser strikes the side of a dye cell exciting dye molecules along the length of the cell. In most cases it is possible to operate the dye laser superradiantly by focusing the pump laser tightly onto the dye cell, but this is not desirable since a dye laser operated in such a way is not tunable, losing the major advantage of this laser over other lasers,[2] so pump light is not focused so tightly that superradiance occurs. Pump light striking the dye cell excites dye molecules in a narrow channel physically just inside the dye cell itself: Penetration of the pump light into the cell is minimal, and essentially all is absorbed within the first few millimeters of dye within the cell.

Optics for a laser-pumped dye laser usually consist of an output coupler and a diffraction grating. Although a simple mirror (HR) could be used in the cavity, a grating is invariably used to allow tuning of the laser across the wide gain bandwidth of the laser, the biggest advantage of a dye laser. In a practical dye laser a beam expander consisting of two lenses is placed between the dye cell and the diffraction grating to utilize more of the grating area. An etalon may also be placed in the optical path to reduce spectral width. Finally, a MOPA configuration may be used in which a separate amplifier is placed in front of the oscillator (complete with wave-

---

[2]The ability to operate the laser superradiantly makes alignment of this type of laser particularly simple. The dye laser is made to operate superradiantly, and cavity optics are aligned with the resulting intracavity beam to boost output and decrease beam divergence of the laser. The pump light is then refocused so that the laser is no longer superradiant.

Figure 14.2.3 Excited dye cell.

length selector). The configuration for a complete dye laser with narrow spectral output width is shown in Figure 14.2.4.

The pump laser is usually a nitrogen or excimer laser emitting in the UV range, but frequency-doubled YAG lasers may also be used. Because the pump laser is pulsed, the dye laser also has a pulsed output, but it is possible to build a CW dye laser pumped by a CW laser source such as an ion laser. In this case the biggest problem becomes heat management and degradation of the dye itself. Both problems are alleviated by forming the dye into a continually flowing sheet of liquid called a *laminar flow*. Flowing dye is pumped through a nozzle to create a broad, flat stream

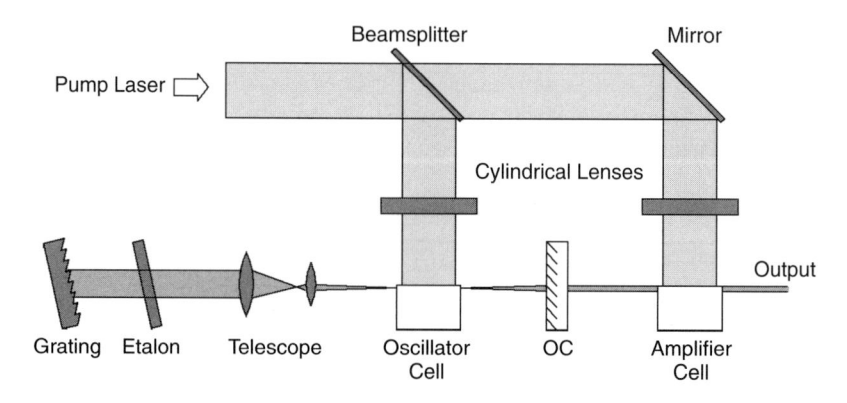

Figure 14.2.4 Optics for a dye laser.

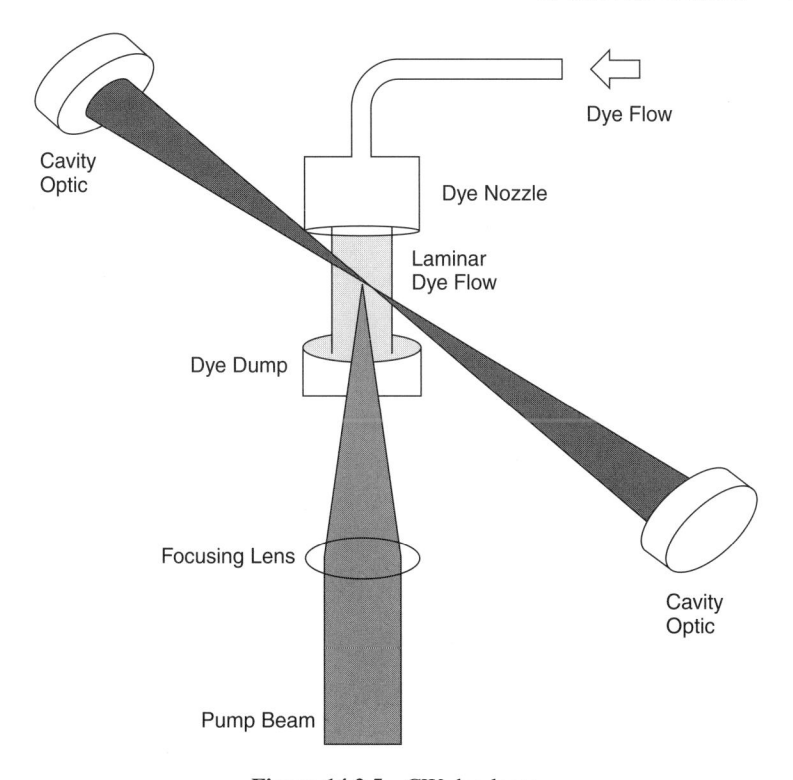

Figure 14.2.5 CW dye laser.

onto which pump laser light, usually from an argon-ion laser given that many laser dyes absorb strongly in the violet-to-green band where the argon-ion laser has the strongest output, is focused by a lens (or a concave mirror) onto the surface of the flowing dye. In a manner similar to that of the laser-pumped arrangement described previously, a column of excited dye molecules serves as the gain medium, which, is off-axis (usually at the Brewster angle) from the axis of the optical resonator. A typical CW dye laser arrangement is shown in Figure 14.2.5. Like most dye lasers, intracavity wavelength selectors are also often included (although they are not shown in the figure).

Aside from heat removal (required when the pump laser is a 10-W CW argon-ion laser!), dye flow helps suppress the effects of triplet absorption in the dye by ensuring a fresh supply of dye in the area irradiated by pump laser light. Fast dye flow is hence required to ensure that dye molecules are exchanged before triplet absorption becomes problematic and affects lasing action.

14.3 OPTICS AND CAVITIES

Most dye lasers are used for their tunability over wide wavelength ranges, and as such, most lasers have integral wavelength selectors in the cavity. Depending on the application, a simple grating may be sufficient to render a spectral output width. Regular diffraction gratings diffract incident light into many orders, so reflectivity of this element as a cavity optic is generally poor. As such, the laser must be operated at a high gain, sometimes close to superradiance, resulting in a broad spectral output. To increase reflectivity, a special type of grating called an echelle grating may be used which is designed to have a reflecting surface as large as possible. Such gratings can reflect up to 70% of incident light into one order. Further enhancement of the laser may be accomplished by using a beam-expanding telescope in the cavity. Aside from allowing the use of a larger grating area than the normally tiny beam emitted from the dye cell, the telescope can also collimate the highly divergent beam exiting from the dye cell. Without collimation the angular spread of light striking the grating limits the resolution possible with the grating alone.

In addition to a diffraction grating, an etalon is frequently included to reduce output bandwidth. A tilted Fabry–Perot etalon between the telescope and the grating greatly reduces the spectral width of the output. The etalon is placed in this manner to ensure that the beam passing through it is collimated, a requirement for proper operation. Etalons may be fine-tuned by changing the angle within the cavity or by changing the pressure of the gas (usually, air or dry nitrogen) between the plates of the etalon. The change of pressure results in a change of refractive index and hence a shift in the resonant frequency of the device. As well, etalons may be fabricated from a solid piece of quartz with reflective coatings on both sides. In this case, angle or temperature tuning may be used. Since the gain of a dye laser is generally large, output couplers are usually plane reflectors with reflectivities below 50%. Another popular use of a dye laser is to produce ultrashort modelocked pulses as covered in Chapter 7. Modelocking requires an intracavity switch such as a passive saturable absorber or an EO modulator, usually a Pockels cell.

14.4 OUTPUT CHARACTERISTICS

The output characteristics of a dye laser are highly dependent on the optics employed. A laser employing broadband optics would feature a naturally broad spectral width, typically spanning almost the entire range of the dye (especially in a laser-pumped dye laser, in which gain is usually large); this can be as large as 100 nm. A laser with such characteristics is not particularly useful, so wavelength selection is invariably employed.

Use of a diffraction grating alone as a wavelength selector (with suitable beam-expanding optics allowing utilization of a large area of the grating surface) renders a spectral width of 0.01 nm, which is suitable for many applications, but to reduce linewidth an intracavity etalon is often included in the optical path. Use of an etalon along with a diffraction grating can render spectral widths as low as 0.0005 nm. Such linewidths are required when a dye laser is used as a tunable laser source for spectroscopy. Etalons may be angle- or pressure-tuned, with pressure tuning preferred for simplicity. As a modelocked laser, dye lasers have produced the shortest pulses produced from any laser source. The wide gain bandwidth of a typical dye allows the production of a series of extremely short pulses, in the femtosecond range.

14.5 APPLICATIONS

As a source for spectroscopy, the dye laser is ideal given the wide range over which tuning may be accomplished and the narrow spectral width of the output. It is used in situations such as atomic absorption spectroscopy, where the beam is passed through a sample in a cell or a hot gas such as exhaust gas from a flame or the gases burning inside the cylinder of an internal combustion engine. Compact flashlamp-pumped dye lasers are occasionally employed in the field of opthalmology for retinal photocoagulation. Tunability allows the output to be optimized for the peak absorption wavelength of hemoglobin, and the pulses are short enough to coagulate blood without generating an explosive shock wave within tissue as Q-switched laser pulses can.

INDEX

Absorber, saturable dye, 195
Absorption, 10–11, 92
 and color, 18
 coefficient, 106–107, 113, 174
 rate, 92
Absorption spectra, YAG, 118
A-coefficient, *see* Einstein coefficients
Acoustic wavelength, 207
Acoustooptic switch, 206–211
 Bragg angle, 207
 Raman–Nath, 209
ADP (ammonium dihydrogen phosphate), 228
Alignment, cavity, 181–183
 HeNe mirrors, 182
 practical, 187–190
 tolerance of confocal cavity, 181
Allowed transitions, 63, 68
Amplifier:
 dye laser, 331
 fiber, *see* EDFA
 helium–neon laser, 109
 ruby laser, 298
Angular momentum, electron, 56
 orbital quantum number, 56
Antireflective coatings, 240
AO modulator, *see* Acoustooptic switch
Arc lamp pumping, 303–305, 309
 power supplies, 307
Argon fluoride, 272
Argon ion (Ar^+) laser, 247–249
 depopulation of lower levels, 129
 doubly ionized, 250–251
 etalon, 173
 gain, 104
 linewidth, 102, 259
 output lines, 249
 wavelength selection, 168–169
ASE (amplified spontaneous emission), 96, 309

Atomic ground state, 12
Autocollimator, 188

Ballast resistance, 244, 257, 291
Balmer series, 21
Bandgap, 41
Bandwidth, *see* Spectral width
B-coefficient, *see* Einstein coefficients
Beam, Gaussian, *see* Gaussian beam
Beamshape, semiconductor laser, 322
Bias voltage, diode, 41
Birefringence:
 in EO modulators, 202–203
 in phase matching, 224
Blackbody, 2
Blackbody radiation, 2–6, 15
 spectrum, 5, 15
Blumlein configuration, 265
Bohr model, 28
Bohr orbit, 28–29
Boltzmann distribution, 13
 in a laser, 86
Boltzman, Ludwig, 3
Bonding, covalent, 73
Boost gas, excimer laser, 280
Bose–Einstein statistics, 8
Bragg angle, 207
Bragg diffraction, AO modulator, 207
Bragg, W.L., 208
Brewster angle, 110, 162
Broadband emission, 15
Broadening, homogeneous, 145
Broadening, lifetime, 176
Buried heterojunction laser, 316

Cadmium (HeCd) ion laser, 250–251
Calcite, as a nonlinear material, 232

Fundamentals of Light Sources and Lasers, by Mark Csele
ISBN 0-471-47660-9 Copyright © 2004 John Wiley & Sons, Inc.

337

338 INDEX

Carbon-dioxide:
 IR absorption spectrum, 78
 vibrational energy in, 74–75
Carbon-dioxide laser, 283–293
 energy levels, 284
 linewidth, 292
 TEA, 269
 water vapor in, 287
Cataphoresis, in a gas laser discharge, 252
Cathode, hot, 253
Cavity, *see* Resonator
Cavity dumping, 211
Cavity modes, 6–7
Cavity radiation, 6
Ceramic plasma tubes, 252
Chromium ion (Cr^{3+}), 296. *See also* Ruby laser
Coaxial flashlamp, 330
Coherence length, 173–174
Coherent light, 84
Coherent photons, rate equations of, 198, 200
Collisions, deexcitation, 130
Collisions, pumping, 88, 236, 262
Color temperature, 16
Concave–convex resonator, 180
Concentric resonator, 180–181
Conduction band, 38
Confocal resonator, 179–181
Continuum, spectrum, 15
Copper vapor laser, 123, 127
Correspondence principle, 55
Coumarin dye, 327
Coupling, l-s, 65–66
Covalent bonding, 73
Cross section, atomic, 131–132, 143
 and gain, 143
 calculating, 144
 in saturation intensity, 148
Cryogenic cooling:
 erbium, 128
 ruby, 297
 solid-state, 315
Cryogenic gas processor, 273
Current regulator, linear passbank, 257
CW (continuous wave) lasers, 124

Damping, critical, 300
DBR (distributed Bragg reflector) laser, 319
DeBroglie, Louis, 50
DeBroglie wavelength, 51
Decay, spontaneous, 34
Degeneracy, 150
Depletion layer, 40
Detector, PIR, 17
Detector, pyroelectric, 17

DFB (distributed feedback) laser, 320
Dielectric mirrors, *see* Mirrors
Diffraction grating, 22
 fiber Bragg, 278
 wavelength selector, 331–332, 334
Diffraction limiting, 176
Diode, 40
Diode laser power supply, 321
Diode laser, threshold, 106
Direct-gap semiconductor, 44
Discharge, gas, 21–22
Dissociation, carbon dioxide, 287
Distribution, energy, in a semiconductor, 45–46
Distributions, probability, 59
Donut mode, 185
Doping, degenerate, 314
Doping, in a semiconductor, 40, 314
Doppler linewidth, 101
Double-pulse ruby lasers, 297, 298
Double refraction, *see* Birefringence
Doublet, 66
 in sodium spectrum, 69–70
Doubly-ionized species, 250
DPSS lasers, 304–306
Dye lasers, 327–335
 CW, 333

EDFA (erbium doped fiber amplifier), 96, 309
Efficiency:
 light sources, 18
 nonlinear conversion, 229
 pumping, 153
 quantum, 153
Einstein coefficients, 92
 solved, 95
Electromagnetic spectrum, 2, 10
Electron:
 conduction band, 38
 collisions, deexcitation, 130
 collisions, pumping, 88, 236, 262
 configuration, 57
 diffraction, 52–53
 emission, 34
 energies, 27, 31–33
 in semiconductors, 38, 40–41
 orbitals, 28–30
 quantum numbers, 56–66
 recombination with hole, 39–40
 transitions, 12
 valence band, 38
 wavefunctions, 53–54
Electronic transitions, 73

Electrooptic effect, 203
Electrooptic switch, 202–206
　as a modelocker, 214
Emission, 10–11
　spontaneous, 92
　stimulated, 90
Emission spectrum:
　flashlamp, 118
　helium, 23
　hydrogen, 23
　hydrogen, hyperfine, 63, 66
　LED, 47
　mercury, 23, 71
　sodium, 29
End pumping, rod, 152
Energy bands:
　in a gas, 76
　in solids, 37–38, 115
Energy distribution, Boltzmann, 13
Energy distribution, in a semiconductor, 45–46
Energy levels:
　argon-ion laser, 249
　carbon dioxide, 76, 284
　copper-vapor laser, 123
　dyes, 328
　excimer laser, 271
　helium–neon laser, 72, 122, 237
　hydrogen, 12, 60, 73
　mercury, 71
　nitrogen, 73
　nitrogen laser, 126, 263
　ruby, 120, 296
　semiconductor, 131, 314
　sodium, 60, 69
　YAG, 121
EO modulators, *see* Electrooptic switch
Equilibrium, thermal, 86
Erbium, 128, 302, 309–311
Etalon, 171
　air-spaced, 173
　in a dye laser, 331, 334
Excimer (excited dimer) laser, 270–271
　gas mix, 272
　wavelengths, 272
Extinction ratio, 206

Fabry–Perot, *see* Etalon; Resonator
Far-IR lasers, 293–294
Fermi levels, 41, 314
Fiber amplifier, *see* EDFA
Finesse, 163
Flashlamps, 118, 299
　pumping, 299–300
　ruby pump, 296

spectra, 118
　YAG pump, 301, 306
Flowing-gas laser, 287
Fluorescence, 35–37
Fluorescent lamp, 35
Fluorine gas, 272, 279
　generation, 279
Forbidden transitions, 63, 68
Forces, kinematic, 219
Franck–Hertz experiment, 31–34
　visual observations, 33
Fresnel equations, 110, 112
FSR (free spectral range), 162
　of an etalon, 172
FWHM, definition, 25

GaAs (gallium-arsenide), 41
　as an optical material, 285
Gain, laser, 98–101
　and cross section, 143
　as a function of wavelength, 103
　as a power increase, 98
　broadening, 101–103, 145
　experimental methods of determining, 108–113
　HeNe transitions, 237, 240
　lineshape, 144
　nitrogen laser, 264
　saturated, 147
　small signal, 145
　switching, 193
　threshold, 104, 145
　threshold, calculated, 106–108
Gas ballast, 254
Gas dynamic laser, 290
Gases, excimer, 272–273
Gas processor, 273
Gas replenishment, 254
Gas return path, 252
Gaussian beam, 176
　profile, 176
　waist size, 177
Germanium, 285
Glass (as a host), 302
Gold mirrors, 285
g-parameters (for a resonator), 179
Graphite plasma tubes, 252
Grating, *see* Diffraction grating
Grating, fiber Bragg, 278
Ground state, 12

Halogen gas, 68, 272
Halogen lamp, 16, 19
Heard, H.G., 261

340 INDEX

Helium:
  line spectrum, 23
  in a carbon-dioxide laser, 284
  in an excimer laser, 273
  in a nitrogen laser, 270
Helium–cadmium (HeCd) laser, 250–251
Helium–neon (HeNe) laser, 83, 235
  depopulation of lower levels, 130
  energy levels, 122
  gain, 111–113
  IR transitions, 236, 238
  linewidth, 102
  superradiance, 238
  visible transitions, 72
Heterodyning, 230
Heterojunction, 316
Holes, in semiconductors, 39
Holmium, 302
Homogeneous broadening, 145
Homojunction, 315
Hydrogen:
  fine structure, 66
  laser, 269
  line spectrum 21, 23
  vibrational energy, 73–74
Hygroscopic materials, 285

Idler beam, 232
Ignitor, plasma, 57–258
  HeNe laser, 41–242
Incandescence, 1, 5
Index-guided lasers, 317, 324
Indirect-gap semiconductor, 44
Intrinsic semiconductors, 40
Inversion, population, 87
Ion, ground state, 248
Ionization, 29
Ion laser, 247
  mercury, 99
IR absorption spectrum, 77–79
IR emission, from objects, 16
IR optics, 285
IR transitions, in a HeNe laser, 236, 238
I–V curve, 43

KDP (potassium dihydrogen phosphate), 228
Kerr effect, 204
Krypton fluoride, 272
Krypton ion (Kr^+) laser, 248–249
  doubly ionized, 250
  output lines, 167, 249
  wavelength selection, 166–168
  white light, 167

KTP (potassium titanyl phosphate), 228

Laminar dye flow, 332
Laser (acronym), 91
Laser-pumped dye laser, 331–333
LASIK surgery, 278
LED (light emitting diode), 41–47
  emission spectrum, 47
  structure, 41–42
  thermal energy in, 45–46
Lenses, collimating, in an output coupler, 177

Lifetime:
  atomic level, 34
  laser level, 93
  molecular nitrogen, 262–263,
  photon, 174
Lifetime broadening, 176
Limelight, 1
Linear passbank, 257
Lineshape, gain, 144
Line spectrum, 12, 21
  analysis, 23
  helium, 23
  hydrogen, 21, 23
  hydrogen, hyperfine, 63, 66
  mercury, 23, 71
  sodium, 29
Linewidth, 25–26, 101
  Doppler broadened, 101
  function of temperature, 103
  He–Ne laser, 102
  in a laser, 101
Lithium niobate, 228
Longitudinal modes, 164
Loss, in a laser, 99, 106
  distributed, 174–175
  effect on output power, 161
  inserted into a laser cavity, 110
l-s coupling, 65–66
Lumen, 18
LWI (lasers without inversion), 87

Macroscopic charge polarization, 221
Magnetic field:
  in a HeNe laser tube, 238
  in an ion laser tube, 252
Magnetic quantum number, 62
MALDI, 269
Mercury, energy levels, 71
Mercury line spectrum, 23
  fine structure, 71
Mercury-vapor laser, 99–100, 250
Metal-vapor lasers, 123, 248, 250

Metastable state, 122, 262
Mirrors, 95
 cleaved, 314
 dielectric, 314
 ion laser tube, 251
 rotating, 195
Mixing, sum and difference, 230
Modelocking:
 dye laser, 334
 frequency domain, 215
 time domain, 212
Modes, cavity, thermal, 6
Modes, longitudinal, 163–166
 semiconductor laser, 322–323
 single-frequency operation, 169
Modes, transverse (TEM), 184–186
 designations, 185
 limiting, 186–187
Molecular laser, 262, 284, 293. *See also* Carbon-dioxide lasers; Nitrogen
Molecules:
 rotations, 76
 vibrations, 73
Molybdenum, 285
Monochromatic light, 85
MOPA (master oscillator power amplifier), 108, 295, 297, 331

Nd:YAG laser, *see* YAG laser
Negative resistance, 244
Neodymium, 301. *See also* YAG laser
Neon-ion (Ne+) laser, 250
Neon (Ne) laser, 90, 269
Neon sign, 36
Nitrogen:
 in a carbon-dioxide laser, 284
 vibrational energy, 73–74
Nitrogen-ion (N^+) laser, 250
Nitrogen (N_2) lasers, 93, 261–270
 level lifetimes, 125–126, 263
 linewidth, 262–263, 269
 molecular ion (N_2^+) species, 270
 TEA, 267, 270
Noble gases, 67
Noise, in an amplifier, 96–98
Nonlinear:
 forces, 219–220
 optics, 219–233
 materials, 227–228
Nonradiative transitions, 34
n-type semiconductor, 40, 314

OC (output coupler), optimizing, 156
OPO (output parametric oscillator), 232–233

Optically-pumped semiconductor lasers, 318
Optical pumping, 87–88, 117–118, 293, 295–311, 330
Orbital, electron, 56–60
Orbital quantum number, 57
Orbit, Bohr, 28–29
Orbit–orbit interaction, 62
Organic dyes, *see* Dye lasers
OSA (optical spectrum analyzer), 96
Oscillator strength, 132, 143
Oxygen ion (O^+) laser, 250

Particle-in-a-box model, 54
Passbank, 257
Passivation, 280–281
Pauli exclusion principle, 67
Periodic table, 67–69
Phase matching, 223
 types, 224–225
Phonon, 34, 130
Phosphor, 35
Photoelectric effect, 27
Photolithography, 278
Photon energy, 10, 30
Photon lifetime, 174
Phototube, 27
Piezoelectric crystals, 228
PIR (passive IR) detector, 17
Planck, Max, 8
Planck's law, 7, 26
Plane mirror resonator, 179–180
p-n junction, 40–41, 313–314
Pockels cell:
 electrooptic modulator, 202–205
 in a ruby laser, 298
Pockels effect, 204
 polarizability, nonlinear material, 221
Polarization:
 Brewster window, 162
 ruby rod, 299
 steady, 223
Polarizer, Brewster, 298
Polarizer, in an EO modulator, 203
Population difference, 143
Population inversion, 87
 in a four-level system, 142
 in a three-level system, 139
Power, output, 155
 of a Q-switched laser, 198–200
Preionization, 275–277
Prism, as a wavelength selector, 168–169
Probability distribution, 59
Probability, pumping, 132

342 INDEX

Probability, transitional, 131–132
Propane, IR absorption spectrum, 79
Proustite, 231
p-type semiconductor, 40, 314
Pulse energy (Q-switched), 201
Pulse power (Q-switched), 200, 202
Pumping, 87–90
 collisional, 89
 flashlamp, 118
 in the HeNe laser, 88–90
 optical, 87
 selective, 117
Pump level, 119
 in the carbon-dioxide laser, 75, 284
 in the HeNe laser, 88, 122
 in the ruby laser, 121
 in the YAG laser, 121

Q, of a laser cavity, 196
Q-switch, 193
 acoustooptic, 195, 206–211, 303
 electrooptic, 195, 202–206, 298
 saturable absorber, 195
Q-switching, 194–211
 output power, 198–200
 pulse energy, 201
 pulse power, 200, 202
 rate equations, 196–200
Quantization, of photon energy, 9–10, 26
Quantum efficiency, 153
Quantum harmonic oscillator, 55
Quantum number, 28
 magnetic, 62, 67
 orbital, 57, 67
 principal, 67
 spin, 65, 67
Quartz:
 nonlinear, 219, 228
 window, 160
Quenching, triplet, 328

Radial probabilities, 58
Radiation trapping, see Resonance trapping
Radiative transitions, 34
Radius, Bohr, 28
Raman–Nath diffraction, 209
Rate equations, 92–96
 at thermal equilibrium, 94
 four-level system, 140–142
 in a Q-switched laser, 196–200
 inversion in a four-level system, 142
 inversion in a three-level system, 139
 pumping, 132
 three-level system, 137–139

 two-level system, 133–135
 ULL populations, 149–151
Rayleigh–Jean's law, 7
Recombination, electron-hole pair, 39
Reflection, at a surface, 110
Resistance, negative, 244
Resistor, ballast, 244, 257, 291
Resistor trimming, 193
Resonance trapping, 127
Resonator, 95, 106, 156
 requirements, 159
RF pumping, 291
Rhodamine-6G dye, 327
Rotational energies, 76
Ruby laser, 120–121, 295–301
 as a four-level laser, 297
 double-pulse, 197, 297
 energy levels, 121, 296
 linewidth, 300
 Q-switched, 297–298
 thermal lensing, 297
Rutherford–Bohr model, 28

Saturable dye absorber, 195
Saturable reactor (electrical), 291
Saturation, gain, 147
Saturation intensity, 148
 cross section for, 144
 measuring, 149
Schroedinger equation, 53–54
Scotch whiskey laser, 293
Sealed carbon-dioxide lasers, 287
Second harmonic term, 223
Selection rules for transitions, 63, 68
Semiconductor, 40
 intrinsic, 40
 junction, 41
 n-type, 40, 314
 p-type, 40, 314
Semiconductor lasers, 313–325
 energy bands, 130
Silicon, 41
Single-frequency operation, 169, 292
Singlet states, 71
 in dye, 328
Small-signal gain, 145
Sodium spectrum, 68–70
 fine structure, 69
Soft-sealed HeNe tubes, 240
Sommerfeld, Arnold, 56
Spark gap, 265–266
Spectral width, 25
Spectroscope, 22
Spectroscopic analysis:

INDEX **343**

infrared, 77
of an unknown gas, 23
Spectroscopic notation, 57
Spectrum, *see* Emission spectrum
Spherical-plane resonator, 180, 182
Spin, electron, 64–66
Spin-spin effect, 66
Spontaneous emission, 34
  as noise, 96–98
  rate, 92, 94–95
Spontaneous lifetime, 34, 124, 132
Spot size, 181
Stability, HeNe output, 245–246
Stability, resonator, 178
Stability, semiconductor laser, 322–324
Stefan–Boltzmann law, 2–3
Stern–Gerlach experiment, 63–65
Stimulated emission, 90
  rate, 92, 95
Stinger electrodes, 253
Strength, transition, 132, 144
Stripe contact, 316, 322
Superradiance, 104
  in a HeNe laser, 238
  in dye lasers, 331
  in nitrogen lasers, 264
Symmetry, in a nonlinear material, 228

Table of elements, 67
TEA lasers, 93, 267, 290
TEM modes, *see* Modes, transverse
Thermal equilibrium, 14, 86–87, 89
Thermal lensing, 297
Thermal light, 1, 12
Thermal population:
  in carbon dioxide, 289
  of levels, 127
Thompson, G.P., 52
Three-level laser, 119, 136–139
Three-phase power, 256
Threshold, gain, 106
  and cross section, 145
  and inversion, 146
Threshold, inversion, 146
Threshold, pumping, 105, 149
  calculated, 150
  diode laser, 106
  four-level laser, 143
  three-level laser, 139
Thyratron, 268, 274–275
Transitional probabilities, 131–132
Transitions:
  allowed, 63, 68
  electronic, 73

forbidden, 63, 68
radiative, 12
selection rules, 63, 68
Transition strength, 132, 143–144
Transmission, EO modulator, 205
Transmission line, 265–266
Transverse electrodes, 93, 265. *See also* TEA lasers
Transverse modes, *see* Modes, transverse
Triplet absorption, 328, 333
Triplet quenching, 328
Triplet states, 71
  in dye, 328, 333
Two-level system, 133

Unstable resonators, 183–184
  in an excimer laser, 274
UV catastrophe, 7
UV lasers, 261–281

Valence band, 38
Vanadate, 302
VCSEL, 317–318
VECSEL, 318
Vernier scale, 24–25
VFD (vacuum fluorescent display), 36
Vibrational energies, 73–76
  in carbon dioxide, 74–75, 284
  in hydrogen, 73
  in nitrogen, 262
Vibronic levels, 74, 76
Vibronic transitions, 262
Voltage, half-wave, 205
Voltage multiplier, 242

Walk-off loss, 172
Wave, electron as a, 52
Wavefunction, 53–54
Waveguide lasers, 290
Wavelength, 10
Wavelength, DeBroglie, 51
Wavelength selection, 166
  using a prism, 168–169
Wavenumber, 78
Wave-particle duality, 52
Wein's law, 4–5
White-light (krypton) laser, 249, 259
Windows, Brewster, 161

Xenon chloride, 272
Xenon flashlamp, 118
  as ruby pump, 296
  as YAG pump, 303
  output spectrum, 118

Xenon fluoride, 272
X-ray laser, 105

YAG (yttrium aluminum garnet) laser, 301
  absorption spectra, 118
  diode-pumped, 225–226, 304–306
  energy levels, 121
  host materials, 302

Q-switched, 303
  thermal effects, 127–128
Young, double slit experiment, 52

Zeeman effect, 63
  in an ion laser plasma, 252
ZnSe (Zinc selenide), 285